AF389489

Antimicrobial Resistance and
Anti-Biofilms

Antimicrobial Resistance and Anti-Biofilms

Editors

Junyan Liu
Ding-Qiang Chen
Yulong Tan
Ren-You Gan
Guanggang Qu
Zhenbo Xu

MDPI • Basel • Beijing • Wuhan • Barcelona • Belgrade • Manchester • Tokyo • Cluj • Tianjin

Editors

Junyan Liu
College of Light Industry
and Food Science
Zhongkai University of
Agriculture and Engineering
Guangzhou
China

Ding-Qiang Chen
Department of Laboratory
Medicine
Southern Medical University
Guangzhou
China

Yulong Tan
Special Food Research
Institute
Qingdao Agricultural
University
Qingdao
China

Ren-You Gan
Agency for Science,
Technology and Research
Singapore Institute of Food
and Biotechnology
Innovation
Nanos
Singapore

Guanggang Qu
Binzhou National Economic
and Technological
Development Zone
Shandong Binzhou Animal
Science & Veterinary
Medicine Academy
Binzhou
China

Zhenbo Xu
School of Food Science and
Engineering
South China University
of Technology
Guangzhou
China

Editorial Office
MDPI
St. Alban-Anlage 66
4052 Basel, Switzerland

This is a reprint of articles from the Special Issue published online in the open access journal *Antibiotics* (ISSN 2079-6382) (available at: www.mdpi.com/journal/antibiotics/special_issues/anti-Biofilms).

For citation purposes, cite each article independently as indicated on the article page online and as indicated below:

LastName, A.A.; LastName, B.B.; LastName, C.C. Article Title. *Journal Name* **Year**, *Volume Number*, Page Range.

ISBN 978-3-0365-6951-2 (Hbk)
ISBN 978-3-0365-6950-5 (PDF)

Contents

Preface to "Antimicrobial Resistance and Anti-Biofilms"

This book is based on a topical collection "Antimicrobial resistance and anti-biofilms" in the journal *Antibiotics* since November of 2020. This book has one editorial article, eight research articles, one review article, and one case report article, covering a time period of 15 months from December of 2021 to February of 2023. Containing three major sub-topics in this book, four manuscripts focus on the prevalence of resistant microbes, the emergence and evolution of resistance, and the molecular investigation of resistance mechanisms fall into the first sub-topic as "Antimicrobial resistance in microorganisms: epidemiology and molecular mechanism"; three manuscripts fall into the second sub-topic as "New antibiofilm strategy against fungal and/or bacterial biofilms"; and three manuscripts regarding the production and characteristics of these functional material-encapsulated/delivered natural compounds, in vitro and in vivo antimicrobial and antivirulent effects, and their potential applications, such as in food and medicine, fall into the third sub-topic as "Influence of Functional Material-Based Encapsulation/Delivery on the Antimicrobial and Antivirulent Effects of Natural Compounds". Articles contained in this book have provided comprehensive knowledge and insight in understanding of antimicrobial resistance epidemiology and molecular mechanism, new antibiofilm strategies, and novel natural compounds on biofilm eradication, covering food and clinical fields.

Junyan Liu, Ding-Qiang Chen, Yulong Tan, Ren-You Gan, Guanggang Qu, and Zhenbo Xu

Editors

Case Report

Antimicrobial Treatment on a Catheter-Related Bloodstream Infection (CRBSI) Case Due to Transition of a Multi-Drug-Resistant *Ralstonia mannitolilytica* from Commensal to Pathogen during Hospitalization

Junyan Liu [1,2,†], Brian M. Peters [3,†], Ling Yang [4,5,*], Hui Yu [4], Donghua Feng [4], Dingqiang Chen [4,5] and Zhenbo Xu [6,7,*]

1 College of Light Industry and Food Science, Guangdong Provincial Key Laboratory of Science and Technology of Lingnan Special Food Science and Technology, Innovation Research Institute of Modern Agricultural Engineering, Zhongkai University of Agriculture and Engineering, Guangzhou 510225, China
2 Key Laboratory of Green Processing and Intelligent Manufacturing of Lingnan Specialty Food, Zhongkai University of Agriculture and Engineering, Guangzhou 510225, China
3 Department of Clinical Pharmacy and Translational Science, University of Tennessee Health Science Center, Memphis, TN 38163, USA
4 Department of Laboratory Medicine, The First Affiliated Hospital of Guangzhou Medical University, Guangzhou Medical University, Guangzhou 510120, China
5 Centre for Translational Medicine, The First Affiliated Hospital of Guangzhou Medical University, Guangzhou Medical University, Guangzhou 510120, China
6 School of Food Science and Engineering, Guangdong Province Key Laboratory for Green Processing of Natural Products and Product Safety, Engineering Research Center of Starch and Vegetable Protein Processing Ministry of Education, South China University of Technology, Guangzhou 510640, China
7 Department of Laboratory Medicine, The Second Affiliated Hospital of Shantou University Medical College, Shantou 515041, China
* Correspondence: jykresearch@126.com (L.Y.); zhenbo.xu@hotmail.com (Z.X.); Tel.: +86-20-87113252 (Z.X.)
† These authors contributed equally to this work.

Citation: Liu, J.; Peters, B.M.; Yang, L.; Yu, H.; Feng, D.; Chen, D.; Xu, Z. Antimicrobial Treatment on a Catheter-Related Bloodstream Infection (CRBSI) Case Due to Transition of a Multi-Drug-Resistant *Ralstonia mannitolilytica* from Commensal to Pathogen during Hospitalization. *Antibiotics* **2022**, *11*, 1376. https://doi.org/10.3390/antibiotics11101376

Academic Editors: Wolf-Rainer Abraham and Nicholas Dixon

Received: 18 August 2022
Accepted: 7 October 2022
Published: 8 October 2022

Publisher's Note: MDPI stays neutral with regard to jurisdictional claims in published maps and institutional affiliations.

Abstract: Despite its commonly overlooked role as a commensal, *Ralstonia mannitolilytica* becomes an emerging global opportunistic human pathogen and a causative agent of various infections and diseases. In respiratory illnesses, including cystic fibrosis and chronic obstructive pulmonary disease (COPD), *R. mannitolilytica* is also identified presumably as colonizer. In this study, one distinctive clone of *R. mannitolilytica* was firstly identified as colonizer for the first 20 days during hospitalization of a patient. It was then identified as a causative agent for catheter-related bloodstream infection with negative identification after effective treatment, verifying its transition from commensal to pathogen. In conclusion, we provide convincing evidence that during hospitalization of a patient, *R. mannitolilytica* transitioned from commensal to pathogen in the respiratory tract leading to catheter-related bloodstream infection (CRBSI).

Keywords: *Ralstonia mannitolilytica*; chronic obstructive pulmonary disease (COPD); catheter-related bloodstream infection (CRBSI); commensal; pathogen

1. Introduction

As a worldwide public health challenge [1], chronic obstructive pulmonary disease (COPD) ranks the third leading cause of death and overall prevalence ranges from 8.6% to 13.6% in China [2,3]. COPD patients present an altered airway microbiome, with *Ralstonia mannitolilytica* identified at a significantly higher rate comparing with a healthy population [4]. Reservoirs of *R. mannitolilytica* in the hospital environment include water [5], saline solutions [6] and oxygen delivery devices [7]. Despite its commonly overlooked role as a commensal, *Ralstonia mannitolilytica* becomes an emerging global opportunistic human pathogen and a causative agent of various infections and diseases, including bacteremia [8–10], meningitis [11], sepsis [12], peritonitis [13], osteomyelitis [14], hemoperitoneum [14] and urinary tract infection [7]. Furthermore, *R. mannitolilytica* is also identified in respiratory illnesses such as cystic fibrosis and COPD, presumably as a colonizer [15–18]. From the first report in 2011, *R. mannitolilytica* isolate G100 was sampled from the sputum of a COPD patient on admission, and the patient was treated with piperacillin/tazobactam [16]. Despite negative identification of *R. mannitolilytica* thereafter, the patient continued to suffer from respiratory failure and eventually died, suggesting this microorganism was a benign colonizer [16,19–21]. Such a finding is in accordance with a recent comprehensive study on airway microbiomes, where the presence of *R. mannitolilytica* was occasionally found colonizing COPD patients and indicating its potential role in recurrent infections and diseases [4,22].

For the first time, we report a catheter-related bloodstream infection (CRBSI) case caused by *R. mannitolilytica* in the respiratory tract of a hospitalized COPD patient.

2. Results and Discussion

2.1. Antimicrobial Agent Treatment

2.1.1. Phase I: From Admission to Day 3

On admission, the patient was diagnosed with acute exacerbations COPD (AECOPD), severe pneumonia and type II respiratory failure, with symptoms of fever and dyspnea on day 1 and high fever and dyspnea on day 2. On day 1, sputum culture identified *Acinetobacter baumannii*, *Chryseobacterium meningosepticum* and *R. mannitolilytica*, with negative culture from a blood sample. During day 1 to 3, antimicrobial treatment had been conducted using meropenem, vancomycin, caspofungin and voriconazole, with peripherally inserted central venous catheters (PICC) used from day 1 onwards.

2.1.2. Phase II: From Day 4 to 18

Despite antimicrobial treatment during day 1 to 3, the patient failed to recover and a blood test on day 4 revealed an uncontrolled infection accompanied by dyspnea and wheezing. In detail, he had suffered from low fever and wheezing on day 4; low fever, dyspnea and wheezing on day 5; and fever and wheezing on day 6. Consequently, the patient was treated with different antimicrobial therapy (Table 1). Nevertheless, he suffered from intermittent fever. Despite negative blood culture, sputum culture on Day 6 identified *A. baumannii*, *C. meningosepticum* and *R. mannitolilytica* (Guangzhou-RMAS10/11, and on day 10, only *R. mannitolilytica* was identified), which are common colonizers in the respiratory tract. As antimicrobial treatment was concerned, the patient had been treated with meropenem, vancomycin, caspofungin and amphotericin B during day 4 to 7; meropenem, linezolid, caspofungin and amphotericin B during day 8 and 9; piperacillin/tazobactam, linezolid, caspofungin and amphotericin B during day 10 to 13; cefoperazone/sulbactam, rifampin, vancomycin and amphotericin B during day 14 to 16; and cefoperazone/sulbactam, meropenem, rifampin, vancomycin and caspofungin during day 17 and 18.

Table 1. Clinical data of the patient.

	Admission Time	Temperature (°C)	WBC ($\times 10^9$ L)	Neutrophils (%)	Detection		
					Blood Gas: pH/HCO$_3$ * (mmol/L) /SpO$_2$ * /PaCO$_2$ * (mm Hg) /PaO$_2$ * (mm Hg)	PCT * (ng/mL)	HCCT *
	D1	38.3	16.6	77.4	7.42/24.5 /98%/51.2/90.6	0.45	RLLLA *
Phase I	D2	38.8–40	20.6	90.1			
	D3	37.5	19.8	94.1			

Table 1. *Cont.*

	Admission Time	Temperature (°C)	WBC (×10⁹ L)	Neutrophils (%)	Detection		
					Blood Gas: pH/HCO$_3$ * (mmol/L) /SpO$_2$ * /PaCO$_2$ * (mm Hg) /PaO$_2$ * (mm Hg)	PCT * (ng/mL)	HCCT *
Phase II	D4		18.3	84.7	7.435/29.1/96.3%/42.8/85.6		
	D5		15.7	82.9			
	D6		15.6	82.3	7.504/26.5/98.8%/33.9/159	4.08	RLLLA
	D7	37.6–38.0	14.6	81.9			
	D8	38.8					
	D9	38.4					
	D10	36.7–38.3					
	D11	38.0					
	D12	38.0	14.3	81.2			
	D13	37.7					
	D14	38.0					
	D15	38.3					
	D16	37.9					
	D17	38.0					
	D18	36.9	16.0	90.7			
Phase III	D19	37.0–38.5			7.346/24.7/98.1%/47.1/152		
	D20	39.4					
	D21	37.9	13.7	75.5			
	D22	37.0–37.9	14.1	88.9			
	D23	38.3					
	D24	37.2–38.3					
	D25	38.0					
	D26	37.1	11.7	89.0	7.365/25.3/98.5%/47.2/119	0.80	Not done
	D27	37.2	9.6	82.7			

* HCCT: High-resolution chest computed tomography. * RLLLA: Right lung lower lobe cavity. * PCT: procalcitonin. * HCO^3: bicarbonate. * SpO$_2$: Saturation of peripheral oxygen. * PaCO$_2$: Partial pressure of carbon dioxide. * PaO$_2$: Partial pressure of oxygen. Sampling on D1, D6, D10, D19 and D26, and results obtained on D3, D8, D12, D20 and D27, respectively. For blood samples, 3 pairs of aerobic and anaerobic blood culture bottles were collected at different fever spikes, including 1 pair drawn through the catheter and another 2 pairs drawn by bilateral peripheral venipuncture.

2.1.3. Phase III: From Day 19 to 27

On day 19, the patient presented with fever, shivering and dyspnea, and on day 20, his condition deteriorated to high fever, severe dyspnea and septic shock. Remarkably, *R. mannitolilytica* had been identified from both sputum and blood samples (Guangzhou-RMAB10/11/12, three isolates, one from the catheter of PICC and two from peripheral blood) on day 19, validating the occurrence of catheter-related bloodstream infection (CRBSI). According to its antimicrobial profile, ciprofloxacin was used from day 21 on. The patient showed decreased body temperature and improvement in dyspnea, with symptoms of low fever and relieved dyspnea during day 21 to 25, then no fever or dyspnea on day 26 and 27. In addition, a blood test on day 26 revealed normal white blood cell count and serum procalcitonin (PCT), with no positive identification of *R. mannitolilytica* from either blood or sputum samples. Also, PICC was removed on day 27 after the result was reported.

2.1.4. Phase IV: After Day 27 to Discharging

Two months later, this patient presented cough with thick yellow sputum and occasional wheezing. On day 82, the sputum culture identified *Pseudomonas aeruginosa*, *Achromobacter xylosoxidans* and *A. baumannii*, and blood culture identified *Enterococcus faecium* and *A. baumannii*. Consequently, routine treatment for these symptoms was further conducted and the patient was eventually discharged on day 132.

2.2. Bacterial Identification

Seven *R. mannitolilytica* isolates were sampled, including four strains (Guangzhou-RMAS10/11/12/13) from sputum and three strains (Guangzhou-RMAB10/11/12) from blood. In detail, *R. mannitolilytica* Guangzhou-RMAS10, Guangzhou-RMAS11, Guangzhou-

RMAS12 and Guangzhou-RMAS13 had been recovered from sputum samples on day 1, day 6, day 10 and day 19, respectively. *R. mannitolilytica* Guangzhou-RMAB10, Guangzhou-RMAB11 and Guangzhou-RMAB12 had been obtained from blood samples on day 19. According to the results of 16S rRNA sequencing, compared with all known sequences in GenBank by BLASTn, the sequencing results of all *R. mannitolilytica* strains were found to be identical and clustered at 100% sequence similarity with the *R. mannitolilytica* strain AU255 (AY043379) and AU428 (AY043378) from a cystic fibrosis patient in the United States, and strains LMG 6866 (NR_025385) from cases of nosocomial recurrent meningitis in Belgium.

2.3. Antimicrobial Susceptibility Testing

Antimicrobial susceptibility testing (AST) using Vitek2™ Automated System had been performed on seven *R. mannitolilytica* strains [23]. All *R. mannitolilytica* strains shared distinctive antimicrobial resistance profiles, and exhibited resistance to 14 antimicrobial agents, including ampicillin/sulbactam, aztreonam, cefazolin, cefepime, cefoperazone/sulbactam, cefotaxime, cefotetan, cefuroxime, gentamicin, imipenem, meropenem, piperacillin, piperacillin/tazobactam and tobramycin, with only a few susceptible exceptions such as ceftriaxone, ciprofloxacin and levofloxacin. The AST results had been used to determine the antimicrobial treatment on *R. mannitolilytica* strains.

2.4. Clonal Relatedness of R. mannitolilytica Strains

According to the results from randomly amplified polymorphic DNA polymerase chain reaction (RAPD-PCR), the acquired indistinguishable fingerprinting patterns had suggested all seven *R. mannitolilytica* belonged to the same genotype and thus were clonally related (patterns from representative strains shown in Figure 1). As further confirmed by genomic comparison, their identical genomes with similarity higher than 99.99% had convincingly verified that all *R. mannitolilytica* strains originated from a distinct clone (the genome sequences of the distinct *R. mannitolilytica* strain are deposited under accession number CP049132 for chromosome 1 and CP049133 for chromosome 2). In connection with the sampling background, the colonizing *R. mannitolilytica* strain had highly likely induced invasive infection in the respiratory tract and subsequently caused bacteremia in the patient, which explained the identification of clonally related bacteria from sputum and blood culture [24].

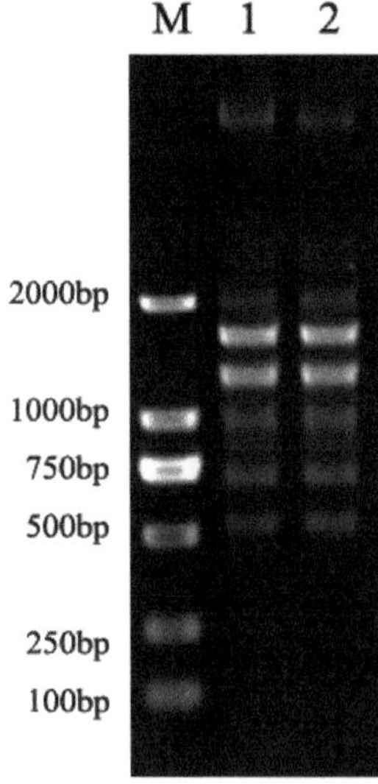

Figure 1. Fingerprinting patterns of two representative strains.

2.5. Extensive Surveillance on R. mannitolilytica Strains

In addition, from an extensive surveillance during 2014 (including the duration of hospitalization), the *R. mannitolilytica* strain was rarely identified as an infectious agent for other patients, nor was it routinely isolated from drinking water, laboratory-based water, saline solutions used for patient care or catheters [25–27].

2.6. Transition from Commensal to Pathogen

In this case, rare identification of *R. mannitolilytica* from patients or environmental samples, no recent hospitalization history for the patient, as well as the immediate detection of *R. mannitolilytica* on admission ruled out the likelihood of nosocomial infection or contamination and, thus, highly suggests a community origin of this microorganism. During hospitalization, one identical *R. mannitolilytica* strain was found colonizing the patient's respiratory tract despite broad antimicrobial therapy (during admission to day 20, shown by four isolates independently sampled from sputum at different time points), and further caused CRBSI (shown by three isolates separately sampled on day 19 including one isolate sampled from the catheter). In addition, the specific treatment of *R. mannitolilytica* (since day 21) correlated with the improvement of symptoms and negative identification of *R. mannitolilytica* (on day 26 and thereafter).

R. mannitolilytica has been implicated in acute exacerbation of chronic obstructive pulmonary disease (AECOPD); however, it is still controversial if the bacterium is the causative agent [4,16]. Mostly importantly, this study reported *R. mannitolilytica* bacteremia in a patient diagnosed with AECOPD, confirming its pathogenic role in this disease. Important evidence for this confirmation and verification were as follows: Firstly, for a patient diagnosed with AECOPD, *R. mannitolilytica* strain Guangzhou-RMAS10 was independently isolated from sputum samples (with identical strains) on day 6, but negative for blood samples. Secondly, *R. mannitolilytica* strain Guangzhou-RMAB10 was independently isolated from all three pairs of blood culture bottles on Day 20. Thirdly, *R. mannitolilytica* strains Guangzhou-RMAS10 and Guangzhou-RMAB10 were identical according to the antimicrobial profile, RAPD and genomic sequences. The above evidence strongly indicates the opportunistic pathogen *R. mannitolilytica* induced invasive infection in the respiratory tract (isolates from sputum) and subsequently caused bacteremia (isolates from blood) for this patient. Fourthly, no *R. mannitolilytica* infection for other patients had been reported in the First Affiliated Hospital of Guangzhou Medical University (FAHGMU) in the whole year of 2014 (this bacterium is extremely rare to be reported), and also, during hospitalization of the patient, none of *R. mannitolilytica* were isolated from drinking water, laboratory-based water, saline solutions used for patient care, or catheters. The above evidence and observation highly suggested the unlikelihood of nosocomial infection or contamination as the infection cause. Fifthly, after the confirmation of *R. mannitolilytica* on day 20, the antimicrobial therapy was changed to ciprofloxacin after day 22. Shortly after this, significant improvement from the blood test and negative identification of *R. mannitolilytica* (on day 25 and thereafter) were found. The above evidence presented strong evidence that *R. mannitolilytica* caused bacteremia during the therapy of AECOPD of a patient, as well as its pathogenic role in this disease. In combination with the above evidence, the *R. mannitolilytica* bacteremia in a patient diagnosed with AECOPD, as well as its pathogenicity, is convincing.

3. Materials and Methods

3.1. Clinical Samples and Bacterial Strains

A total of 16 strains were isolated during the hospitalization of a patient, including 7 *Acinetobacter baumannii* and *R. mannitolilytica* (4 from sputum and 3 from blood samples, sampling on day 1 and 6, and on day 82 (June 22) from both sputum and blood), 2 *Chryseobacterium meningosepticum* (sampling on day 1 and 6), 1 *Pseudomonas aeruginosa* (sampling on day 82 (22 June) from sputum), 1 *Achromobacter xylosoxidans* (sampling on day 82 (22 June) from sputum), and 1 *Enterococcus faecium* (sampling on day 82 (22 June) from blood) strains.

3.2. Clinical Case

On 2 April 2014, a 72-year-old male patient presenting with fever and dyspnea was admitted to the First Affiliated Hospital of Guangzhou Medical University (FAHGMU) in Guangzhou, China. With a COPD history dating back to 2004, this patient was treated to relieve symptoms of dyspnea. On admission, the patient was diagnosed with acute exacerbations COPD (AECOPD), severe pneumonia and type II respiratory failure, with symptoms of fever and dyspnea on day 1 and high fever and dyspnea on day 2.

Concerning the clinical history, in January of 2014, the patient was diagnosed with pulmonary aspergillosis and was treated with voriconazole during hospitalization. On 26 February 2014, the patient was hospitalized again due to gastrointestinal bleeding and cured by an antiacid drug. Considering the pulmonary aspergillosis was not cured, the patient was successively treated with voriconazole, mezlocillin sodium/sulbactam sodium, and meropenem in combination with continuous positive airway pressure. On 14 March 2014, the patient was transferred to ICU and successively treated with biapenem, vancomycin hydrochloride, voriconazole and moxifloxacin hydrochloride.

3.3. Species Identification

For all samples, bacterial identification was performed with a Vitek2^{TM} Automated System (bioMerieux, Saint-Louis, MO, USA). In addition, all *R. mannitolilytica* isolates were further assessed by sequencing the *16S rRNA* gene [28] using universal primers 27F and 1492R to amplify and then sequence the *16S rRNA* gene, and the obtained sequences were compared to all known sequences in GenBank by BLASTn against the Nr database.

3.4. Antimicrobial Susceptibility Testing

Antimicrobial susceptibility testing (AST) was performed with a Vitek2^{TM} Automated System [23]. For *R. mannitolilytica* isolates, 17 antibiotics were used including ampicillin/sulbactam, aztreonam, cefazolin, cefepime, cefoperazone/sulbactam, cefotaxime, cefotetan, ceftriaxone, cefuroxime, ciprofloxacin, gentamicin, imipenem, levofloxacin, meropenem, piperacillin, Piperacillin/Tazobactam and tobramycin (BBI Solutions, Crumlin, UK).

3.5. Analysis on Clonal Relatedness of R. mannitolilytica Strains

Random amplification of polymorphic DNA (RAPD) PCR was applied to characterize the genetic relatedness of all *R. mannitolilytica* strains with different origin. The RAPD primer (RM270: 5′–TGC GCG CGG G–3′) was used for random amplification as described previously for *Ralstonia* spp. [29]. Obtaining identical DNA fingerprinting from all 7 *R. mannitolilytica* strains, additionally, genomic sequencing by Illumina HiSeq 2500 (Illumina, San Diego, CA, USA) and genome comparison were further performed for confirmation [30,31].

3.6. Extensive Surveillance on R. mannitolilytica Strains

Extensive surveillance on *R. mannitolilytica* was conducted at the same medical setting (FAHGMU) during the whole year of hospitalization (2014).

4. Conclusions

In conclusion, we provide convincing evidence that during hospitalization of a patient, *R. mannitolilytica* transitioned from commensal to pathogen in the respiratory tract leading to CRBSI.

Author Contributions: Conceptualization, L.Y.; Data curation, J.L.; Formal analysis, J.L.; Funding acquisition, Z.X.; Investigation, J.L. and D.F.; Methodology, L.Y. and H.Y.; Project administration, Z.X.; Resources, L.Y.; Software, H.Y.; Supervision, D.C.; Validation, D.C.; Visualization, D.F.; Writing—original draft, J.L.; Writing—review & editing, B.M.P. All authors have read and agreed to the published version of the manuscript.

Funding: This work was supported by the Guangdong Major Project of Basic and Applied Basic Research (2020B0301030005), Guangdong International S&T Cooperation Programme (2021A0505030007), Guangdong Provincial Key Laboratory of Lingnan Specialty Food Science and Technology (No. 2021B1212040013), Young S&T Talent Training Program of Guangdong Provincial Association for S&T, China (SKXRC202207), Young Talent Support Project of Guangzhou Association for Science and Technology (QT20220101076), Collaborative grant with AEIC (KEO-2019-0624-001-1), the 111 Project (B17018).

Institutional Review Board Statement: Not applicable.

Informed Consent Statement: The research has been approved by the medical ethics committee of the First Affiliated Hospital of Guangzhou Medical University.

Data Availability Statement: Accession number: CP049132 and CP049133.

Acknowledgments: We express high gratitude to the staff in the First Affiliated Hospital of Guangzhou Medical University (FAHGMU) in Guangzhou representative of Southern China for collecting the *R. mannitolilytica* isolates.

Conflicts of Interest: The authors declare no conflict of interest.

References

1. Adeloye, D.; Song, P.; Zhu, Y.; Campbell, H.; Sheikh, A.; Rudan, I. Global, regional, and national prevalence of, and risk factors for, chronic obstructive pulmonary disease (COPD) in 2019: A systematic review and modelling analysis. *Lancet Respir. Med.* **2022**, *10*, 447–458. [CrossRef]
2. Fang, L.; Gao, P.; Bao, H.; Tang, X.; Wang, B.; Feng, Y.; Cong, S.; Juan, J.; Fan, J.; Lu, K.; et al. Chronic obstructive pulmonary disease in China: A nationwide prevalence study. *Lancet Respir. Med.* **2018**, *6*, 421–430. [CrossRef]
3. Wang, C.; Xu, J.; Yang, L.; Xu, Y.; Zhang, X.; Bai, C.; Kang, J.; Ran, P.; Shen, H.; Wen, F.; et al. Prevalence and risk factors of chronic obstructive pulmonary disease in China (the China Pulmonary Health [CPH] study): A national cross-sectional study. *Lancet* **2018**, *391*, 1706–1717. [CrossRef]
4. Wang, Z.; Liu, H.; Wang, F.; Yang, Y.; Wang, X.; Chen, B.; Stampfli, M.R.; Zhou, H.; Shu, W.; Brightling, C.E.; et al. A refined view of airway microbiome in chronic obstructive pulmonary disease at species and strain-levels. *Front. Microbiol.* **2020**, *30*(11), 1758. [CrossRef] [PubMed]
5. Baird, R.M.; Elhag, K.M.; Shaw, E.J. *Pseudomonas Thomasii* in a hospital distilled-water supply. *J. Med. Microbiol.* **1976**, *9*, 493–495. [CrossRef] [PubMed]
6. Pan, H.J.; Teng, L.-J.; Tzeng, M.S.; Chang, S.-C.; Ho, S.W.; Luh, K.T.; Hsieh, W.C. Identification and typing of *Pseudomonas pickettii* during an episode of nosocomial outbreak. *Zhonghua Minguo Wei Sheng Wu Ji Mian Yi Xue Za Zhi* **1992**, *25*, 115–123. [PubMed]
7. Jhung, M.A.; Sunenshine, R.H.; Noble-Wang, J.; Coffin, S.E.; John, K.S.; Lewis, F.M.; Jensen, B.; Peterson, A.; LiPuma, J.; Arduino, M.J.; et al. A National Outbreak of *Ralstonia mannitolilytica* Associated With Use of a Contaminated Oxygen-Delivery Device Among Pediatric Patients. *Pediatrics* **2007**, *119*, 1061–1068. [CrossRef]
8. Boattini, M.; Bianco, G.; Biancone, L.; Cavallo, R.; Costa, C. *Ralstonia mannitolilytica* bacteraemia: A case report and literature review. *Infez. Med.* **2018**, *26*, 374–378.
9. Basso, M.; Venditti, C.; Raponi, G.; Navazio, A.S.; Alessandri, F.; Giombini, E.; Nisii, C.; Di Caro, A.; Venditti, M. A case of persistent bacteraemia by *Ralstonia mannitolilytica* and *Ralstonia pickettii* in an intensive care unit. *Infect. Drug Resist.* **2019**, *12*, 2391. [CrossRef] [PubMed]
10. De Souza, D.C.; Palmeiro, J.K.; Maestri, A.C.; Cogo, L.L.; Rauen, C.H.; Graaf, M.E.; Grein, F.L.; Nogueira, K.D.S. *Ralstonia mannitolilytica* bacteremia in a neonatal intensive care unit. *Rev. da Soc. Bras. de Med. Trop.* **2018**, *51*, 709–711. [CrossRef] [PubMed]
11. Vaneechoutte, M.; De Baere, T.; Wauters, G.; Steyaert, S.; Claeys, G.; Vogelaers, D.; Verschraegen, G. One Case Each of Recurrent Meningitis and Hemoperitoneum Infection with *Ralstonia mannitolilytica*. *J. Clin. Microbiol.* **2001**, *39*, 4588–4590. [CrossRef]
12. Owusu, M.; Acheampong, G.; Annan, A.; Marfo, K.S.; Osei, I.; Amuasi, J.; Sarpong, N.; Im, J.; Mogeni, D.; Chiang, H.-Y.; et al. *Ralstonia mannitolilytica* sepsis: A case report. *J. Med. Case Rep.* **2019**, *13*, 318. [CrossRef]
13. Dotis, J.; Printza, N.; Orfanou, A.; Papathanasiou, E.; Papachristou, F. Peritonitis due to *Ralstonia mannitolilytica* in a pediatric peritoneal dialysis patient. *New Microbiol.* **2012**, *35*, 503–506. [PubMed]
14. Ryan, M.P.; Adley, C.C. *Ralstonia* spp.: Emerging global opportunistic pathogens. *Eur. J. Clin. Microbiol. Infect. Dis.* **2013**, *33*, 291–304. [CrossRef] [PubMed]
15. Coenye, T.; Vandamme, P.; LiPuma, J.J. Infection by *Ralstonia* Species in Cystic Fibrosis Patients: Identification of *R. pickettii* and *R. mannitolilytica* by Polymerase Chain Reaction. *Emerg. Infect. Dis.* **2002**, *8*, 692–696. [CrossRef] [PubMed]
16. Zong, Z.Y.; Peng, C.H. *Ralstonia mannitolilytica* and COPD: A case report. *Eur. Respir. J.* **2011**, *38*, 1482–1483. [CrossRef] [PubMed]
17. Coman, I.; Bilodeau, L.; Lavoie, A.; Carricart, M.; Tremblay, F.; Zlosnik, J.; Berthiaume, Y. *Ralstonia mannitolilytica* in cystic fibrosis: A new predictor of worse outcomes. *Respir. Med. Case Rep.* **2016**, *20*, 48–50. [CrossRef] [PubMed]

18. Mukhopadhyay, C.; Bhargava, A.; Ayyagari, A. *Ralstonia mannitolilytica* infection in renal transplant recipient: First report. *Indian J. Med. Microbiol.* **2003**, *21*, 284–286. [CrossRef]
19. Shankar, M.; Rampure, S.; Siddini, V.; Ballal, H. Outbreak of *Ralstonia mannitolilytica* in hemodialysis unit: A case series. *Indian J. Nephrol.* **2018**, *28*, 323–326. [CrossRef] [PubMed]
20. Gröbner, S.; Heeg, P.; Autenrieth, I.B.; Schulte, B. Monoclonal outbreak of catheter-related bacteraemia by *Ralstonia mannitolilytica* on two haemato-oncology wards. *J. Infect.* **2007**, *55*, 539–544. [CrossRef]
21. Said, M.; Dangor, Y.; Van Hougenhouck-Tulleken, W.; Mbelle, N.; Strydom, K.-A.; Ismail, F. First outbreak of *Ralstonia mannitolilytica* bacteraemia in patients undergoing haemodialysis at a tertiary hospital in Pretoria, South Africa. *Int. J. Infect. Dis.* **2018**, *73*, 93. [CrossRef]
22. Saiman, L. Providing a Safety Net for Children: Investigating a Multistate Outbreak of *Ralstonia mannitolilytica* Related to a Contaminated Reusable Device. *Pediatrics* **2007**, *119*, 1207–1209. [CrossRef] [PubMed]
23. Bayot, M.L.; Bragg, B.N. *Antimicrobial Susceptibility Testing*; StatPearls Publishing: Treasure Island, FL, USA, 2022.
24. Vaneechoutte, M.; Kampfer, P.; De Baere, T.; Falsen, E.; Verschraegen, G. Wautersia gen. nov., a novel genus accommodating the phylogenetic lineage including Ralstonia eutropha and related species, and proposal of *Ralstonia [Pseudomonas] syzygii* (Roberts et al. 1990) comb. nov. *Int. J. Syst. Evol. Microbiol.* **2004**, *54 Pt 2*, 317–327. [CrossRef]
25. Daroy, M.L.G.; Lopez, J.S.; Torres, B.C.L.; Loy, M.J.; Tuaño, P.M.C.; Matias, R.R. Identification of unknown ocular pathogens in clinically suspected eye infections using ribosomal RNA gene sequence analysis. *Clin. Microbiol. Infect.* **2011**, *17*, 776–779. [CrossRef]
26. Ferro, P.; Vaz-Moreira, I.; Manaia, C.M. Association between gentamicin resistance and stress tolerance in water isolates of *Ralstonia pickettii* and *R. mannitolilytica*. *Folia Microbiol.* **2019**, *64*, 63–72. [CrossRef]
27. Lim, C.T.S.; Lee, S.E. A rare case of *Ralstonia mannitolilytica* infection in an end stage renal patient on maintenance dialysis during municipal water contamination. *Pak. J. Med. Sci.* **2017**, *33*, 1047–1049. [CrossRef] [PubMed]
28. Stackebrant, E.; Goodfellow, M.; Lane, D.J. 16S/23S rRNA sequencing. In *Nucleic Acid Tech-Niques in Bacterial Systematics*; Stackebrant, E., Goodfellow., M., Eds.; John Wiley & Sons: New York, NY, USA, 1991; pp. 115–175.
29. Daxboeck, F.; Stadler, M.; Assadian, O.; Marko, E.; Hirschl, A.M.; Koller, W. Characterization of clinically isolated *Ralstonia mannitolilytica* strains using random amplification of polymorphic DNA (RAPD) typing and antimicrobial sensitivity, and comparison of the classification efficacy of phenotypic and genotypic assays. *J. Med. Microbiol.* **2005**, *54*, 55–61. [CrossRef]
30. Coenye, T.; Spilker, T.; Reik, R.; Vandamme, P.; LiPuma, J.J. Use of PCR Analyses To Define the Distribution of *Ralstonia* Species Recovered from Patients with Cystic Fibrosis. *J. Clin. Microbiol.* **2005**, *43*, 3463–3466. [CrossRef]
31. Suzuki, M.; Nishio, H.; Asagoe, K.; Kida, K.; Suzuki, S.; Matsui, M.; Shibayama, K. Genome Sequence of a Carbapenem-Resistant Strain of *Ralstonia mannitolilytica*. *Genome Announc.* **2015**, *3*, e00405-15. [CrossRef] [PubMed]

Article

Occurrence of *Serratia marcescens* Carrying *bla*$_{IMP-26}$ and *mcr-9* in Southern China: New Insights in the Evolution of Megaplasmid IMP-26

Yuxia Zhong [1,†], Wanting Liu [1,†], Peibo Yuan [1], Ling Yang [2], Zhenbo Xu [3,4,5,*] and Dingqiang Chen [1,*]

[1] Microbiome Medicine Center, Department of Laboratory Medicine, Zhujiang Hospital, Southern Medical University, Guangzhou 510282, China; yuxia0301@163.com (Y.Z.); lwtjyk2022@163.com (W.L.); pbyuanpb@163.com (P.Y.)

[2] Department of Laboratory Medicine, The First Affiliated Hospital of Guangzhou Medical University, Guangzhou 510120, China; jykresearch@126.com

[3] School of Food Science and Engineering, Guangdong Province Key Laboratory for Green Processing of Natural Products and Product Safety, Engineering Research Center of Starch and Vegetable Protein Processing Ministry of Education, South China University of Technology, Guangzhou 510640, China

[4] Department of Civil and Environmental Engineering, University of Maryland, College Park, MD 20742, USA

[5] Research Institute for Food Nutrition and Human Health, Guangzhou 510640, China

* Correspondence: zhenbo.xu@hotmail.com (Z.X.); jyksys@126.com (D.C.)

† These authors contributed equally to this work.

Citation: Zhong, Y.; Liu, W.; Yuan, P.; Yang, L.; Xu, Z.; Chen, D. Occurrence of *Serratia marcescens* Carrying *bla*$_{IMP-26}$ and *mcr-9* in Southern China: New Insights in the Evolution of Megaplasmid IMP-26. *Antibiotics* **2022**, *11*, 869. https://doi.org/10.3390/antibiotics11070869

Academic Editors: Krisztina M. Papp-Wallace and Marc Maresca

Received: 26 May 2022
Accepted: 25 June 2022
Published: 28 June 2022

Publisher's Note: MDPI stays neutral with regard to jurisdictional claims in published maps and institutional affiliations.

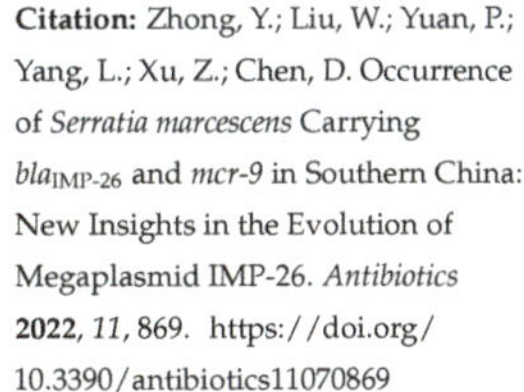

Abstract: The spread of multidrug-resistant enterobacteria strains has posed a significant concern in public health, especially when the strain harbors metallo-beta-lactamase (MBL)-encoding and mobilized colistin resistance (*mcr*) genes as such genetic components potentially mediate multidrug resistance. Here we report an IncHI2/2A plasmid carrying *bla*$_{IMP-26}$ and *mcr-9* in multidrug-resistant *Serratia marcescens* human isolates YL4. Antimicrobial susceptibility testing was performed by the broth microdilution method. According to the results, *S. marcescens* YL4 was resistant to several antimicrobials, including β-lactams, fluorquinolones, sulfanilamide, glycylcycline, and aminoglycosides, except for amikacin. To investigate the plasmid further, we conducted whole-genome sequencing and sequence analysis. As shown, *S. marcescens* YL4 possessed a circular chromosome with 5,171,477 bp length and two plasmids, pYL4.1 (321,744 bp) and pYL4.2 (46,771 bp). Importantly, sharing high similarity with plasmids pZHZJ1 and pIMP-26, pYL4.1 has an IncHI2/2A backbone holding a variable region containing *bla*$_{IMP-26}$, *mcr-9*, and two copies of *bla*$_{TEM-1B}$. After comprehensively comparing relevant plasmids, we proposed an evolutionary pathway originating from ancestor pZHZJ1. Then, via an acquisition of the *mcr-9* element and a few recombination events, this plasmid eventually evolved into pYL4.1 and pIMP-26 through two different pathways. In addition, the phage-like plasmid pYL4.2 also carried a *bla*$_{TEM-1B}$ gene. Remarkably, this study first identified a multidrug-resistant *S. marcescens* strain co-harboring *bla*$_{IMP-26}$ and *mcr-9* on a megaplasmid pYL4.1 and also included a proposed evolutionary pathway of epidemic megaplasmids carrying *bla*$_{IMP-26}$.

Keywords: *Serratia marcescens*; multidrug resistance; *mcr-9*; *bla*$_{IMP-26}$; megaplasmids IMP-26; evolutionary pathway

1. Introduction

Serratia marcescens is a bacterium of the *Enterobacteriaceae* family that thrives in damp environments such as water and soil and can survive for months on inanimate surfaces. The bacteria were thought to be nonpathogenic for a long time before the infections caused by this microorganism were confirmed [1]. Its threat to health remained unclear until the outbreak of nosocomial *S. marcescens* infections in the late 20th century [2,3]. Even though this organism exhibits a wide range of virulence factors and is relatively weak in virulence, it can infect critically ill or immunocompromised patients, as well as infants and newborns [4–6]. Researchers have pointed out that this microorganism could cause multiple infections, including meningitis, pneumonia, septicemia, and urinary tract infections, associated with poor clinical outcomes [7–10]. Due to their capability of adherence to invasive hospital equipment and forming biofilm, nosocomial infections caused by *S. marcescens* were difficult to treat [11,12]. What is more, these strains were usually resistant to ampicillin, ampicillin-sulbactam, amoxicillin, amoxicillin-clavulanate, narrow-spectrum cephalosporins, cephamycins, cefuroxime, nitrofurantoin, and colistin according to the Clinical and Laboratory Standards Institute (CLSI) guidelines. The increasing multidrug-resistant *S. marcescens* in the nosocomial environment has become a major concern. The situation worsens when this species acquires the ability to resist the last-resort antibiotics, such as carbapenems, due to the transfer of resistance genes, for example, bla_{KPC} was reported [13–15]. Now, the World Health Organization (WHO) has designated it as a research priority for developing alternative antibacterial strategies.

Beta-Lactam antibiotics have been popular as therapeutic drugs for over 70 years, which has led to an abundance of β-lactam-inactivating β-lactamases. Beta-lactamases, including plasmid-mediated extended-spectrum β-lactamases, AmpC cephalosporinases, and carbapenemases, are now present globally with some variants preferred in particular regions. A significant challenge in modern medicine has been the presence and dissemination of carbapenemases in *Enterobacteriaceae* since this hydrolase can decompose penicillins, cephalosporins, monobactams, and carbapenems [16,17]. IMP-26 was first identified and characterized as one kind of metallo-beta-lactamases (MBLs) from a *Pseudomonas aeruginosa* isolated in Singapore in 2010 [18]. The protein resembles IMP-4 but differs by one amino acid (Phe49Val). It displayed higher carbapenem hydrolysis activity toward meropenem than IMP-1 [19]. Since then, such MBLs have been reported in a variety of microbes worldwide, including *Enterobacteria* [20,21], *P. aeruginosa* [19,22], and *Klebsiella pneumoniae* [23]. As reported previously, the bla_{IMP-26} gene was found on various microbial chromosomes; only four articles ever described its location on a plasmid [24–27].

Colistin is one of the last therapeutic options for infections caused by multidrug-resistant Gram-negative bacteria [28]. However, a plasmid-mediated colistin resistance gene, *mcr-1*, was first identified in Chinese *Escherichia coli* isolates in 2016 [29]. In the years following the first description, several reports have described the emergence of *mcr-1* to *mcr-10* in different host species and geographic locations [30–32]. The *mcr-9* gene was found in a colistin-exposed *Salmonella* Typhimurium in 2019 [33]. It shared 65% and 63% amino acid identities with the closest relatives, MCR-3 and MCR-7, and between 33% and 45% with the other MCRs. Since then, the *mcr-9* gene has been identified in 40 countries across six continents [34]. These genes encode phosphoethanolamine transferase enzymes responsible for adding phosphoethanolamine to lipid A, which leads to a diminished affinity for colistin and antibiotic resistance [31].

Mobile elements are common in prokaryotic genomes and crucial for the evolution of plasmids and bacteria [35]. Typically, the evolution of a plasmid driven by transposons or IS elements usually results in structural changes through homologous recombination [36]. Porse et al. discovered that IS26 mediated a large-scale deletion of a plasmid's conjugation machinery, leading to a reduction in the plasmid fitness cost in *E. coli* hosts, improving the plasmid–host adaptation [37]. IS26 disseminated antibiotic resistance genes in two distinct ways which differ from other non-IS6 family members, disseminating antibiotic resistance genes in two specific ways. The first way is achieved through an IS26-flanked structure creating a cointegrate formation with duplication of the IS26 and generation of a target duplication. The second way is forming a non-replicating circular intermediate containing a single IS26 named a translocatable unit (TU). The IS26 in TU targets an existing copy of IS26 on the receptor's sequence and adjacent to it without increasing the number of IS26 copies or making a further duplication of the target [38,39]. Another noteworthy mobile element was Tn3 [40]. Tn3 transposons are a big and widespread transposon family allowing assembly, diversification, and redistribution of antimicrobial resistance genes, contributing to the transport of bacterial resistance genes among bacteria. They transpose in a "copy-and-paste" way in which the donor and target molecules are fused by repeated transposon copies [41].

Here, we characterized an IncHI2/2A plasmid harboring $bla_{\text{IMP-26}}$ and *mcr-9* in a multidrug-resistant *S. marcescens* which displayed resistance to carbapenems and further studied the evolutionary pathway of such epidemic megaplasmids carrying $bla_{\text{IMP-26}}$ mediated by IS26 and Tn3.

2. Results

2.1. Antimicrobial Susceptibility Profiles

MICs obtained by the broth micro-dilution method are shown in Table 1. The isolate was susceptible only to amikacin according to CLSI breakpoints of 2021 (Table 1).

Table 1. Antimicrobial susceptibility of *S. marcescens* YL4 isolate.

Antimicrobial Class	Antimicrobial Agents	MIC (µg/mL)	S	I	R
Cephalosporins	Cefepime	16	≤2	4–8	≥16
	Ceftazidime	≥64	≤4	8	≥16
	Cefatriaxone	≥64	≤1	2	≥4
β-lactam inhibitor combinations	Ticarcillin/clavulanate	≥128/2	≤16/2	32/2–64/2	≥128/2
Carbapenems	Imipenem	≥16	≤1	2	≥4
	Meropenem	≥16	≤1	2	≥4
	Ertapenem	≥8	≤0.5	1	≥2
Aminoglycosides	Tobramycin	≥16	≤4	8	≥16
	Amikacin	16	≤16	32	≥64
Fluorquinolones	Levofloxacin	4	≤0.5	1	≥2
	Ciprofloxacin	2	≤0.25	0.5	≥1
Sulfanilamides	Trimethoprim/sulfamethoxazole	≥16/304	≤2/38	-	≥4/76
Glycylcycline	Tigecycline	≥8	≤0.5	-	-
Monobactams	Aztreonam	≥64	≤4	8	≥16
Tetracyclines	Minocycline	≥16	≤4	8	≥16
	Doxycycline	≥16	≤4	8	≥16

2.2. Genome Sequencing of S. marcescens YL4

S. marcescens YL4 possessed a circular chromosome with 5,171,477 bp and two plasmids (pYL4.1 and pYL4.2) with 316,459 bp and 46,771 bp, respectively. The GC content of chromosomes and plasmids was 59.30%, 47.64%, and 53.11%. On the YL4's chromosome, 88 tRNA, 22 rRNA, 38 sRNA, and 4762 open reading frames were annotated using the Prokaryotic Genomes Annotation Pipeline server. In addition, 2 prophages, 1 CRISPR array, and 27 insertion elements were detected on the chromosome. ResFinder server analysis showed that the YL4 strain chromosome contains three resistance genes, including one beta-lactam resistance gene ($bla_{\text{SRT-1}}$), aminoglycoside resistance gene ($aac(6')\text{-}Ic$), and Tetracycline resistance gene (*tet(41)*) (Table 2).

2.3. Characteristics of the IncHI2/2A Plasmid pYL4.1

The IncHI2/2A plasmid pYL4.1's DNA sequence comprises 316,459 bp with a G+C content of 47.64%. There are two replicons on the plasmid. One was 876 bp (291,621..292,496) in size; the other was 1056 bp (227,123..278,178) in length. BLAST search showed the backbone regions of pYL4 similar to pIMP-26 (Genbank ID: MH399264) [25], pEHZJ1 (Genbank ID: CP033103) [26], pEC-IMPQ (Genbank ID: EU855788) [42], pGMI14-002 (Genbank ID: CP028197), and p505108-MDR (Genbank ID: KY978628) (Figure 1).

Table 2. Antibiotic resistance genes in *S. marcescens* YL4.

Location	Antimicrobial Agents	Resistant Genes	Identity	Alignment Length/Gene Length	Start	End
Chromosome of YL4 strain (access No. CP083754)	Beta-lactam	*bla*$_{SRT-1}$	96.22	1137/1137	775,741	776,877
	Tetracyline	*tet(41)*	93.57	1151/1182	1,054,107	1,055,257
	Aminoglycoside	*aac(6′)-Ic*	94.33	441/441	2,918,180	2,918,620
Plasmid pYL4.1 (access No. CP083755)	Aminoglycoside	*aph(6)-Id*	100.0	837/837	168,510	169,346
		aph(3″)-Ib	99.88	804/804	169,346	170,149
		aac(6′)-Ib3	100.0	555/555	241,940	242,494
	Polymyxin	*mcr-9*	100.0	1620/1620	158,061	159,680
	Fosfomycin	*fosA5*	100.0	420/420	181,153	181,572
	Macrolide	*mph(A)*	100.0	906/906	66,782	67,687
	Folate pathway antagonist	*sul1*	100.0	840/840	189,399	190,238
	Tetracycline	*dfrA19*	100.0	570/570	195,200	195,769
		tet(D)	100.0	1185/1185	70,462	71,646
	Beta-lactam	*bla*$_{TEM-1B}$	100.0	861/861	175,551	176,411
		bla$_{IMP-26}$	100.0	741/741	192,691	193,431
		bla$_{TEM-1B}$	100.0	861/861	205,714	206,574
		bla$_{SHV-12}$	100.0	861/861	245,958	246,818
	Quaternary ammonium compound	*qacE*	100.0	282/333	190,298	190,579
	Amphenicol	*catA2*	96.11	642/642	231,688	232,329
Phage-like plasmid pYL4.2 (access No. CP083756)	Beta-lactam	*bla*$_{TEM-1B}$	100.0	861/861	23,202	24,062

2.4. Gene Environments of bla$_{IMP-26}$

The genetic environment of *bla*$_{IMP-26}$ reveals that it was partitioned into the class 1 integron cassette, sequentially arranged as *sul1-qacEΔ1-ItrA-bla*$_{IMP-26}$*-Int1*. It contained a 5′-conserved segment (5′-CS) adjacent to IS6100, InsB, ISVsa5, *fosA5*, Tn3, *bla*$_{TEM-1B}$, Tn3, and the 3′-CS was abutted to *drfA19*, mobile elements (IS26, couple of ISEc63, Tn2, the last two belong to Tn3 family), *bla*$_{TEM-1B}$. The surrounding of *bla*$_{IMP-26}$ in pYL4.1 was similar to but opposite to that of pEHZJ1 from *E. hormaechei* ST1103 in Zhejiang (accession: CP033103) and pIMP26 from *E. cloacae* RJ702 in Shanghai (accession: MH399264). The three highly similar plasmids share most genetic structures outside the class 1 integron cassette, for example, insertion elements (IS26, ISEc63, IS4, IS1, Tn3) and resistance genes (*fosA5, bla*$_{TEM-1B}$). Two Tn2 transposons flank the multiple resistant regions, one on either end, and could be combined to create a composite transposon capable of moving as a single unit. Additionally, the Tn3 family transposon at the 5′-CS of pEHZJ1 was ISEc63, a transposon of the Tn3 family, but not Tn2. Moreover, we found a *bla*$_{IMP-26}$ containing plasmid pIMP1572 from *K. pneumoniae* KP-1572 (accession: MH464586) which differed from pYL4.1. The two plasmids only share the class 1 integron cassette region, and they do not share the other resistance genes and other elements. Importantly, although there is more or less a difference between the four plasmids, the *bla*$_{IMP-26}$ is always located in the class 1 integron cassette, *IntI1-bla*$_{IMP-26}$*-ORF1-qacEΔ1-sul1*, which consists of a complete 5′-conserved sequence (5′-CS, integrase *intI1*) and 3′-CS (*qacEΔ1-sul1*). It infers that class 1 integron may be significant for the transmission of *bla*$_{IMP-26}$ between plasmids (Figure 2).

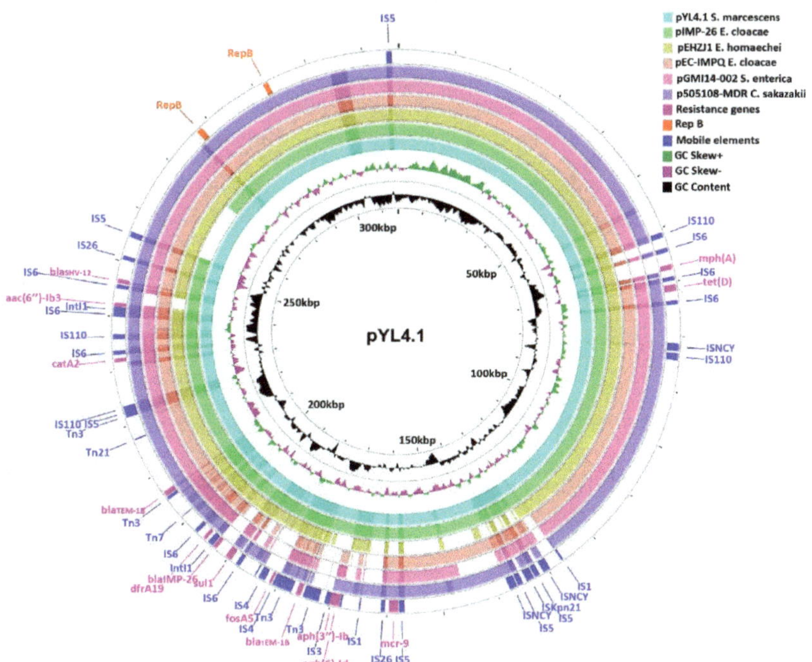

Figure 1. Genetic features of the pYL4.1 plasmid. Inner and outer circles correspond to average G+C content (black circle) and GC skew information (green and purple circles, respectively). The colored circles represent different plasmids (details are in the legend), and the Genbank numbers are as follows: pIMP26 (MH399264), pEHZJ1 (CP033103), pEC-IMPQ (EU855788), pGMI14-002 (CP028197), p505108-MDR (KY978628). An outer cyan-blue circle shows where resistance genes and *IntI* are located. The location of the resistance genes (magenta), insert elements (blue), and replication origins (orange) are also shown outside the round and tagged by different colors.

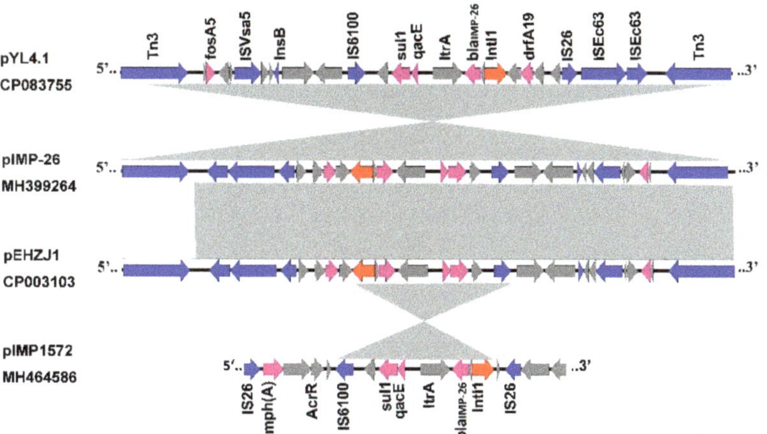

Figure 2. Genetic content of *bla*IMP-26. Comparison of the genetic environment of *bla*IMP-26 between different plasmids. Magenta arrows indicate resistance genes, and blue arrows indicate insertion sequences. Shading regions denote homologous regions (>95% nucleotide identity).

2.5. Gene Environments of Mcr-9

We compared three *mcr-9* harboring plasmids reported in the last two years with our target plasmids. The four plasmids were pIMP-26 from *Enterobacter cloacae*, p1575-1 from *E. hormaechei* (accession: CP068288) [43], pEC3 from *E. coli* (accession: MW509820) [44], and pK714029-2 from *K. pneumoniae* (accession: CP073658) [45]. Comparing *mcr-9*-possessing regions among pYL4.1, pIMP-26, and p1575-1 revealed that they shared the same areas flanked by two mobile elements (from IS*CNY* to *InsB*) with 100% coverage and 98.88% identity. In all five plasmids, the upstream sequences of *mcr-9* were highly homologous, except for pK710429.2, which contained an insertion of the *InsB*. The structure of *rcnR-rcnA-pcoE-pcoS*-IS*903-mcr-9-wbuC* is present on all four plasmids, providing further evidence that these core elements perform a significant role in the conjugation and recombination processes of *mcr-9*. There are some unique sequences evident in the downstream line. Regulatory genes *qseC* and *qseB*, crucial to colistin resistance induction, were typically absent among the four observed plasmids, excluding pEC (Figure 3). However, the *mcr-9* gene expression did not increase after pretreatment with colistin in *qseC-qseB* carrying pEC [44]. In contrast, up-regulation of the *mcr-9* gene following colistin treatment was observed in pK714029-2 [45], the plasmid lacking *qseC-qseB*. The evidence suggests that the effect of *qseC-qseB* on *mcr-9* induction might differ among isolates with diverse genetic backgrounds [46]. Some additional genes may play a prominent role in *mcr-9* stimulation [34,47]. Further studies will be necessary to verify whether the *qseC-qseB* module or the other genes are essential for *mcr-9* induction.

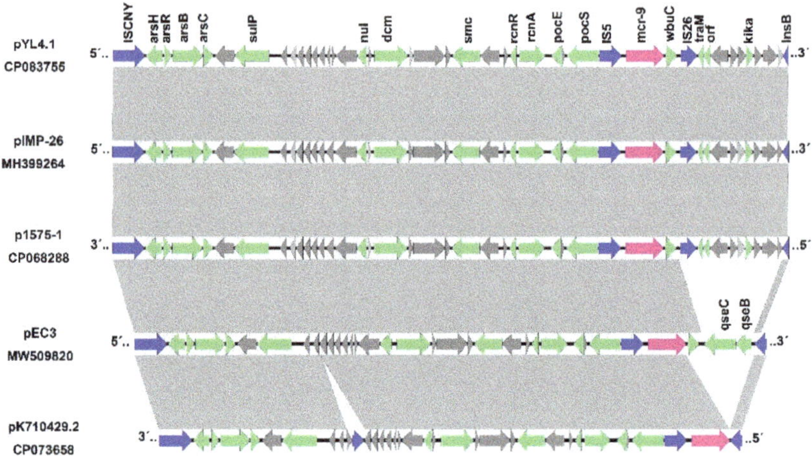

Figure 3. Comparison of the genetic environment of *mcr-9* in different plasmids. Magenta arrows indicate resistance genes, blue arrows indicate insertion sequences, green arrows indicate identified ORFs, and gray arrows represent hypothetical genes. Shaded regions denote homologous regions (>95% nucleotide identity).

2.6. Evolutionary Pathway of Megaplasmids IMP-26

We compared pYL4.1's whole sequence to the two most similar plasmids, pIMP-26 with 95% query coverage and 99.99% nucleotide identity and pEHZJ1 with 79% query coverage and 99.99% nucleotide identity. After a comprehensive comparison with all megaplasmids with *bla*$_{IMP-26}$, the evolutionary path was hypothesized as below (Figure 4). For the pEHZJ1-derived module, a deletion of a 54 kb region followed by the acquisition of the *mcr-9* harboring multi-resistant region (MRR) and another acquisition of the *IntI1* area that occurred in pEHZJ1 generated a hypothetical module A plasmid. In the second step, a 179 kb MRR flanked by IS*26* and Tn3 on module A reversed and developed a critical theoretical module B, from which pYL4 and pIMP-26 evolved in different pathways. A series of restructuring events occurred in the evolution from module B to pYL4.1. First, an IS*26* flanking MRR shear from module B formed a circle by connecting IS*26* on each end of the fragment. Second, deletion of a class 1 integron also flanked by IS*26* was conducted, creating a new component. Next, this fragment was reinserted between Tn*3* and IS*26*, located downstream of module B. The recombination was followed by a reversion of the *bla*IMP-26 harboring region backed by Tn*3*. In contrast, another evolutionary pathway from module B to pIMP-26 was simple, with a 13 kb hypothetical protein deletion behind IS*5* on the tail of module B and evolving into pIMP-26.

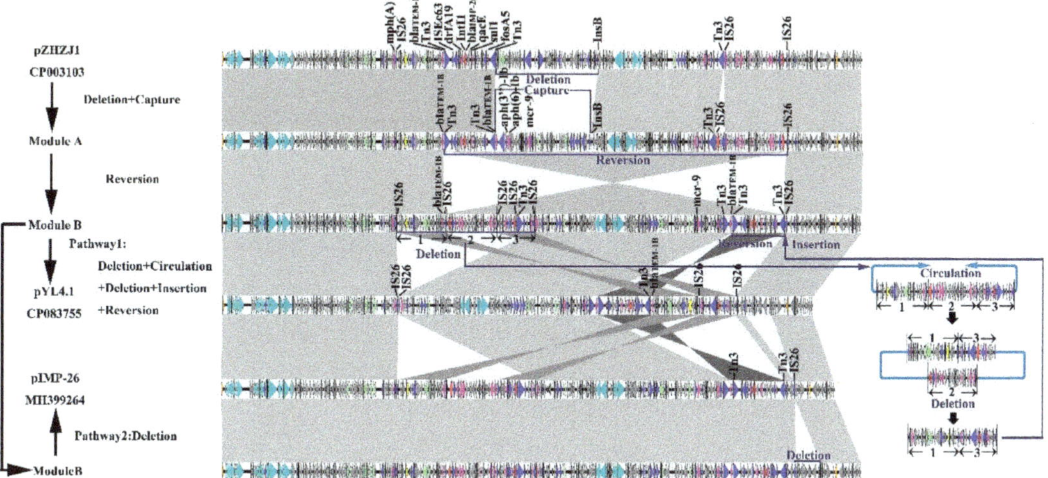

Figure 4. Evolutionary path between pEHZJ1 (CP033103), pYL4 (CP083755), and pIMP-26 (MH399264). The arrows indicate ORFs that have been confirmed or putatively confirmed and their orientations. Length of the arrow corresponds to the predicted ORF. The color code represents: replication initiation protein genes (orange), integrase genes (red), transposase genes and insertion elements (blue), antibiotic resistance genes (magenta), toxin-antitoxin system genes (yellow), heavy metal resistance genes (green), conjugal transfer genes (cray), putative, hypothetical, and other genes are represented by gray arrows. Shaded gray areas denote homologous regions with up to 95% nucleotide identity. When the shadow areas overlap, the color of the shaded region is darkened.

2.7. A Phage-Like Plasmid pYL4.2 Carrying bla*TEM-1B*

Interestingly, the *S. marcescens* YL4 harbors another small plasmid, pYL4.2, a phage-like plasmid containing a large proportion of phage-related sequences, 46,771 bp in the plasmid sequence, with a 53.10% GC content. A fragment of ~8 kb in the plasmid was identified as the genome sequence of *S. marcescens* strain CBS12 isolated from contaminated platelet concentrates obtained from a Canadian donor. No more fragments are found similar to pYL4.2. Plasmid pYL4.2 belongs to the IncFII group and contains 74 ORFs. Noteworthy, the pYL4.2 genome encodes 47 proteins related to phages, such as coat protein, tail fiber protein, portal protein, terminator, outer membrane lytic protein, and others (Table 1). It suggests that pYL4.2 is a phage-like plasmid that carries many phage-like elements. A beta-lactamase encoding gene, *bla*TEM-1B, was identified on the pYL4.2, located downstream of a complete transposon Tn*3* and the insertion elements IS*26*. No additional antibiotic resistance genes were found in pYL4.2 (Figure 5).

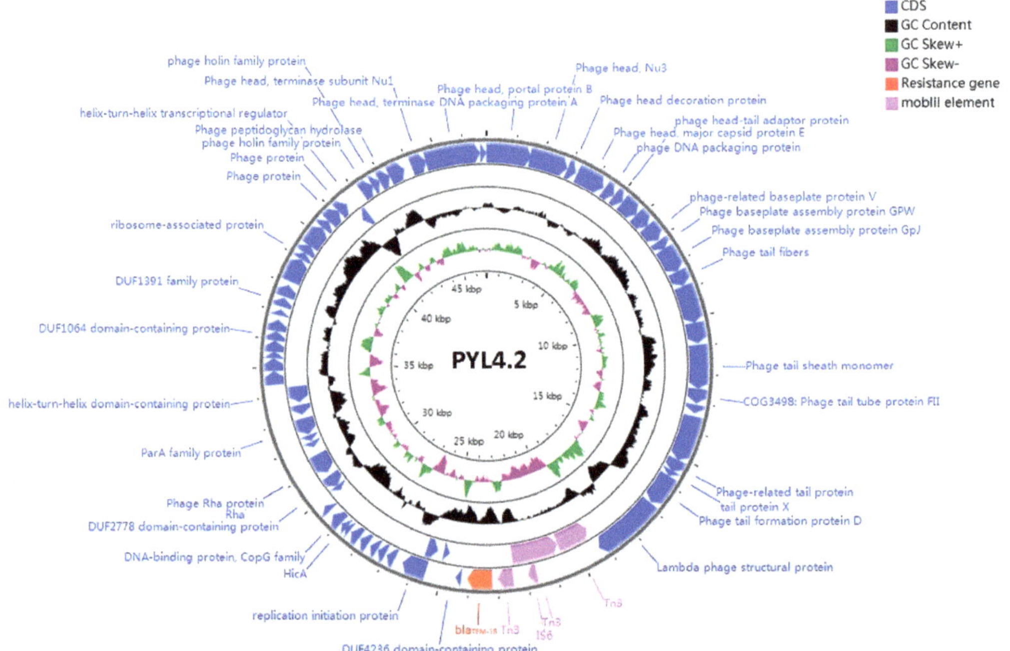

Figure 5. Structure of phage-like plasmid pYL4.2 carrying *bla*$_{TEM-1B}$ from *S. marcescens* isolate YL4. The arrows with different colors represent open reading frames (ORFs), with purple, orange, and blue representing mobile elements, resistance genes, and ORFs, respectively. Inner and outer circles correspond to average G+C content (black circle) and GC skew information (green and purple circles), respectively.

3. Discussion

S. marcescens is an opportunistic nosocomial pathogen and is intrinsically multidrug resistant, mainly due to the presence of a large number of efflux pump genes [48,49], with the ability to produce myriad extracellular enzymes as well as metabolites which make it capable of adapting to both hostile and changing environments [50–52]. In *S. marcescens*, the PhoPQ system can regulate the expression of the *arn* operon by sensing polymyxin B (PB) and Mg^{2+}, which results in the LPS being modified and leads to PB resistance [53,54]. Another important feature of *S. marcescens* is that pathogens can acquire antibiotic resistance genes from the surroundings rapidly, mainly due to the acquisition of plasmids and resistance genes [52], for example, *bla*$_{KPC}$, as reported [13–15].

Here, we found the *S. marcescens* isolate YL4 possessed three drug-resistance genes on the chromosome, including beta-lactam resistance gene (*bla*$_{SRT-1}$), aminoglycoside resistance gene (*aac(6′)-Ic*), and tetracycline resistance gene (*tet(41)*), together with two plasmids harboring a resistance gene. In addition, we found a plasmid-mediated colistin resistance gene, *mcr-9*, from a plasmid captured by the YL4 strain. This gene encodes a phosphoethanolamine transferase, contributing to narrowing the negative charge of the outer membrane of bacteria, and attenuates the affinity for colistin via the phosphoethanolamine incorporation into lipid A's phosphate group, which results in colistin resistance [31]. It was suspected that *mcr-9* gene entered the *S. marcescens* YL4 by plasmid transfer, which is unnecessary for the survival of *S. marcescens*, bacteria intrinsically resistant to colistin [52]. The phenomenon portends a worrying prospect that the *mcr-9* gene may induce colistin resistance in other non-colistin resistant bacteria by horizontal gene transfer as reported [45].

pYL4.1 is a megaplasmid, which belongs to IncHI2/2A plasmid. The entire gene analysis revealed that the plasmid pYL4.1 was a close match to two bla_{IMP-26} harboring plasmids, pIMP-26 from *E. cloacae* RJ702 in Shanghai (accession: MH399264) [25] and pEHZJ1 from *E. hormaeechei* in Zhejiang (accession: CP033103) [26] reported before. An evolutionary path from pEHZJ1 to pYL4.1 or from pEHZJ1 to pIMP-26 was hypothesized. We assumed an IS26-flanked MRR experienced a process with shear, circulation, deletion, and insertion from module B to pYL4.1. This hypothesis is built on the theory posted by Harmer CJ et al. IS26 transmits drug-resistance genes in two distinct forms, a couple of IS26s flanked cointegrate formation or single IS26 containing a translocatable unit (TU) [38]. Past reports have confirmed that IS26 is associated with the transfer of antibiotic resistance genes for its capability of forming translatable units. The translatable unit can be excised from the chromosome and reinserted into it. As a result, a tandem array is created and the number of copies of the resistant gene increases, which does not involve any fitness cost but does increase resistance to drugs [55,56]. Other deletions and reversion are closely related to IS26 and another mobile element, Tn3, a transposon using a "copy-and-paste" mechanism to transfer gene fragments [41]. It is speculated that IS26 and Tn3 participation in the plasmid reorganization from clinical strains are likely to promote the spread of resistance genes among *S. marcescens* and other *Enterobacterales*.

Furthermore, the genetic content of bla_{IMP-26} was analyzed, and the results show that the bla_{IMP-26} was closely followed by *IntI1*. The bla_{IMP-26} harboring a class 1 integron cassette, sequentially arranged as $sul1-qacE\Delta1-ItrA-bla_{IMP-26}-Int1$, were identical to the other three plasmids reported before (pEHZJ1 from *E. hormaechei* ST1103 (accession: CP033103) [26], pIMP-26 from *E. cloacae RJ702* (accession: MH399264) [25], pIMP1572 from *K. pneumoniae* KP-1572 (accession: MH464586)) [27]. It suggests that the transmission of bla_{IMP-26} between plasmids may be mediated by and dependent on a class 1 integron, a mobile genetic component responsible for the transmission of multiple drug resistance [57]. These elements are capable of capturing, mobilizing, and integrating antibiotic-resistant gene cassettes [58]. Through lateral DNA transfer, they gained access to a variety of commensal and pathogenic bacteria and subsequently accumulate diverse antibiotic resistance genes [59,60]. The bla_{IMP-26} transfer mediated by class 1 integron might lead to the increase in carbapenem-resistant isolates, which are a risk to healthcare systems.

4. Materials and Methods

4.1. Bacterial Strain and Clinical Data

A 66-year-old female was hospitalized with pneumonia in Guangzhou, China, in February 2021. An injury to the central nervous system (CNS), type II respiratory failure, and type II diabetes were observed in the patient. *S. marcescens* was isolated from sputum samples collected during this patient's hospitalization.

4.2. Bacterial Identification and Antimicrobial Susceptibility Testing

Antimicrobial susceptibility tests and biochemical identifications were performed using the Vitek 2 Compact system. The MIC values were determined using the broth micro-dilution methods. Results were interpreted according to guidelines set by the Clinical and Laboratory Standards Institute (CLSI) in 2021. The *S. marcescens* strain YL4 was used for further determinant analysis and antimicrobial element testing.

4.3. Whole-Genome Sequencing and Bioinformatics Analysis

According to the manufacturer's instructions, genomic DNA was extracted using HiPure Bacterial DNA Kits (Magen, Guangzhou, China). Our libraries were constructed from a 350 bp small fragment genomic DNA library and a 10 kb fragment library. Genomic sequencing was performed using the Illumina Novaseq 6000 and Pacific Biosciences Sequel platforms (Guangzhou Gene Denovo Bioinformatics Technology Company, Guangzhou, China) to obtain short-read data and long-read data, respectively. Hybrid assembly was performed using Falcon (version 0.3.0) for long-read de novo assembly. The filtered short reads were utilized to correct the genome sequences to improve the quality of the assembly and determine the final genome sequences using Pilon (version 1.23). Genomic sequences were annotated using RAST 2.0 (http://rast.nmpdr.org/, accessed on 4 November 2021) and BLAST (https://blast.ncbi.nlm.nih.gov/Blast.cgi, accessed on 4 November 2021) to confirm the annotation results. PlasmidFinder was used to identify the plasmid (https://cge.food.dtu.dk/services/PlasmidFinder, accessed on 4 November 2021). ISFinder (https://www-is.biotoul.fr, accessed on 4 November 2021) and ResFinder (https://cge.cbs.dtu.dk/services/ResFinder, accessed on 4 November 2021) were used to identify resistance genes and insertion elements. Prophages were predicted by PHASTER (http://phaster.ca/, accessed on 15 November 2021). Plasmid sequence alignment to the GenBank database was performed using BLASTn (https://blast.ncbi.nlm.nih.gov/Blast.cgi, accessed on 4 November 2021).

4.4. GenBank Accession Numbers

The complete genome sequence of *S. marcescens* YL4 was submitted to GenBank under the accession number CP083754 (chromosome of *S. marcescens* YL4 strain), CP083755 (Plasmid pYL4.1), CP083756 (Phage like plasmid pYL4.2).

5. Conclusions

In summary, we first identified an IncHI2/2A plasmid carrying bla_{IMP-26} and *mcr-9* in a multidrug-resistant *Serratia marcescens* human isolate. After comprehensively comparing this plasmid and other relevant similar plasmids, we proposed an evolutionary pathway originating from ancestor pZHZJ1, including the acquisition of the *mcr-9* element and a few recombination events mediated by mobile factors, which resulted in the capture and loss of some drug-resistance genes. As widely known, mobile genetic elements play a crucial part in the transmission of antibiotic resistance genes between plasmids or between plasmids and strains' chromosomes. Focusing on the evolutionary pathways between structurally similar plasmids is necessary in order to fully understand how multidrug-resistant plasmids evolve and move through microbial populations in diverse settings.

Author Contributions: Conceptualization, D.C. and Z.X.; methodology, Y.Z., W.L., L.Y. and P.Y.; writing—original draft preparation, Y.Z. and W.L.; writing—review and editing, Y.Z. and W.L.; funding acquisition, D.C. and Z.X. All authors have read and agreed to the published version of the manuscript.

Funding: This work is supported by the research grants from the Natural Science Foundation of China (No. 81974318), Natural Science Foundation of Guangdong Province (No. 2018A030310170 and 2018A030313279), Guangdong Province Science and Technology Innovation Strategy Special Fund (No. 2019B020209001), Guangdong International S&T Cooperation Programme (2021A0505030007), and the 111 Project (B17018).

Institutional Review Board Statement: Not applicable.

Informed Consent Statement: Not applicable.

Data Availability Statement: The complete genome sequence of *S. marcescens* YL4 was submitted to GenBank under the accession numbers CP083754 (chromosome of *S. marcescens* YL4 strain), CP083755 (Plasmid pYL4.1), CP083756 (Phage like plasmid pYL4.2).

Conflicts of Interest: The authors declare that there is no conflict of interest.

References

1. Hejazi, A.; Falkiner, F.R. Serratia marcescens. *J. Med. Microbiol.* **1997**, *46*, 903–912. [CrossRef] [PubMed]
2. Fleisch, F.; Zimmermann-Baer, U.; Zbinden, R.; Bischoff, G.; Arlettaz, R.; Waldvogel, K.; Nadal, D.; Ruef, C. Three consecutive outbreaks of *Serratia marcescens* in a neonatal intensive care unit. *Clin. Infect. Dis.* **2002**, *34*, 767–773. [CrossRef] [PubMed]
3. Al Jarousha, A.M.; El Qouqa, I.A.; El Jadba, A.H.; Al Afifi, A.S. An outbreak of *Serratia marcescens* septicaemia in neonatal intensive care unit in Gaza City, Palestine. *J. Hosp. Infect.* **2008**, *70*, 119–126. [CrossRef] [PubMed]
4. Daoudi, A.; Benaoui, F.; El Idrissi Slitine, N.; Soraa, N.; Rabou Maoulainine, F.M. An Outbreak of *Serratia marcescens* in a Moroccan Neonatal Intensive Care Unit. *Adv. Med.* **2018**, *2018*, 4867134. [CrossRef] [PubMed]

5. Cristina, M.L.; Sartini, M.; Spagnolo, A.M. *Serratia marcescens* Infections in Neonatal Intensive Care Units (NICUs). *Int. J. Environ. Res. Public Health* **2019**, *16*, 610. [CrossRef]

6. Redondo-Bravo, L.; Gutiérrez-González, E.; San Juan-Sanz, I.; Fernández-Jiménez, I.; Ruiz-Carrascoso, G.; Gallego-Lombardo, S.; Sánchez-García, L.; Elorza-Fernández, D.; Pellicer-Martínez, A.; Omeñaca, F.; et al. *Serratia marcescens* outbreak in a neonatology unit of a Spanish tertiary hospital: Risk factors and control measures. *Am. J. Infect. Control.* **2019**, *47*, 271–279. [CrossRef]

7. da Silva, K.E.; Rossato, L.; Jorge, S.; de Oliveira, N.R.; Kremer, F.S.; Campos, V.F.; da Silva Pinto, L.; Dellagostin, O.A.; Simionatto, S. Three challenging cases of infections by multidrug-resistant *Serratia marcescens* in patients admitted to intensive care units. *Braz. J. Microbiol.* **2021**, *52*, 1341–1345. [CrossRef]

8. Bes, T.; Nagano, D.; Martins, R.; Marchi, A.P.; Perdigao-Neto, L.; Higashino, H.; Prado, G.; Guimaraes, T.; Levin, A.S.; Costa, S. Bloodstream Infections caused by Klebsiella pneumoniae and *Serratia marcescens* isolates co-harboring NDM-1 and KPC-2. *Ann. Clin. Microbiol. Antimicrob.* **2021**, *20*, 57. [CrossRef]

9. Stewart, A.G.; Paterson, D.L.; Young, B.; Lye, D.C.; Davis, J.S.; Schneider, K.; Yilmaz, M.; Dinleyici, R.; Runnegar, N.; Henderson, A.; et al. Meropenem Versus Piperacillin-Tazobactam for Definitive Treatment of Bloodstream Infections Caused by AmpC beta-Lactamase-Producing Enterobacter spp, Citrobacter freundii, Morganella morganii, Providencia spp, or *Serratia marcescens*: A Pilot Multicenter Randomized Controlled Trial (MERINO-2). *Open Forum. Infect. Dis.* **2021**, *8*, ofab387. [CrossRef]

10. Madide, A.; Smith, J. Intracranial complications of *Serratia marcescens* infection in neonates. *S. Afr. Med. J.* **2016**, *106*, 36–38. [CrossRef]

11. Rodrigues, A.P.; Holanda, A.R.; Lustosa, G.P.; Nóbrega, S.M.; Santana, W.J.; Souza, L.B.; Coutinho, H.D. Virulence factors and resistance mechanisms of *Serratia marcescens*. A short review. *Acta Microbiol. Immunol. Hung.* **2006**, *53*, 89–93. [CrossRef] [PubMed]

12. Merkier, A.K.; Rodriguez, M.C.; Togneri, A.; Brengi, S.; Osuna, C.; Pichel, M.; Cassini, M.H.; Serratia marcescens Argentinean Collaborative, G.; Centron, D. Outbreak of a cluster with epidemic behavior due to Serratia marcescens after colistin administration in a hospital setting. *J. Clin. Microbiol.* **2013**, *51*, 2295–2302. [CrossRef] [PubMed]

13. Huang, X.; Shen, S.; Shi, Q.; Ding, L.; Wu, S.; Han, R.; Zhou, X.; Yu, H.; Hu, F. First Report of bla IMP-4 and bla SRT-2 Coproducing *Serratia marcescens* Clinical Isolate in China. *Front. Microbiol.* **2021**, *12*, 743312. [CrossRef] [PubMed]

14. Ferreira, R.L.; Rezende, G.S.; Damas, M.S.F.; Oliveira-Silva, M.; Pitondo-Silva, A.; Brito, M.C.A.; Leonardecz, E.; de Goes, F.R.; Campanini, E.B.; Malavazi, I.; et al. Characterization of KPC-Producing *Serratia marcescens* in an Intensive Care Unit of a Brazilian Tertiary Hospital. *Front. Microbiol.* **2020**, *11*, 956. [CrossRef]

15. Batah, R.; Loucif, L.; Olaitan, A.O.; Boutefnouchet, N.; Allag, H.; Rolain, J.M. Outbreak of *Serratia marcescens* Coproducing ArmA and CTX-M-15 Mediated High Levels of Resistance to Aminoglycoside and Extended-Spectrum Beta-Lactamases, Algeria. *Microb. Drug Resist.* **2015**, *21*, 470–476. [CrossRef]

16. Queenan, A.M.; Bush, K. Carbapenemases: The versatile beta-lactamases. *Clin. Microbiol. Rev.* **2007**, *20*, 440–458. [CrossRef]

17. Sawa, T.; Kooguchi, K.; Moriyama, K. Molecular diversity of extended-spectrum beta-lactamases and carbapenemases, and antimicrobial resistance. *J. Intensive Care* **2020**, *8*, 13. [CrossRef]

18. Koh, T.H.; Khoo, C.T.; Tan, T.T.; Arshad, M.A.; Ang, L.P.; Lau, L.J.; Hsu, L.Y.; Ooi, E.E. Multilocus sequence types of carbapenem-resistant Pseudomonas aeruginosa in Singapore carrying metallo-beta-lactamase genes, including the novel bla(IMP-26) gene. *J. Clin. Microbiol.* **2010**, *48*, 2563–2564. [CrossRef]

19. Tada, T.; Nhung, P.H.; Miyoshi-Akiyama, T.; Shimada, K.; Tsuchiya, M.; Phuong, D.M.; Anh, N.Q.; Ohmagari, N.; Kirikae, T. Multidrug-Resistant Sequence Type 235 Pseudomonas aeruginosa Clinical Isolates Producing IMP-26 with Increased Carbapenem-Hydrolyzing Activities in Vietnam. *Antimicrob. Agents Chemother.* **2016**, *60*, 6853–6858. [CrossRef]

20. Xia, Y.; Liang, Z.; Su, X.; Xiong, Y. Characterization of carbapenemase genes in *Enterobacteriaceae* species exhibiting decreased susceptibility to carbapenems in a university hospital in Chongqing, China. *Ann. Lab. Med.* **2012**, *32*, 270–275. [CrossRef]

21. Beyrouthy, R.; Barets, M.; Marion, E.; Dananché, C.; Dauwalder, O.; Robin, F.; Gauthier, L.; Jousset, A.; Dortet, L.; Guérin, F.; et al. Novel Enterobacter Lineage as Leading Cause of Nosocomial Outbreak Involving Carbapenemase-Producing Strains. *Emerg. Infect. Dis.* **2018**, *24*, 1505–1515. [CrossRef] [PubMed]

22. Kim, M.J.; Bae, I.K.; Jeong, S.H.; Kim, S.H.; Song, J.H.; Choi, J.Y.; Yoon, S.S.; Thamlikitkul, V.; Hsueh, P.R.; Yasin, R.M.; et al. Dissemination of metallo-β-lactamase-producing Pseudomonas aeruginosa of sequence type 235 in Asian countries. *J. Antimicrob. Chemother.* **2013**, *68*, 2820–2824. [CrossRef] [PubMed]

23. Lascols, C.; Peirano, G.; Hackel, M.; Laupland, K.B.; Pitout, J.D. Surveillance and molecular epidemiology of Klebsiella pneumoniae isolates that produce carbapenemases: First report of OXA-48-like enzymes in North America. *Antimicrob. Agents Chemother.* **2013**, *57*, 130–136. [CrossRef] [PubMed]

24. Matsumura, Y.; Peirano, G.; Bradford, P.A.; Motyl, M.R.; DeVinney, R.; Pitout, J.D.D. Genomic characterization of IMP and VIM carbapenemase-encoding transferable plasmids of *Enterobacteriaceae*. *J. Antimicrob. Chemother.* **2018**, *73*, 3034–3038. [CrossRef]

25. Wang, S.; Zhou, K.; Xiao, S.; Xie, L.; Gu, F.; Li, X.; Ni, Y.; Sun, J.; Han, L. A Multidrug Resistance Plasmid pIMP26, Carrying blaIMP-26, fosA5, blaDHA-1, and qnrB4 in Enterobacter cloacae. *Sci. Rep.* **2019**, *9*, 10212. [CrossRef]

26. Gou, J.J.; Liu, N.; Guo, L.H.; Xu, H.; Lv, T.; Yu, X.; Chen, Y.B.; Guo, X.B.; Rao, Y.T.; Zheng, B.W. Carbapenem-Resistant Enterobacter hormaechei ST1103 with IMP-26 Carbapenemase and ESBL Gene bla (SHV-178). *Infect. Drug Resist.* **2020**, *13*, 597–605. [CrossRef]

27. Yao, H.; Cheng, J.; Li, A.; Yu, R.; Zhao, W.; Qin, S.; Du, X.D. Molecular Characterization of an IncFII(k) Plasmid Co-harboring bla (IMP-26) and tet(A) Variant in a Clinical Klebsiella pneumoniae Isolate. *Front. Microbiol.* **2020**, *11*, 1610. [CrossRef]

28. Gogry, F.A.; Siddiqui, M.T.; Sultan, I.; Haq, Q.M.R. Current Update on Intrinsic and Acquired Colistin Resistance Mechanisms in Bacteria. *Front. Med.* **2021**, *8*, 677720. [CrossRef]

29. Liu, Y.Y.; Wang, Y.; Walsh, T.R.; Yi, L.X.; Zhang, R.; Spencer, J.; Doi, Y.; Tian, G.; Dong, B.; Huang, X.; et al. Emergence of plasmid-mediated colistin resistance mechanism MCR-1 in animals and human beings in China: A microbiological and molecular biological study. *Lancet Infect. Dis.* **2016**, *16*, 161–168. [CrossRef]

30. Ling, Z.; Yin, W.; Shen, Z.; Wang, Y.; Shen, J.; Walsh, T.R. Epidemiology of mobile colistin resistance genes *mcr-1* to *mcr-9*. *J. Antimicrob. Chemother.* **2020**, *75*, 3087–3095. [CrossRef]

31. Sun, J.; Zhang, H.; Liu, Y.-H.; Feng, Y. Towards Understanding MCR-like Colistin Resistance. *Trends Microbiol.* **2018**, *26*, 794–808. [CrossRef] [PubMed]

32. Wang, C.; Feng, Y.; Liu, L.; Wei, L.; Kang, M.; Zong, Z. Identification of novel mobile colistin resistance gene mcr-10. *Emerg. Microbes. Infect.* **2020**, *9*, 508–516. [CrossRef] [PubMed]

33. Carroll, L.M.; Gaballa, A.; Guldimann, C.; Sullivan, G.; Henderson, L.O.; Wiedmann, M. Identification of Novel Mobilized Colistin Resistance Gene *mcr-9* in a Multidrug-Resistant, Colistin-Susceptible Salmonella enterica Serotype Typhimurium Isolate. *mBio* **2019**, *10*, e00853-19. [CrossRef] [PubMed]

34. Li, Y.; Dai, X.; Zeng, J.; Gao, Y.; Zhang, Z.; Zhang, L. Characterization of the global distribution and diversified plasmid reservoirs of the colistin resistance gene *mcr-9*. *Sci. Rep.* **2020**, *10*, 8113. [CrossRef]

35. Romaniuk, K.; Golec, P.; Dziewit, L. Insight Into the Diversity and Possible Role of Plasmids in the Adaptation of Psychrotolerant and Metalotolerant *Arthrobacter* spp. to Extreme Antarctic Environments. *Front. Microbiol.* **2018**, *9*, 3144. [CrossRef]

36. Wein, T.; Dagan, T. Plasmid evolution. *Curr. Biol.* **2020**, *30*, R1158–R1163. [CrossRef]

37. Porse, A.; Schønning, K.; Munck, C.; Sommer, M.O. Survival and Evolution of a Large Multidrug Resistance Plasmid in New Clinical Bacterial Hosts. *Mol. Biol. Evol.* **2016**, *33*, 2860–2873. [CrossRef]

38. Harmer, C.J.; Moran, R.A.; Hall, R.M. Movement of IS26-associated antibiotic resistance genes occurs via a translocatable unit that includes a single IS26 and preferentially inserts adjacent to another IS26. *mBio* **2014**, *5*, e01801–e01814. [CrossRef]

39. Harmer, C.J.; Hall, R.M. Targeted conservative formation of cointegrates between two DNA molecules containing IS26 occurs via strand exchange at either IS end. *Mol. Microbiol.* **2017**, *106*, 409–418. [CrossRef]

40. Szuplewska, M.; Czarnecki, J.; Bartosik, D. Autonomous and non-autonomous Tn3-family transposons and their role in the evolution of mobile genetic elements. *Mob. Genet. Elem.* **2014**, *4*, 1–4. [CrossRef]

41. Nicolas, E.; Lambin, M.; Dandoy, D.; Galloy, C.; Nguyen, N.; Oger, C.A.; Hallet, B. The Tn3-family of Replicative Transposons. *Microbiol. Spectr.* **2015**, *3*, 4. [CrossRef] [PubMed]

42. Chen, Y.T.; Liao, T.L.; Liu, Y.M.; Lauderdale, T.L.; Yan, J.J.; Tsai, S.F. Mobilization of qnrB2 and ISCR1 in plasmids. *Antimicrob. Agents Chemother.* **2009**, *53*, 1235–1237. [CrossRef] [PubMed]

43. Ai, W.; Zhou, Y.; Wang, B.; Zhan, Q.; Hu, L.; Xu, Y.; Guo, Y.; Wang, L.; Yu, F.; Li, X. First Report of Coexistence of bla (SFO-1) and bla (NDM-1) β-Lactamase Genes as Well as Colistin Resistance Gene *mcr-9* in a Transferrable Plasmid of a Clinical Isolate of Enterobacter hormaechei. *Front. Microbiol.* **2021**, *12*, 676113. [CrossRef] [PubMed]

44. Simoni, S.; Mingoia, M.; Brenciani, A.; Carelli, M.; Lleò, M.M.; Malerba, G.; Vignaroli, C. First IncHI2 Plasmid Carrying *mcr-9.1*, bla(VIM-1), and Double Copies of bla(KPC-3) in a Multidrug-Resistant Escherichia coli Human Isolate. *mSphere* **2021**, *6*, e0030221. [CrossRef] [PubMed]

45. Liu, Z.; Hang, X.; Xiao, X.; Chu, W.; Li, X.; Liu, Y.; Li, X.; Zhou, Q.; Li, J. Co-occurrence of bla (NDM-1) and *mcr-9* in a Conjugative IncHI2/HI2A Plasmid From a Bloodstream Infection-Causing Carbapenem-Resistant Klebsiella pneumoniae. *Front. Microbiol.* **2021**, *12*, 756201. [CrossRef] [PubMed]

46. Tyson, G.H.; Li, C.; Hsu, C.H.; Ayers, S.; Borenstein, S.; Mukherjee, S.; Tran, T.T.; McDermott, P.F.; Zhao, S. The *mcr-9* Gene of *Salmonella* and Escherichia coli Is Not Associated with Colistin Resistance in the United States. *Antimicrob. Agents Chemother.* **2020**, *64*, e00573-20. [CrossRef]

47. Chavda, K.D.; Westblade, L.F.; Satlin, M.J.; Hemmert, A.C.; Castanheira, M.; Jenkins, S.G.; Chen, L.; Kreiswirth, B.N. First Report of bla (VIM-4)- and *mcr-9*-Coharboring Enterobacter Species Isolated from a Pediatric Patient. *mSphere* **2019**, *4*, e00629-19. [CrossRef]

48. Shirshikova, T.V.; Sierra-Bakhshi, C.G.; Kamaletdinova, L.K.; Matrosova, L.E.; Khabipova, N.N.; Evtugyn, V.G.; Khilyas, I.V.; Danilova, I.V.; Mardanova, A.M.; Sharipova, M.R.; et al. The ABC-Type Efflux Pump MacAB Is Involved in Protection of *Serratia marcescens* against Aminoglycoside Antibiotics, Polymyxins, and Oxidative Stress. *mSphere* **2021**, *6*, e00033-21. [CrossRef]

49. Dalvi, S.D.; Worobec, E.A. Gene expression analysis of the SdeAB multidrug efflux pump in antibiotic-resistant clinical isolates of *Serratia marcescens*. *Indian J. Med. Microbiol.* **2012**, *30*, 302–307. [CrossRef]

50. Sandner-Miranda, L.; Vinuesa, P.; Cravioto, A.; Morales-Espinosa, R. The Genomic Basis of Intrinsic and Acquired Antibiotic Resistance in the Genus Serratia. *Front. Microbiol.* **2018**, *9*, 828. [CrossRef]

51. Soo, P.C.; Wei, J.R.; Horng, Y.T.; Hsieh, S.C.; Ho, S.W.; Lai, H.C. Characterization of the dapA-nlpB genetic locus involved in regulation of swarming motility, cell envelope architecture, hemolysin production, and cell attachment ability in *Serratia marcescens*. *Infect. Immun.* **2005**, *73*, 6075–6084. [CrossRef] [PubMed]

52. Mahlen, S.D. Serratia infections: From military experiments to current practice. *Clin. Microbiol. Rev.* **2011**, *24*, 755–791. [CrossRef] [PubMed]

53. Lin, Q.Y.; Tsai, Y.L.; Liu, M.C.; Lin, W.C.; Hsueh, P.R.; Liaw, S.J. *Serratia marcescens* arn, a PhoP-regulated locus necessary for polymyxin B resistance. *Antimicrob. Agents Chemother.* **2014**, *58*, 5181–5190. [CrossRef]
54. Mariscotti, J.F.; Garcia Vescovi, E. *Serratia marcescens* RamA Expression Is under PhoP-Dependent Control and Modulates Lipid A-Related Gene Transcription and Antibiotic Resistance Phenotypes. *J. Bacteriol.* **2021**, *203*, e0052320. [CrossRef] [PubMed]
55. Hubbard, A.T.M.; Mason, J.; Roberts, P.; Parry, C.M.; Corless, C.; van Aartsen, J.; Howard, A.; Bulgasim, I.; Fraser, A.J.; Adams, E.R.; et al. Piperacillin/tazobactam resistance in a clinical isolate of *Escherichia coli* due to IS26-mediated amplification of blaTEM-1B. *Nat. Commun.* **2020**, *11*, 4915. [CrossRef] [PubMed]
56. Hansen, K.H.; Andreasen, M.R.; Pedersen, M.S.; Westh, H.; Jelsbak, L.; Schønning, K. Resistance to piperacillin/tazobactam in Escherichia coli resulting from extensive IS26-associated gene amplification of blaTEM-1. *J. Antimicrob. Chemother.* **2019**, *74*, 3179–3183. [CrossRef] [PubMed]
57. Mazel, D. Integrons: Agents of bacterial evolution. *Nat. Rev. Microbiol.* **2006**, *4*, 608–620. [CrossRef]
58. Harms, K.; Starikova, I.; Johnsen, P.J. Costly Class-1 integrons and the domestication of the the functional integrase. *Mob. Genet. Elem.* **2013**, *3*, e24774. [CrossRef]
59. Karimi Dehkordi, M.; Halaji, M.; Nouri, S. Prevalence of class 1 integron in Escherichia coli isolated from animal sources in Iran: A systematic review and meta-analysis. *Trop. Med. Health* **2020**, *48*, 16. [CrossRef]
60. Li, W.; Ma, J.; Sun, X.; Liu, M.; Wang, H. Antimicrobial Resistance and Molecular Characterization of Gene Cassettes from Class 1 Integrons in Escherichia coli Strains. *Microb. Drug Resist.* **2022**, *28*, 413–418. [CrossRef]

Article

NDM Production as a Dominant Feature in Carbapenem-Resistant Enterobacteriaceae Isolates from a Tertiary Care Hospital

Fakhur Uddin [1], Syed Hadi Imam [2], Saeed Khan [3], Taseer Ahmed Khan [4], Zulfiqar Ahmed [4], Muhammad Sohail [5,*], Ashraf Y. Elnaggar [6], Ahmed M. Fallatah [7] and Zeinhom M. El-Bahy [8]

[1] Jinnah Postgraduate Medical Center (JPMC), Department of Microbiology, Basic Medical Sciences Institute (BMSI), Karachi 75510, Pakistan; fakharindhar@hotmail.com
[2] Basildon University Hospital Essex, Basildon SS16 5NL, UK; dr_hadiimam@hotmail.com
[3] Department of Pathology, Dow University of Health Sciences (DUHS), Karachi 74200, Pakistan; saeed.khan@duhs.edu.pk
[4] Department of Physiology, University of Karachi, Karachi 75270, Pakistan; takhan@uok.edu.pk (T.A.K.); zuahmed@uok.edu.pk (Z.A.)
[5] Department of Microbiology, University of Karachi, Karachi 75270, Pakistan
[6] Department of Food Nutrition Science, College of Science, Taif University, P.O. Box 11099, Taif 21944, Saudi Arabia; aynaggar@tu.edu.sa
[7] Department of Chemistry, College of Science, Taif University, P.O. Box 11099, Taif 21944, Saudi Arabia; a.fallatah@tu.edu.sa
[8] Department of Chemistry, Faculty of Science, Al-Azhar University, Nasir City, Cairo 11884, Egypt; zeinelbahy@azhar.edu.eg
* Correspondence: msohail@uok.edu.pk

Citation: Uddin, F.; Imam, S.H.; Khan, S.; Khan, T.A.; Ahmed, Z.; Sohail, M.; Elnaggar, A.Y.; Fallatah, A.M.; El-Bahy, Z.M. NDM Production as a Dominant Feature in Carbapenem-Resistant Enterobacteriaceae Isolates from a Tertiary Care Hospital. *Antibiotics* **2022**, *11*, 48. https://doi.org/10.3390/antibiotics11010048

Academic Editors: Ding-Qiang Chen, Yulong Tan, Ren-You Gan, Zhenbo Xu, Junyan Liu and Michele Bartoletti

Received: 18 November 2021
Accepted: 27 December 2021
Published: 31 December 2021

Publisher's Note: MDPI stays neutral with regard to jurisdictional claims in published maps and institutional affiliations.

Abstract: The worldwide spread and increasing prevalence of carbapenem-resistant Enterobacteriaceae (CRE) is of utmost concern and a problem for public health. This resistance is mainly conferred by carbapenemase production. Such strains are a potential source of outbreaks in healthcare settings and are associated with high rates of morbidity and mortality. In this study, we aimed to determine the dominance of NDM-producing Enterobacteriaceae at a teaching hospital in Karachi. A total of 238 Enterobacteriaceae isolates were collected from patients admitted to Jinnah Postgraduate Medical Centre (Unit 4) in Karachi, Pakistan, a tertiary care hospital. Phenotypic and genotypic methods were used for detection of metallo-β-lactamase. Out of 238 isolates, 52 (21.8%) were CRE and 50 isolates were carbapenemase producers, as determined by the CARBA NP test; two isolates were found negative for carbapenemase production by CARB NP and PCR. Four carbapenemase-producing isolates phenotypically appeared negative for metallo-β-lactamase (MBL). Of the 52 CRE isolates, 46 (88.46%) were bla_{NDM} positive. Most of the NDM producers were *Klebsiella pneumoniae*, followed by *Enterobacter cloacae* and *Escherichia coli*. In all the NDM-positive isolates, the bla_{NDM} gene was found on plasmid. These isolates were found negative for the VIM and IPM MBLs. All the CRE and carbapenem-sensitive isolates were sensitive to colistin. It is concluded that the NDM is the main resistance mechanism against carbapenems and is dominant in this region.

Keywords: carbapenem-resistant Enterobacteriaceae; metallo-β-lactamases; NDM producers

1. Introduction

Many Enterobacteriaceae species are the pathogens involved in hospital-associated and community-acquired infections, especially in urinary and respiratory tracts, the blood stream, and intra-abdominal and surgical sites [1]. The most commonly encountered pathogens of the family Enterobacteriaceae are *Escherichia coli*, *Klebsiella pneumoniae*, *Proteus*, *Salmonella*, *Shigella* and *Enterobacter* spp. These genera are reportedly very susceptible to carbapenems [2]. Hence, carbapenems are considered as a good option to treat the infections caused by extended-spectrum β-lactamase (ESBL) producing strains and other multidrug-resistant (MDR) bacteria [3,4]. Owing to the emergence (up to 32%) of carbapenem-resistant Enterobacteriaceae (CRE) and their inclusion in the list of priority pathogens, they have received attention globally [4–9]. CRE strains can also spread resistance markers by horizontal transfer to other strains in hospitals and augment problems in healthcare sectors [4]. In the case of bloodstream infections, CRE significantly increase the mortality rate up to 65.4% in comparison to carbapenemase-susceptible Enterobacteriaceae pathogens [8]. CRE strains have a common mechanism of resistance against carbapenem antibiotics by producing carbapenemases [4].

Many types and subtypes of carbapenemases (bla_{IMP}, bla_{VIM}, bla_{SIM}, bla_{SPM}, bla_{GIM}, bla_{KPC}, bla_{SME}) have been recognized among Enterobacteriaceae; the arrival of NDM-1 is the 'final straw' in this increasing antimicrobial resistance problem [10]. Rapid and easy determination of carbapenemases in Enterobacteriaceae is needed for effective clinical practices and infection control measures. Different phenotypic approaches are employed for the diagnosis of CPE, including the modified Hodge test (MHT), Blue-Carba test and Rapidec CARBA NP test. The Rapidec CARBA NP test is a biochemical assay with a good sensitivity and specificity, but it cannot differentiate among types of carbapenemases [7,11]. This test is also recommended by the Clinical and Laboratory Standards Institute (CLSI) [12]. Likewise, another inhibitory test, EDTA with imipenem or meropenem, is a phenotypic test for the detection of metallo-β-lactamases in Enterobacteriaceae. Both tests are cost-effective and are feasible for the screening of carbapenemases and metallo-β-lactamases prior to performing costly and sophisticated genotypic tests in low-income countries.

This cross-sectional, single-center study was conducted on the admitted patients in the medical unit of Jinnah Postgraduate Medical Centre in Karachi, Pakistan. The different clinical specimens were collected according to the site of infection. Traditional microbiological techniques and some cost-effective assays were used for the detection of the carbapenemases, with specific emphasis on metallo-β-lactamases. The goal of this research was three-fold: to monitor antimicrobial resistance among the members of Enterobacteriaceae, to detect the presence of CRE and to study the prevalence of metallo-β-lactamases producing Enterobacteriaceae by phenotypic and genotypic methods.

2. Results

Among 238 Enterobacteriaceae isolates, *E. cloacae* (40; 16.80%), *K. pneumonia* (69; 28.99%) and *E. coli* (84; 35.29%) were the most common pathogens (Table 1). *Klebsiella aerogenes*, *K. oxytoca*, *P. mirabilis*, *P. vulgaris* and *S. typhi* were isolated less frequently. Out of 238 isolates, 52 (21.84%) were CRE. The carbapenem resistance was higher in *Klebsiella aerogenes* (3; 23.1%), *K. pneumoniae* (20; 28.9%), *E. cloacae* (9; 22.5%) and *E. coli* (18; 21.4%) in comparison to other species of Enterobacteriaceae (Table 2). The isolates of different species are small in number, and for the validation of these results, the large scale studies are required. The resistance against the ampicillin, cefazolin and cefuroxime was 90–100%. A large number of isolates (46.2% to 83.3%) exhibited resistance against β-lactamase inhibitors and cephalosporin. The resistance to aminoglycosides was 45–71% (Figure 1), and all the isolates were susceptible to colistin, except the *Proteus* spp. having intrinsic resistance. Out the 52 CRE isolates, 46 (88.5%) found carbapenemase producers by the phenotypic colorimetric assay, Rapedic CARBA NP. The metallo-β-lactamase detection was determined by the phenotypic inhibitor based EDTA+ IMP and MEM discs. A total of 43 (82.7%) showed metallo-β-lactamase producers using this method. For the confirmation by the PCR assay, 41 (78.8%) were positive for bla_{NDM}. The bla_{VIM} and bla_{IMP} could not be detected in any isolate, and out of 52 CRE, 11(21%) isolates were negative for bla_{NDM}, bla_{VIM} and bla_{IMP}. The bla_{NDM} positive isolates were analyzed for the plasmid extraction, and PCR was performed; all the isolates showed the presence of bla_{NDM} on plasmids. The conjugation results revealed that bla_{NDM} was transferred to *E. coli* J53 recipient successfully from 38 bla_{NDM} positive isolates, and 03 *E. coli* isolates failed to show the transmissibility by conjugation on several attempts.

Table 1. Proportion of different species of Enterobacteriaceae (*n* = 238).

Species	Number (%) of CR *	Number (%) of CS **	Total No. of Isolates (%)
E. coli	18 (21.42)	66 (78.57)	84 (35.29)
K. pneumoniae	20 (28.98)	49 (71.01)	69 (28.99)
Enterobacter cloacae	9 (22.50)	31 (77.50)	40 (16.80)
Klebsiella aerogenes	3 (23.07)	10 (76.92)	13 (5.46)
S. typhi	-	10 (100)	10 (4.20)
P. mirabilis	2 (20.0)	8 (80.00)	10 (4.20)
K. oxytoca	-	6 (100)	6 (2.52)
P. vulgaris	-	4 (100)	4 (1.68)
S. marcescens	-	2 (100)	2 (0.84)
Total	52 (21.84)	186 (78.15)	238 (100)

* CR, carbapenem-resistant isolate (resistant to both imipenem and meropenem); ** CS, carbapenem susceptible.

Table 2. Phenotypic detection of carbapenemases by Rapedic CARBA NP test, MBL detection by EDTA synergy with carbapenems (double disc diffusion test, (DDST)) and carbapenemases by PCR in CRE isolates (*n* = 52).

Carbapenem-Resistant (CR) Species	Rapedic CARBA NP Positive No. (%), *n* = 52	MBL Positive No. (%), *n* = 52	PCR bla_{NDM} No. (%), *n* = 52
E. coli	16 (30.76)	15 (28.84)	15 (28.84)
K. pneumoniae	19 (36.53)	17 (32.69)	16 (30.76)
E. cloacae	8 (15.38)	8 (15.38)	7 (13.46)
Klebsiella aerogenes	3 (5.76)	3 (5.76)	3 (5.76)
Total No. (%)	46 (88.46)	43 (82.69)	41 (78.84)

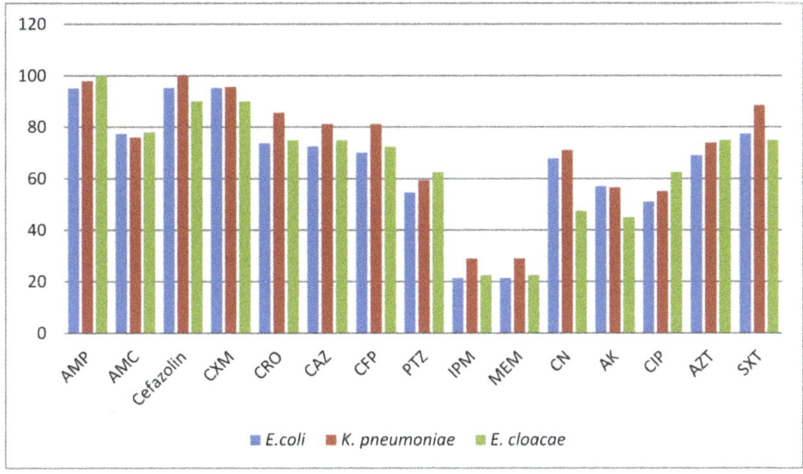

Figure 1. Resistance pattern of major isolated pathogens *Escherichia coli*, *Klebsiella pneumoniae* and *Enterobacter cloacae*. AMP: ampicillin, AMC: amoxicillin/clavulanic acid, CXM: cefuroxime, CRO: ceftriaxone, CAZ: ceftazidime, CFP: Cefepime, PTZ: piperacillin/tazobatam, IPM: imipenem, MEM: meropenem, CN: gentamicin, AK: amikacin, CIP: ciprofloxacin, AZT: aztreonam, SXT: sulfamethoxazole-trimethoprim.

3. Discussion

The incidence of infection by the carbapenem-resistant Enterobacteriaceae (CRE) is increasing worldwide and poses a threat to public health and a challenge for physicians [2]. Among CREs, *K. pneumoniae* is a leading pathogen, followed by *E. coli* and *E. cloacae* [13]. The common mode for carbapenem resistance in Enterobacteriaceae is the production of carbapenemases, particularly New Delhi metallo-β-lactamase (NDM-1), on the Indian subcontinent [14]. The carbapenem resistance and carbapenemase production in Enterobacteriaceae vary geographically with respect to prevalence, pathogen and type of carbapenemase, so it is necessary to continuously monitor the presence and prevalence of these bugs [15]. Although, *K. pneumoniae* Carbapenemase-2 (KPC-2) has been found dominantly in China, NDM is the most frequently detected (71.4%) metallo-β-lactamase in Hunan among CRE isolates, as reported by Chinese network for CRE surveillance [13]. The prevalence of carbapenem resistance in Enterobacteriaceae isolates was 21.84% in the present study. The resistance to other classes of β-lactams, including cephalosporins, penicillins and aztreonam, was higher (61–100%) in comparison to ciprofloxacin and aminoglycosides. This higher resistance to penicillins and cephalosporins may be attributed to the higher prevalence (60–80%) of ESBLs, modification in the outer membrane porins or AmpC overexpression or other mechanisms as deciphered for the isolates from Asia [16].

Comparing carbapenem resistance among the members of Enterobacteriaceae, most of the *K. pneumoniae* isolates appeared resistant, followed by *Enterobacter* spp. and *E. coli*. In the majority of CRE isolates (88.5%), carbapenemase production was the main mode of resistance to carbapenems by the phenotypic Rapidec CARBA NP test. This is a colorimetric, phenotypic test that is easy to perform without any special requirements. A simple color change can be read by the technician, so it does not require highly skilled or specially trained staff. This provides rapid identification of carbapenemase-producing strains within 30 min to 2 h, at a lower cost in comparison to the molecular assays [17]. This is helpful for the screening of carbapenemase-producing isolates, especially in low- and middle-income countries, including Pakistan, where molecular assays are not very common. The prevalence of metallo-β-lactamases was higher (82.69%) in CRE by phenotypic assay in the present study. These results were in accordance with the PCR results; therefore, this cost-effective technique may be used to screen metallo-β-lactamases in CRE isolates and provide information regarding the selection of therapeutic options.

NDM production is the main mode of resistance against the carbapenems in enterobactericeae, as found in this study. NDM was initially reported in *K. pneumoniae* and *E. coli* strains isolated from a Swedish patient, having history of seeking medical care in New Delhi, India, in 2009. It has since been spread all over the world and has been detected in different species of Enterobacteriaceae as well as in other Gram-negative bacilli. Unlike in North America and Europe, NDM, IMP and VIM are the most common carbapenemases in CRE in Southeast Asia and were similarly observed in the present work in Karachi, Pakistan. The previous data revealed that the NDM is endemic in Pakistan, Bangladesh and India, while KPCs are endemic in Colombia, Brazil, Argentina and USA [7,13]. Surveillance reports from India and neighboring countries summarized that, in Enterobacteriaceae, the most predominant carbapenemase is NDM [7]. These findings coincide with the present work. The higher prevalence of NDM producers in this study revealed that the NDM-harboring isolates are dominant in this region. A previous study from Pakistan reported that the major carbapenemase among the carbapenemase-producing Enterobacteriaceae is NDM [18].

The results of the plasmid DNA extraction and conjugation assay indicated that the bla_{NDM} is plasmid mediated, although a more sensitive assay needs to be performed to confirm this conclusion. Similar findings were reported in an earlier report [15]. In the present research work, the bla_{VIM} and bla_{IPM} could not be detected in the CRE isolates. However, these are reported in the members of Enterobacteriaceae isolates in the regions and health care settings where bla_{VIM} and bla_{IPM} carrying *P. aeruginosa* and other glucose non-fermenter Gram-negative bacilli are common [19,20]. In the present work six CRE isolates were negative for carbapenemase production by the Rapidec CARBA NP test, and 11 isolates were negative for bla_{NDM}, bla_{IPM} and bla_{VIM} by PCR. These isolates may carry any AmpC, KPC or OXA carbapenemses in combination with other carbapenem-resistant mechanisms, including overexpression of efflux pumps and the decreased permeability to carbapenems.

4. Materials and Methods

4.1. Setting

The study was conducted in the Department of Microbiology, Basic Medical Sciences Institute (BMSI) and Medical Unit-4 of Jinnah Postgraduate Medical Centre (JPMC) Karachi, Pakistan, in collaboration with the department of Microbiology and the department of Physiology, University of Karachi.

4.2. Sample Size and Collection

A total of 238 Enterobacteriaceae isolates were included in this study. The clinical specimens were collected from January to June 2018 from medical Unit-4 (Department of Medicine and Critical Care), which comprises a general ward and an ICU. The samples were collected from the urine patients suspecting of having urinary tract infections, from respiratory secretions including endotracheal tubes, tracheal aspirates and sputum for respiratory tract infections, and from blood in the case of septicemia. One isolate per patient was included, and the repeated samples of the same patient were excluded in the present study according to the adjusted criteria.

Inclusion criteria: During this period all the admitted patients presenting urinary tract infection, hospital acquired pneumonia and bacteremia were included.

Exclusion criteria: Isolates other than Enterobacteriaceae and growth-negative specimens were excluded.

4.3. Identification of the Isolates

Bacterial isolates were identified by the routine techniques, which comprised cultural and morphological features and a battery of biochemical and motility tests. The species identification was further confirmed by API 20E (BioMérieux, Lyon, France). The antimicrobial susceptibility testing (AST) was determined by the disc diffusion technique following the CLSI recommendations and protocol [12]. Colistin susceptibility testing was performed by the broth microdilution method [12]. The quality control strains included *E. coli* (ATCC 25922) for AST and *Pseudomonas aeruginosa* (ATCC 27853) for carbapenem resistance. The break points for imipenem (IPM) and meropenem (MEM) against Enterobacteriaceae were interpreted as sensitive, intermediate and resistant at the MICs of ≤ 1 μg mL^{-1}, 2 μg mL^{-1} and ≥ 4 μg mL^{-1}, respectively, as recommended by the CLSI [12]. The MICs were performed by the Etest strip (BioMérieux, Lyon, France). The Rapedic CARBA NP (RCNP) kit was used for the phenotypic detection of carbapenemases. The test was performed according to the manufacturer's instructions and standard operating procedures (BioMérieux, Lyon, France). The metallo-β-lactamases were phenotypically screened and detected by the double disc diffusion method using EDTA as a metallo-β-lactamase inhibitor and imipenem (IPM, 10 μg) and meropenem (MEM, 10 μg) disc, (Oxoid, Basingstoke, Hampshire, UK). After lawn formation on Muller Hinton agar (MHA) surface, IPM and MEM discs were placed 30 mm apart from one another, and a filter paper disc having 10 μL of 0.5 M EDTA solution was placed at the center of both discs. Inoculated plates were incubated at 37 °C overnight [21]. The synergistic effect of EDTA and carbapenems against metallo-β-lactamase producers appeared when the zone of inhibition due to the carbapenem discs increased with EDTA-containing discs.

4.4. Manual PCR Method

The RCNP positive isolates were selected to analyze metallo-β-lactamase production by targeting common genes, including *bla*$_{VIM}$ (F-GGTGTTTGGTCGCATATCGC R-CCATTCAGCCAGATCGGCATC), *bla*$_{NDM}$ (F-CACCTCATGTTTGAATTCGCC R- CTCTGT CACATCGAAATCGC) and *bla*$_{IPM}$ (F-GGAATAGAGTGGCTTAATTC R-CAACCAGTTTTG CCTTACC), with 503bp, 984bp and 327bp, respectively. PCR reaction (25 μL) was prepared in according to the master mix protocol (Promega, Madison, WI, USA). The PCR conditions were maintained as previously described [19,22].

4.5. Plasmid Extraction and Conjugation

The GeneJET plasmid miniprep kit (Thermo Scientific™, Waltham, MA, USA, #K0502) was used for the plasmid extraction, according to the manufacturer's instructions and the protocol for amplification of bla_{VIM}, bla_{IMP} and bla_{NDM}. The PCR mixture and conditions were the same as for the whole DNA sample amplification. Conjugation was performed as described by Borgia et al. [23]. The PCR positive bla_{NDM} CRE strains were used as the donors with the recipient *E. coli* J53 (a sodium azide resistant) strain. The transconjugants were selected by inoculating a mixture of donor and recipient fresh cultures on Mueller-Hinton (MH) agar containing sodium azide (100 μg mL^{-1}) and imipenem (1 μg mL^{-1}). The bla_{NDM} gene was detected in transconjugants by using PCR, with the same primers and conditions used for the whole genome DNA and plasmid of clinical isolates.

5. Conclusions

In conclusion, this study supports the dominance of NDM in this setting and supports continuous monitoring to control outbreaks and infection mitigating measures against the spread of these bugs.

Author Contributions: Conceptualization, F.U., M.S., S.H.I., S.K. and A.Y.E.; methodology, F.U., M.S., A.M.F., T.A.K. and Z.A.; software, Z.M.E.-B. and S.H.I.; validation, M.S., A.Y.E., S.K., Z.A. and A.M.F.; formal analysis, A.M.F., T.A.K. and Z.A.; investigation, F.U., M.S., T.A.K. and Z.A.; resources, A.M.F., Z.M.E.-B. and S.H.I.; data curation, F.U., M.S. and S.H.I.; writing—original draft preparation, F.U. and M.S.; writing—review and editing, all authors; supervision, M.S., S.H.I. and S.K.; project administration, A.Y.E., A.M.F. and Z.M.E.-B.; funding acquisition, A.Y.E., A.M.F. and Z.M.E.-B. All authors have read and agreed to the published version of the manuscript.

Funding: This research was funded by Taif research Supporting Project (TURSP-2020/46) Taif University Taif, Saudi Arabia.

Institutional Review Board Statement: Informed consent was obtained from the patient or attendant, and ethical approval was received from the institutional review board (IRB), JPMC, Karachi (No. F.2-18/2014-GENL/31649577/JPMC). This work was the extended part of research conducted on "MBLs in *Pseudomonas* and *Acinetobacter*".

Informed Consent Statement: Informed consent was obtained from all subjects involved in the study.

Data Availability Statement: Data associated with this work are given in the manuscript. Raw data can be obtained from the corresponding author upon a reasonable request.

Acknowledgments: The authors acknowledge Taif University Research Supporting Project (TURSP-2020/46) Taif University Taif, Saudi Arabia.

Conflicts of Interest: The authors declare no conflict of interest.

References

1. Wong, M.H.; Chan, E.W.; Chen, S. Isolation of carbapenem-resistant *Pseudomonas* spp. from food. *J. Glob. Antimicrob. Resist.* **2015**, *5*, 109–114. [CrossRef]
2. Doi, Y. Treatment options for carbapenem-resistant gram-negative bacterial infections treatment options for carbapenem-resistant gram-negative bacterial infections. *Clin. Infect. Dis.* **2018**, *69*, S565–S575. [CrossRef] [PubMed]
3. Marco, C.; Alberto, A.; Tommaso, G.; Teresa, S.; Marzia, L.F.; Carla, F.; Mirandola, W.; Gargiulo, R.; Barozzi, A.; Mauri, C.; et al. Multicenter evaluation of the RAPIDEC! CARBA NP test for rapid screening of carbapenemase-producing Enterobacteriaceae and Gram-negative nonfermenters from clinical specimens. *Diagn. Microbiol. Infect. Dis.* **2017**, *88*, 207–213. [CrossRef]
4. Nordmann, P.; Poirel, L.; Walsh, T.R.; Livermore, D.M. The emerging NDM carbapenemases. *Trends. Microbiol.* **2011**, *19*, 588–595. [CrossRef]
5. Rossolini, G.M. Extensively drug-resistant carbapenemase-producing Enterobacteriaceae: An emerging challenge for clinicians and healthcare systems. *J. Intern. Med.* **2015**, *277*, 528–531. [CrossRef] [PubMed]
6. Suwantarat, N.; Carroll, K.C. Epidemiology and molecular characterization of multidrug-resistant Gram-negative bacteria in Southeast Asia. *Antimicrob. Resist. Infect. Control* **2016**, *4*, 15. [CrossRef]
7. Nordmann, P.; Poirel, L. Epidemiology and diagnostics of carbapenem resistance in gram-negative bacteria. *Clin. Infect. Dis.* **2019**, *69*, S521–S528. [CrossRef]

8. Logan, L.K.; Weinstein, R.A. The Epidemiology of carbapenem-resistant Enterobacteriaceae: The impact and evolution of a global menace. *J. Infect. Dis.* **2017**, *215*, S28–S36. [CrossRef]
9. Li, X.; Ye, H. Clinical and mortality risk factors in bloodstream infections with carbapenem-resistant Enterobacteriaceae. *Can. J. Infect. Dis. Med. Microbiol.* **2017**, 6212910. [CrossRef]
10. Mitra, S.; Mukherjee, S.; Naha, S.; Chattopadhyay, P.; Dutta, S.; Basu, S. Evaluation of co-transfer of plasmid-mediated fluoro-quinolone resistance genes and *bla*$_{NDM}$ gene in Enterobacteriaceae causing neonatal septicaemia. *Antimicrob. Resist. Infect. Control* **2019**, *8*. [CrossRef]
11. Rao, M.R.; Chandrashaker, P.; Mahale, R.P.; Shivappa, S.G.; Gowda, R.S.; Chitharagi, V.B. Detection of carbapenemase production in Enterobacteriaceae and *Pseudomonas* species by carbapenemase Nordmann–Poirel test. *J. Lab. Physicians* **2019**, *11*, 107–110. [CrossRef]
12. Clinical & Laboratory Standards Institute (CLSI). *Performance Standards for Antimicrobial Susceptibility Testing*; 28th Informational Supplement. M100-S28; CLSI: Wayne, PA, USA, 2018.
13. Zhang, Y.; Wang, Q.; Yin, Y.; Chen, H.; Jin, L.; Gu, B.; Xie, L.; Yang, C.; Ma, X.; Li, H.; et al. Epidemiology of carbapenem-resistant Enterobacteriaceae infections: Report from the China CRE Network. *Antimicrob. Agents Chemother.* **2018**, *62*, e01882-17. [CrossRef]
14. Dortet, L.; Poirel, L.; Nordmann, P. Worldwide dissemination of the NDM-type carbapenemases in Gram-negative bacteria. *BioMed Res. Int.* **2014**, *2014*, 249856. [CrossRef]
15. Yang, Q.; Fang, L.; Fu, Y.; Du, X.; Shen, Y.; Yu, Y. Dissemination of NDM-1-producing Enterobacteriaceae mediated by the IncX3-type plasmid. *PLoS ONE* **2012**, *5*, e0129454. [CrossRef]
16. Abrar, S.; Hussain, S.; Khan, R.A.; Ul-Ain, N.; Haider, H.; Riaz, S. Prevalence of extended-spectrum-β-lactamase-producing Enterobacteriaceae: First systematic meta-analysis report from Pakistan. *Antimicrob. Resist. Infect. Control* **2018**, *7*, 26. [CrossRef]
17. Garg, A.; Garg, J.; Upadhyay, G.C.; Agarwal, A.; Bhattacharjee, A. Evaluation of the Rapidec Carba NP test kit for detection of carbapenemase-producing Gram-Negative Bacteria. *Antimicrob. Agents Chemother.* **2015**, *59*, 7870–7872. [CrossRef]
18. Masseron, A.; Poirel, L.; Ali, B.J.; Syed, M.A.; Nordmann, P. Molecular characterization of multidrug-resistance in Gram-negative bacteria from the Peshawar teaching hospital, Pakistan. *New Microbes New Infect.* **2019**, *32*, 100605. [CrossRef]
19. Kaase, M.; Nordmann, P.; Wichelhaus, T.A.; Gatermann, S.G.; Bonnin, R.A.; Poirel, L. NDM-2 carbapenemase in Acinetobacter baumannii from Egypt. *J. Antimicrob. Chemother.* **2011**, *66*, 1260–1262. [CrossRef]
20. Manenzhe, R.I.; Zar, H.J.; Nicol, M.P.; Kaba, M. The spread of carbapenemase-producing bacteria in Africa: A systematic review. *J. Antimicrob. Chemother.* **2015**, *70*, 23–40. [CrossRef]
21. Pandya, N.P.; Prajapati, S.B.; Mehta, S.J.; Kikani, K.M.; Joshi, P.J. Evaluation of various methods for detection of metallo-β-lactamase (MBL) production in gram-negative bacilli. *Int. J. Biol. Med. Res.* **2011**, *2*, 775–777.
22. Wolter, D.J.; Khalaf, N.; Robledo, I.E.; Vázquez, G.J.; Santé, M.I.; Aquino, E.E.; Goering, R.V.; Hanson, N.D. Surveillance of carbapenem-resistant Pseudomonas aeruginosa isolates from Puerto Rican Medical Center Hospitals: Dissemination of KPC and IMP-18 ß-lactamases. *Antimicrob. Agents Chemother.* **2009**, *53*, 1660–1664. [CrossRef]
23. Borgia, S.; Lastovetska, O.; Richardson, D.; Eshaghi, A.; Xiong, J.; Chung, C.; Baqi, M.; McGeer, A.; Ricci, G.; Sawicki, R.; et al. Outbreak of carbapenem-resistant Enterobacteriaceae containing blaNDM-1, Ontario, Canada. *Clin. Infect. Dis.* **2012**, *55*, e109–e117. [CrossRef]

antibiotics

Article

The 30-Day Economic Burden of Newly Diagnosed Complicated Urinary Tract Infections in Medicare Fee-for-Service Patients Who Resided in the Community

Thomas P. Lodise [1,*], **Michael Nowak** [2] **and Mauricio Rodriguez** [2]

1   Albany College of Pharmacy and Health Sciences, Albany, NY 12208, USA
2   Spero Therapeutics, Inc., Cambridge, MA 02139, USA; mnowak@sperotherapeutics.com (M.N.); mrodriguez@sperotherapeutics.com (M.R.)
*   Correspondence: thomas.lodise@acphs.edu; Tel.: +1-518-694-7292

Abstract: Introduction: Scant data are available on the 30-day financial burden associated with incident complicated urinary tract infections (cUTIs) in a cohort of predominately elderly patients. This study sought to examine total and cUTI-related 30-day Medicare spending (MS), a proxy for healthcare costs, among Medicare fee-for-service (FFS) beneficiaries who resided in the community with newly diagnosed cUTIs. Methods: A retrospective multicenter cohort study of adult beneficiaries in the Medicare FFS database with a cUTI between 2017 and 2018 was performed. Patients were included if they were enrolled in Medicare FFS and Medicare Part D from 2016 to 2019, had a cUTI first diagnosis in 2017–2018, no evidence of any UTI diagnoses in 2016, and residence in the community between 2016 and 2018. Results: During the study period, 723,324 cases occurred in Medicare beneficiaries who met the study criteria. Overall and cUTI-related 30-day MS were $7.6 and $4.5 billion, respectively. The average overall and cUTI-related 30-day MS per beneficiary were $10,527 and $6181, respectively. The major driver of cUTI-related 30-day MS was acute care hospitalizations ($3.2 billion) and the average overall and cUTI-related 30-day MS per hospitalizations were $16,431 and $15,438, respectively. Conclusion: Overall 30-day MS for Medicare FSS patients who resided in the community with incident cUTIs was substantial, with cUTI-related MS accounting for 59%. As the major driver of cUTI-related 30-day MS was acute care hospitalizations, healthcare systems should develop well-defined criteria for hospital admissions that aim to avert hospitalizations in clinically stable patients and expedite the transition of patients to the outpatient setting to complete their care.

Keywords: complicated urinary tract infections; outcomes; costs; burden of illness; Medicare

Citation: Lodise, T.P.; Nowak, M.; Rodriguez, M. The 30-Day Economic Burden of Newly Diagnosed Complicated Urinary Tract Infections in Medicare Fee-for-Service Patients Who Resided in the Community. *Antibiotics* **2022**, *11*, 578. https:// doi.org/10.3390/antibiotics11050578

Academic Editors: Ding-Qiang Chen, Yulong Tan, Ren-You Gan, Guanggang Qu, Zhenbo Xu and Junyan Liu

Received: 23 March 2022
Accepted: 20 April 2022
Published: 26 April 2022

Publisher's Note: MDPI stays neutral with regard to jurisdictional claims in published maps and institutional affiliations.

1. Introduction

Complicated urinary tract infections (cUTI) are among the most frequent bacterial infections in the community and were the 14th-ranked principal diagnosis for hospital admissions in the 2018 Healthcare Cost and Utilization Project [1–3]. Recently, a U.S. national database study indicated that there are over 2.8 million cases of cUTI per year, resulting in annual 30-day total costs of more than $6 billion [4]. Complicated urinary tract infections are also commonplace in elderly patients [5–7], with a spectrum of disease severity ranging from a mild illness with limited or no systemic symptoms to severe sepsis [8–11]. Despite the frequency of cUTIs in elderly patients, most U.S. burden of illness studies have focused on younger patient cohorts [4,12] and scant data are available on the financial burden associated with incident cUTIs episodes in a cohort of predominately elderly patients. Given this gap in the literature, this study sought to examine total and cUTI-related 30-day Medicare spending (MS), a proxy for healthcare costs, among Medicare beneficiaries who resided in the community (i.e., non-long-term care facility) with newly diagnosed cUTIs. As part of the study, we were most interested in defining the number of incident cUTI cases, 30-day cUTI-related costs, and drivers of cUTI-related costs. Given that Medicare represents approximately 15% of total federal spending in the US [13], we believed a comprehensive understanding of the overall and drivers of 30-day cUTI MS would provide insights into the treatment approaches required to potentially reduce the financial burden associated with cUTIs in Medicare beneficiaries.

2. Methods

A retrospective multicenter cohort study of adult beneficiaries in the Medicare fee-for-service (FFS) database with a cUTI between 2017 and 2018 was performed. Patients were included in the study if they: (1) were enrolled in Medicare FFS and Medicare Part D (Medicare pharmacy benefits) from 2016 to 2019, (2) were not enrolled in Medicare Advantage, (3) had a cUTI first diagnosis (Supplemental Table S1) [4,14] in 2017–2018, (4) had no evidence of any UTI diagnoses in 2016, and (4) had no residence in a long-term care facility in 2016–2018. For patients with ≥ 1 cUTI between 2017 and 2018, the first cUTI was examined.

Baseline demographics and covariates included age, sex, race, dual eligibility status (enrolled in Medicare and Medicaid), and low-income subsidy (LIS) status. Charlson Comorbidity Index (CCI) score [15] and other diagnoses in the 30-day post-cUTI diagnosis period were recorded [15]. Overall and cUTI-related healthcare resource utilization (HRU), MS, and Medicare Spending per beneficiary (MSPB) were also collected in the 30-day post-cUTI index date period for the entire study population and among beneficiaries with an encounter in the following service categories: acute care inpatient facilities, non-acute/long-term care inpatient facilities, physician offices, and outpatient. Acute care inpatient facilities included acute care hospitals, critical access hospitals, and other inpatient facilities. Non-acute/long-term care inpatient facilities included skilled nursing facilities (SNF), inpatient rehabilitation facilities (IRF), long-term acute care hospitals (LTACH), and inpatient psychiatric facilities. Physician classifications included primary care, physician specialist, and carrier non-physicians. Outpatient included hospital outpatient, home health, hospices, renal dialysis, outpatient nursing homes, outpatient pharmacy (Medicare Part D), and other outpatient. To quantify the 30-day economic burden of newly diagnosed cUTIs in Medicare beneficiaries who resided in the community, the following overall and cUTI-related outcomes were reported: (1) MS (total sum of MS in the 30-day follow-up period across all beneficiaries included in the study), (2) MS in each service category (total sum of MS in each service category in the 30-day follow-up period across all beneficiaries included in the study), (3) average MSPB (total sum of MS divided by the number of beneficiaries included in the study), (4) average number of encounters per beneficiary (total number of encounters in the study population divided by number of beneficiaries included in the study) (5), average number of encounters per beneficiary in each service category (total number of encounters in the study population in each service category divided by the number of beneficiaries included in the study), and (6) average MS per encounter in each service category (total sum of MS in the study population in each service category divided by the number of total encounters in the service category). Medicare spending was considered related to the cUTI if the claim(s) included any diagnosis (i.e., primary or secondary) of a UTI (Supplemental Table S1) [4]. Data aggregation, analysis, and visualization were performed using Tableau 2021.4 and Microsoft Excel.

3. Results

Among the beneficiaries in Medicare FFS only and Medicare Part D between 2016 and 2019 with no evidence of any UTI diagnoses in 2016, 2,330,123 had a newly diagnosed UTI in 2017–2018. Of the 2,330,123 beneficiaries with a UTI, 772,896 had a cUTI and 723,324 (93.5%) occurred in beneficiaries who resided in the community (i.e., non-long-term care facility). Among the beneficiaries in the final study population, most were male (61%), and age distribution was as follows: 13% were <65 years, 29% were 65–74 years, 35% were 75–84 years, and 23% were ≥ 85 years. Seventeen percent of beneficiaries in the study population had partial or full dual Medicare and Medicaid coverage and 19% had a partial or full LIS status (Table 1). The average CCI score was 2.2 in the 30-day post-cUTI index follow-up period. Diagnosed conditions identified in the 30-day post-cUTI index follow-up period are listed in Supplemental Table S2.

Table 1. Baseline demographic and clinical characteristics.

Characteristics	Patients	
	(N = 723,324)	
Age Distribution	*n*	(%)
<65	93,459	(12.9%)
65–74	211,279	(29.2%)
75–84	253,524	(35.0%)
85+	165,062	(22.8%)
Sex		
Male	441,452	(61.0%)
Female	256,878	(35.5%)
Unknown	24,994	(3.5%)
Race		
Non-Hispanic White	575,316	(79.5%)
Black	49,713	(6.9%0
Hispanic	39,770	(5.5%)
Asian/Pacific Islander	17,163	(2.4%)
Other/Unknown	41,362	(5.7%)
Dual eligible status		
Full dual	95,311	(13.2%)
Partial dual	27,145	(3.8%)
Non-dual	600,868	(83.1%)
Low-income subsidy (LIS) status		
Full LIS	124,354	(17.2%)
Partial LIS	14,604	(2.0%)
No LIS	584,366	(80.8%)

Overall and cUTI-related 30-day MS were $7.6 and $4.5 billion, respectively (Figure 1). The average overall and cUTI-related 30-day MSPB were $10,527 and $6181, respectively. Acute care hospitalizations represented the highest proportion of overall and cUTI-related MS and MSPB. Non-acute/long-term care facilities were the second highest, followed by physician, and outpatient services. On average, beneficiaries had 4.4 and 2.2 overall and cUTI-related service encounters, respectively. The mean number of encounters per beneficiary in each service category and average MS per encounter in each service category are displayed in Figure 2. The mean number of encounters per beneficiary for cUTI-related physician specialists and primary care visits were 0.8 and 0.3, respectively. Receipt of care in acute care and outpatient hospitals were also commonplace in the 30-day follow-up period (mean number of cUTI-related encounters per beneficiary was 0.3 for both service categories). The mean number of cUTI-related encounters per beneficiary was ≤0.05 for all other service categories. Average MS for cUTI-related encounters for physician specialist, primary care, and outpatient hospitals were $269, $310, and $869, respectively. Among beneficiaries with acute care hospitalizations, the average overall and cUTI-related 30-day MS per encounter were $16,431 and $15,438, respectively.

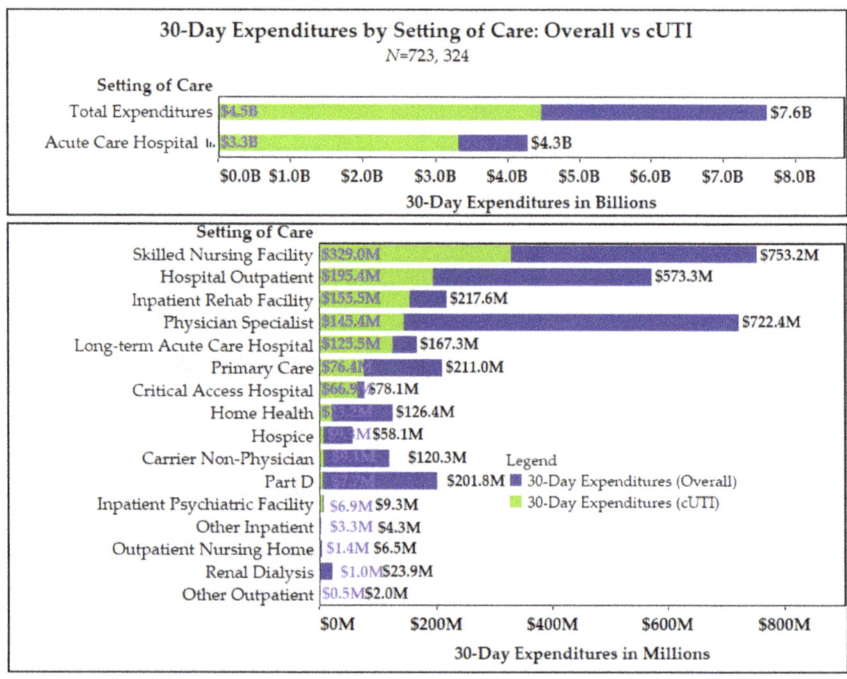

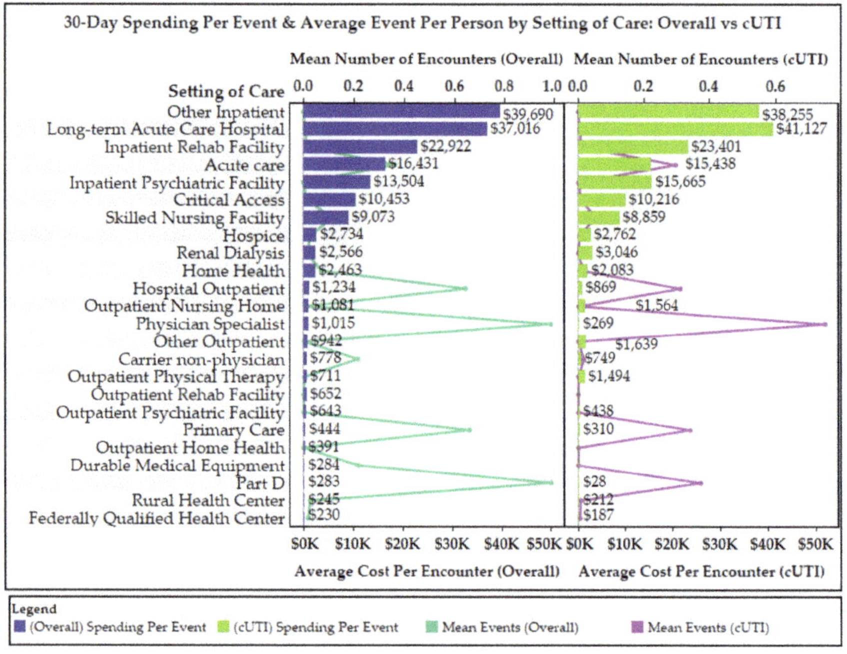

Figure 1. Overall and cUTI-related 30-day Medicare spending by setting of care.

Figure 2. Mean number of overall and cUTI-related 30-day encounters per beneficiary in each service category and average 30-day Medicare spending per encounter in each service category.

4. Discussion

Thirty-day MS for beneficiaries who resided in the community (i.e., non-long-term care facility) with incident cUTIs between 2017 and 2018 was substantial (overall: $7.6 billion), with cUTI-related MS accounting for 59% (cUTI-related: $4.5 billion) of 30-day MS. On average, 30-day cUTI-related MSPB was $6181. The major driver of 30-day cUTI-related MS was acute care hospitalizations, which totaled at $3.2 billion. Approximately 30% of beneficiaries in the study cohort had an acute care hospitalization and the average cost of a cUTI-related hospitalization was $15,438. Although the proportion of beneficiaries with encounters for non-acute long-term care facilities in the 30-day follow-up period was low, it was the second major contributor to 30-day cUTI-related MS, followed by MS on physician visits and other outpatient medical services. Medicare Part D spending (cUTI-related: 7.6 million) was only a small component of 30-day cUTI-related MS, reflecting the low costs of the generic antibiotics used to treat cUTIs patients.

To put these 30-day cUTI-related MS statistics in proper perspective, the annual Medicare FFS payments for Parts A and B benefits were $403 billion, and $95 billion for Part D, in 2018 [13]. Annual average MS per enrollee in Medicare Parts A and/or B was $10,229 in 2018 [16]. Thus, the cUTI-related MS in the 30-day post-cUTI follow-up period alone consumed ~0.5% of the annual Medicare budget FFS for Parts A and B, but was only a small fraction of Part D spending. More importantly, 30-day cUTI-related MS for each beneficiary who experienced a cUTI was considerable when compared with the total average amount spent on each beneficiary per year. Given the aging of the population and growth in Medicare enrollment, the future number of cUTIs and total cUTI-related MS is expected to be even greater. These findings highlight the critical need for educating patients on ways to mitigate the occurrence of cUTIs (i.e., maintain adequate hydration, proper hygiene, urinate as soon as the need arises, etc.). They also indicate that clinicians need to proactively implement cUTI prevention measures [17,18], especially among patients with indwelling urinary catheters, and create individualized patient care plans [19] to reduce the clinical and economic sequelae associated with cUTIs. Since 70% of cUTI-related 30-day MS was for acute care hospitalizations, healthcare systems also need to develop institutional site-of-care clinical pathways that aim to expedite the transition of patients to the outpatient setting once they are stabilized in the emergency department, observation unit, or hospital [20–24]. Given the vulnerability of many elderly patients and observed diagnosed conditions identified in the 30-day post-cUTI index follow-up period (Supplemental Table S2), it is likely that most elderly patients will require initial care in the hospital. However, data indicates that admission patterns were highly variable among cUTI patients and many hospitalized cUTI patients $\geq$65 years had low disease acuity and were potential candidates for outpatient care or early hospital discharge [4,12,25]. For instance, if hospitals were able to avert acute care hospitalizations by 5–10% among elderly patients, it would reduce 30-day cUTI MS by upwards of $200–400 million.

Outpatient treatment options for appropriate elderly cUTI patients that may potentially lessen the financial burden of cUTI-related hospitalization on the Medicare program include oral antibiotics and outpatient parenteral therapy (OPAT). While it is always preferred to treat with oral antibiotics when possible, their use is somewhat limited for many elderly cUTI patients due to the high resistance rates to first-line oral cUTI agents among common uropathogens [26–28]. An alternative to oral antibiotics is OPAT, which can be administered at home, in a physician's office or infusion suite, or in a non-acute long-term care facility. There are several considerations with the use of OPAT across these settings. Medicare currently provides limited coverage of home infusions, limiting their use for many elderly cUTI patients [29]. It is challenging for many elderly patients, especially those homebound, to travel daily to physicians' offices/infusion centers for receipt of OPAT. Furthermore, data indicate that elderly patients are at an increased risk for OPAT-related adverse events and subsequent hospital admissions/readmissions [30–38]. Cost of delivering care in non-acute long-term care facilities is comparable to those associated with acute care hospitalizations [39]. Although their use was limited in the 30-day post-cUTI follow-up period, the average MS for a SNF, IRF, and LTACH encounter were \$8859, \$23,401, and \$41,127, respectively. Fortunately, there are several pharmacokinetic and pharmacodynamic dose-optimized oral antibiotics in development with reliable activity against highly resistant common uropathogens [40] and these have the potential to lessen the financial burden of cUTIs on the Medicare program. Like all new treatments, the ability of these new agents to improve the efficiency of healthcare delivery for elderly cUTI patients will need to be demonstrated before they can be adopted in clinical practice.

Several things should be noted when interpreting study findings. This study was subject to the limits inherent in all administrative claims database analyses. Diagnoses of cUTIs were based on diagnostic and procedure codes, but many of the claims used have been shown to have high positive predictive values [41,42]. The study population included all Medicare FFS beneficiaries with part D coverage, including those $\leq$65 years who qualified for Medicare due to a permanent disability. We included Medicare beneficiaries <65 years because they represented 12.9% of the study population and their rates of chronic conditions, functional limitations, and cognitive impairments are consistent with Medicare beneficiaries $\geq$65 years, which minimizes the potential for any downward bias on observed results due to younger age [43]. Due to the nature of the study, we could not determine if beneficiaries presented to the hospital with a cUTI or developed their cUTI during a hospitalization. However, it is likely that a fair proportion of patients presented to the hospital with a cUTI given that this report was limited to beneficiaries who resided in the community. Lastly, MS may have been a conservative estimate of the true costs as data shows that the amount reimbursed by Medicare is often less than the accrued healthcare costs [44]. Future studies are needed to determine the actual cost of cUTI-related hospitalizations. Since cUTIs have different pathologies, future studies should analyze costs across the different cUTI diagnosis-related groups (DRGs).

In summary, 30-day MS for beneficiaries who resided in the community (i.e., non-long-term care facility) with incident cUTIs was substantial, with cUTI-related MS accounting for 56% of the total. While a variety of service categories contributed to 30-day cUTI-related MS, the major driver was acute care hospitalizations. Healthcare systems should develop well-defined criteria for hospital admissions that aim to expedite the transition of stable patients to the outpatient setting to complete their care. Although acute care hospitalization will still be required for most elderly cUTI patients, even modest reductions in hospitalization rates will have a major impact on cUTI-related MS. The findings also highlight the need for additional treatment options that maximize the efficiency of healthcare delivery for beneficiaries who can be safely and effectively managed in the outpatient setting.

Supplementary Materials: The following supporting information can be downloaded at: https://www.mdpi.com/article/10.3390/antibiotics11050578/s1, Table S1: Algorithm for Identifying Medicare Beneficiaries with cUTIs Based on International Classification of Diseases, Tenth Revision (ICD-10) Diagnostic Codes and Procedure Codes and Current Procedural Terminology (CPT) Codes; Table S2: Distribution of Diagnoses in 30-Day Post Index-cUTI period.

Author Contributions: Conceptualization, T.P.L. and M.R.; methodology, T.P.L., M.R. and M.N.; writing—original draft preparation, T.P.L.; writing—review and editing, T.P.L., M.R. and M.N.; funding acquisition, M.R. All authors have read and agreed to the published version of the manuscript.

Funding: This study was supported by Spero Therapeutics, Inc.

Conflicts of Interest: T.P.L. reports the following disclosures in the past 24 months: Spero Therapeutics, Inc.: consultant; Merck & Co.: consultant and investigator-initiated grant support; Shionogi, Inc.: consultant and speaker. M.N. and M.R. are employees of Spero Therapeutics, Inc.

References

1. Agency for Healthcare Research and Quality (AHRQ); Healthcare Cost and Utilization Project (HCUP); National Inpatient Sample (NIS). Most Frequent Principal Diagnoses for Inpatient Stays in U.S. Hospitals. 2018. Available online: https://www.hcup-us.ahrq.gov/reports/statbriefs/sb277-Top-Reasons-Hospital-Stays-2018.pdf (accessed on 24 February 2022).
2. Foxman, B. Urinary tract infection syndromes: Occurrence, recurrence, bacteriology, risk factors, and disease burden. *Infect. Dis. Clin. N. Am.* **2014**, *28*, 1–13. [CrossRef]
3. Flores-Mireles, A.L.; Walker, J.N.; Caparon, M.; Hultgren, S.J. Urinary tract infections: Epidemiology, mechanisms of infection and treatment options. *Nat. Rev. Microbiol.* **2015**, *13*, 269–284. [CrossRef]
4. Carreno, J.J.; Tam, I.M.; Meyers, J.L.; Esterberg, E.; Candrilli, S.D.; Lodise, T.P., Jr. Longitudinal, Nationwide, Cohort Study to Assess Incidence, Outcomes, and Costs Associated with Complicated Urinary Tract Infection. *Open. Forum. Infect. Dis.* **2019**, *6*, ofz446. [CrossRef]
5. Shallcross, L.; Rockenschaub, P.; Blackburn, R.; Nazareth, I.; Freemantle, N.; Hayward, A. Antibiotic prescribing for lower UTI in elderly patients in primary care and risk of bloodstream infection: A cohort study using electronic health records in England. *PLoS Med.* **2020**, *17*, e1003336. [CrossRef]
6. Gharbi, M.; Drysdale, J.H.; Lishman, H.; Goudie, R.; Molokhia, M.; Johnson, A.P.; Holmes, A.H.; Aylin, P. Antibiotic management of urinary tract infection in elderly patients in primary care and its association with bloodstream infections and all cause mortality: Population based cohort study. *BMJ* **2019**, *364*, l525. [CrossRef]
7. Linhares, I.; Raposo, T.; Rodrigues, A.; Almeida, A. Frequency and antimicrobial resistance patterns of bacteria implicated in community urinary tract infections: A ten-year surveillance study (2000–2009). *BMC Infect. Dis.* **2013**, *13*, 19. [CrossRef]
8. Wagenlehner, F.M.; Lichtenstern, C.; Rolfes, C.; Mayer, K.; Uhle, F.; Weidner, W.; Weigand, M.A. Diagnosis and management for urosepsis. *Int. J. Urol.* **2013**, *20*, 963–970. [CrossRef]
9. Martin, G.S.; Mannino, D.M.; Moss, M. The effect of age on the development and outcome of adult sepsis. *Crit. Care Med.* **2006**, *34*, 15–21. [CrossRef]
10. Tal, S.; Guller, V.; Levi, S.; Bardenstein, R.; Berger, D.; Gurevich, I.; Gurevich, A. Profile and prognosis of febrile elderly patients with bacteremic urinary tract infection. *J. Infect.* **2005**, *50*, 296–305. [CrossRef]
11. Rodriguez-Manas, L. Urinary tract infections in the elderly: A review of disease characteristics and current treatment options. *Drugs Context* **2020**, *9*, 13. [CrossRef]
12. Turner, R.M.; Wu, B.; Lawrence, K.; Hackett, J.; Karve, S.; Tunceli, O. Assessment of Outpatient and Inpatient Antibiotic Treatment Patterns and Health Care Costs of Patients with Complicated Urinary Tract Infections. *Clin. Ther.* **2015**, *37*, 2037–2047. [CrossRef]
13. Henry J Kaiser Family Foundation (KFF). Analysis of Medicare Spending Data from the 2009–2019 Annual Report of the Boards of Trustees of the Federal Hospital Insurance and Federal Supplementary Medical Insurance Trust Funds, Table II.B1. Available online: https://files.kff.org/attachment/Issue-Brief-Facts-on-Medicaid-Spending-and-Financing (accessed on 26 February 2022).
14. Lodise, T.; Ye, M.J.; Zhao, Q. Prevalence of Invasive Infections due to Carbapenem-Resistant Enterobacteriaceae among Adult Patients in U.S. Hospitals. *Antimicrob. Agents Chemother.* **2017**, *61*, e00228-17. [CrossRef]
15. Deyo, R.A.; Cherkin, D.C.; Ciol, M.A. Adapting a clinical comorbidity index for use with ICD-9-CM administrative databases. *J. Clin. Epidemiol.* **1992**, *45*, 613–619. [CrossRef]
16. Centers for Medicare & Medicaid Services (CMS). CMS Program Statistics. Available online: https://www.kff.org/medicare/state-indicator/per-enrollee-spending-by-residence/?currentTimeframe=1&selectedRows=%7B%22wrapups%22:%7B%22united-states%22:%7B%7D%7D%7D&sortModel=%7B%22colId%22:%22Location%22,%22sort%22:%22asc%22%7D (accessed on 7 January 2021).
17. Rowe, T.A.; Juthani-Mehta, M. Urinary tract infection in older adults. *Aging Health* **2013**, *9*. [CrossRef]
18. Gould, C.V.; Umscheid, C.A.; Agarwal, R.K.; Kuntz, G.; Pegues, D.A.; Healthcare Infection Control Practices Advisory Committee. Guideline for prevention of catheter-associated urinary tract infections 2009. *Infect. Control Hosp. Epidemiol.* **2010**, *31*, 319–326. [CrossRef]
19. Dumkow, L.E.; Kenney, R.M.; MacDonald, N.C.; Carreno, J.J.; Malhotra, M.K.; Davis, S.L. Impact of a Multidisciplinary Culture Follow-up Program of Antimicrobial Therapy in the Emergency Department. *Infect. Dis. Ther.* **2014**, *3*, 45–53. [CrossRef]
20. Schrock, J.W.; Reznikova, S.; Weller, S. The effect of an observation unit on the rate of ED admission and discharge for pyelonephritis. *Am. J. Emerg. Med.* **2010**, *28*, 682–688. [CrossRef]
21. Kim, K.; Lee, C.C.; Rhee, J.E.; Suh, G.J.; Lee, H.J.; Kim, H.B.; Singer, A.J. The effects of an institutional care map on the admission rates and medical costs in women with acute pyelonephritis. *Acad. Emerg. Med.* **2008**, *15*, 319–323. [CrossRef]
22. Elkharrat, D.; Chastang, C.; Boudiaf, M.; Le Corre, A.; Raskine, L.; Caulin, C. Relevance in the emergency department of a decisional algorithm for outpatient care of women with acute pyelonephritis. *Eur. J. Emerg. Med.* **1999**, *6*, 15–20.
23. Ward, G.; Jorden, R.C.; Severance, H.W. Treatment of pyelonephritis in an observation unit. *Ann. Emerg. Med.* **1991**, *20*, 258–261. [CrossRef]
24. Dumkow, L.E.; Beuschel, T.S.; Brandt, K.L. Expanding Antimicrobial Stewardship to Urgent Care Centers through a Pharmacist-Led Culture Follow-up Program. *Infect. Dis. Ther.* **2017**, *6*, 453–459. [CrossRef] [PubMed]
25. Lodise, T.P.; Chopra, T.; Nathanson, B.H.; Sulham, K. Hospital admission patterns of adult patients with complicated urinary tract infections who present to the hospital by disease acuity and comorbid conditions: How many admissions are potentially avoidable? *Am. J. Infect. Control* **2021**, *49*, 1528–1534. [CrossRef] [PubMed]

26. Critchley, I.A.; Cotroneo, N.; Pucci, M.J.; Mendes, R. The burden of antimicrobial resistance among urinary tract isolates of Escherichia coli in the United States in 2017. *PLoS ONE* **2019**, *14*, e0220265. [CrossRef] [PubMed]

27. Rank, E.L.; Lodise, T.; Avery, L.; Bankert, E.; Dobson, E.; Dumyati, G.; Hassett, S.; Keller, M.; Pearsall, M.; Lubowski, T.; et al. Antimicrobial Susceptibility Trends Observed in Urinary Pathogens Obtained from New York State. *Open Forum Infect. Dis.* **2018**, *5*, ofy297. [CrossRef]

28. Sanchez, G.V.; Babiker, A.; Master, R.N.; Luu, T.; Mathur, A.; Bordon, J. Antibiotic Resistance among Urinary Isolates from Female Outpatients in the United States in 2003 and 2012. *Antimicrob. Agents Chemother.* **2016**, *60*, 2680–2683. [CrossRef]

29. Mansour, O.; Keller, S.; Katz, M.; Townsend, J.L. Outpatient Parenteral Antimicrobial Therapy in the Time of COVID-19: The Urgent Need for Better Insurance Coverage. *Open Forum.Infect. Dis.* **2020**, *7*, ofaa287. [CrossRef]

30. Underwood, J.; Marks, M.; Collins, S.; Logan, S.; Pollara, G. Intravenous catheter-related adverse events exceed drug-related adverse events in outpatient parenteral antimicrobial therapy. *J. Antimicrob. Chemother.* **2019**, *74*, 787–790. [CrossRef]

31. Means, L.; Bleasdale, S.; Sikka, M.; Gross, A.E. Predictors of Hospital Readmission in Patients Receiving Outpatient Parenteral Antimicrobial Therapy. *Pharmacotherapy* **2016**, *36*, 934–939. [CrossRef]

32. Lane, M.A.; Marschall, J.; Beekmann, S.E.; Polgreen, P.M.; Banerjee, R.; Hersh, A.L.; Babcock, H.M. Outpatient parenteral antimicrobial therapy practices among adult infectious disease physicians. *Infect. Control Hosp. Epidemiol.* **2014**, *35*, 839–844. [CrossRef]

33. Krah, N.M.; Bardsley, T.; Nelson, R.; Esquibel, L.; Crosby, M.; Byington, C.L.; Pavia, A.T.; Hersh, A.L. Economic Burden of Home Antimicrobial Therapy: OPAT versus Oral Therapy. *Hosp. Pediatr.* **2019**, *9*, 234–240. [CrossRef]

34. Keller, S.C.; Wang, N.Y.; Salinas, A.; Williams, D.; Townsend, J.; Cosgrove, S.E. Which Patients Discharged to Home-Based Outpatient Parenteral Antimicrobial Therapy Are at High Risk of Adverse Outcomes? *Open Forum Infect. Dis.* **2020**, *7*, ofaa178. [CrossRef] [PubMed]

35. Keller, S.C.; Cosgrove, S.E.; Arbaje, A.I.; Chang, R.H.; Krosche, A.; Williams, D.; Gurses, A.P. It's Complicated: Patient and Informal Caregiver Performance of Outpatient Parenteral Antimicrobial Therapy-Related Tasks. *Am. J. Med. Qual.* **2020**, *35*, 133–146. [CrossRef] [PubMed]

36. Hale, C.M.; Steele, J.M.; Seabury, R.W.; Miller, C.D. Characterization of Drug-Related Problems Occurring in Patients Receiving Outpatient Antimicrobial Therapy. *J. Pharm. Pract.* **2017**, *30*, 600–605. [CrossRef] [PubMed]

37. Durojaiye, O.C.; Kritsotakis, E.I.; Johnston, P.; Kenny, T.; Ntziora, F.; Cartwright, K. Developing a risk prediction model for 30-day unplanned hospitalization in patients receiving outpatient parenteral antimicrobial therapy. *Clin. Microbiol. Infect.* **2019**, *25*, 905.e1–905.e7. [CrossRef]

38. Allison, G.M.; Muldoon, E.G.; Kent, D.M.; Paulus, J.K.; Ruthazer, R.; Ren, A.; Snydman, D.R. Prediction model for 30-day hospital readmissions among patients discharged receiving outpatient parenteral antibiotic therapy. *Clin. Infect. Dis.* **2014**, *58*, 812–819. [CrossRef]

39. Tian, W. An All-Payer View of Hospital Discharge to Postacute Care, 2013. HCUP Statistical Brief #205; Agency for Healthcare Research and Quality, : Rockville, MD, USA, May 2016. Available online: http://www.hcup-us.ahrq.gov/reports/statbriefs/sb2 05-Hospital-Discharge-Postacute-Care.pdf (accessed on 31 March 2022).

40. Veeraraghavan, B.; Bakthavatchalam, Y.D.; Sahni, R.D. Oral Antibiotics in Clinical Development for Community-Acquired Urinary Tract Infections. *Infect. Dis. Ther.* **2021**, *10*, 1815–1835. [CrossRef]

41. Wiese, A.D.; Griffin, M.R.; Stein, C.M.; Schaffner, W.; Greevy, R.A.; Mitchel, E.F., Jr.; Grijalva, C.G. Validation of discharge diagnosis codes to identify serious infections among middle age and older adults. *BMJ Open.* **2018**, *8*, e020857. [CrossRef]

42. Schneeweiss, S.; Robicsek, A.; Scranton, R.; Zuckerman, D.; Solomon, D.H. Veteran's affairs hospital discharge databases coded serious bacterial infections accurately. *J. Clin. Epidemiol.* **2007**, *60*, 397–409. [CrossRef]

43. Henry, J.; Kaiser Family Foundation. A Primer on Medicare March 2015 Key Facts about the Medicare Program and the People It Covers. Available online: https://files.kff.org/attachment/report-a-primer-on-medicare-key-facts-about-the-medicare-program-and-the-people-it-covers (accessed on 22 February 2022).

44. American Hospital Association. Underpayment by Medicare and Medicaid Fact Sheet—January 2019. Available online: https://www.aha.org/factsheet/2019-01-02-underpayment-medicare-and-medicaid-fact-sheet-january-2019 (accessed on 31 March 2022).

Review

A Review of Biofilm Formation of *Staphylococcus aureus* and Its Regulation Mechanism

Qi Peng [1], Xiaohua Tang [1], Wanyang Dong [1], Ning Sun [2,*] and Wenchang Yuan [1,*]

1 Guangzhou Key Laboratory for Clinical Rapid Diagnosis and Early Warning of Infectious Diseases, KingMed School of Laboratory Medicine, Guangzhou Medical University, Guangzhou 510180, China
2 Guangzhou First People's Hospital, School of Medicine, South China University of Technology, Guangzhou 510180, China
* Correspondence: ning.sun@connect.polyu.hk (N.S.); yuanwenchang95@163.com (W.Y.)

Abstract: Bacteria can form biofilms in natural and clinical environments on both biotic and abiotic surfaces. The bacterial aggregates embedded in biofilms are formed by their own produced extracellular matrix. *Staphylococcus aureus* (*S. aureus*) is one of the most common pathogens of biofilm infections. The formation of biofilm can protect bacteria from being attacked by the host immune system and antibiotics and thus bacteria can be persistent against external challenges. Therefore, clinical treatments for biofilm infections are currently encountering difficulty. To address this critical challenge, a new and effective treatment method needs to be developed. A comprehensive understanding of bacterial biofilm formation and regulation mechanisms may provide meaningful insights against antibiotic resistance due to bacterial biofilms. In this review, we discuss an overview of *S. aureus* biofilms including the formation process, structural and functional properties of biofilm matrix, and the mechanism regulating biofilm formation.

Keywords: *Staphylococcus aureus*; biofilms; extracellular matrix; quorum sensing; antibiofilm; antibiotic resistance

Citation: Peng, Q.; Tang, X.; Dong, W.; Sun, N.; Yuan, W. A Review of Biofilm Formation of *Staphylococcus aureus* and Its Regulation Mechanism. *Antibiotics* **2023**, *12*, 12. https://doi.org/10.3390/antibiotics12010012

Academic Editors: Wolf-Rainer Abraham and Zhenbo Xu

Received: 15 October 2022
Revised: 23 November 2022
Accepted: 25 November 2022
Published: 22 December 2022

1. Introduction

Biofilm is an organized bacterial population and refers to the membrane-like extracellular matrix (ECM) formed by the adhesion of bacterial colonies and extracellular polymeric substances (EPS) such as polysaccharides, nucleic acids, and proteins secreted by bacteria during the growth process [1]. The interaction between EPS and bacterial aggregates endows biofilm with cohesion and viscoelasticity [2]. As a result, bacteria can attach to both biotic and abiotic surfaces. The formation of pathogenic biofilm plays an important role in causing chronic persistent infection [3]. Currently, researchers generally believe that more than 80% of chronic infections are mediated by bacterial biofilms [4]. *Staphylococcus aureus* (*S. aureus*) is prevalent in hospital environments. It attaches to and persists on host tissues and indwelling medical devices. This may cause skin and soft tissue infection, osteomyelitis, endocarditis, pneumonia, bacteremia, etc. [5–8]. These infections are difficult to cure due to the biofilm formed that enhances the resistance of *S. aureus* to antibiotics [9]. Additionally, biofilm formation is considered to be a protected growth mode for bacteria to adapt to harsh environments [10]. The biofilm acts as a barrier to create a stable internal environment for bacterial cell activity and protects bacterial cells from adverse conditions including extreme temperature, nutritional restriction and dehydration, and even antibacterial drugs [11]. Consequently, bacteria can settle quickly and protect themselves from host defense mechanisms and then promote long-term infection by enhancing adhesion to the host surface. Biofilm is therefore the first self-protection line of bacteria. It has been known that biofilm-forming bacteria are resistant to most antibiotics [12]. Most clinical antibiotics are developed targeting planktonic microbial cells. Antibiotics targeting planktonic cells may exert selective pressure on microorganisms, thus giving them a survival advantage over susceptible competitors [13]. Therefore, antibiotic therapy against biofilm usually requires long-term use antibiotics at high doses [14]. However, chronic treatment with such antibiotics may lead to increase the risk of antibiotic resistance and drug toxicity [15]. Due to the highly complex and rapid adaptability of biofilm population [16], an in-depth understanding of biofilm formation mechanism may provide new insights for the development of effective infection control strategy against biofilms [17,18].

2. Biofilm Formation Process

The formation of three-dimensional biofilm by bacteria is a complex process. It is generally divided into four stages: adhesion, aggregation, maturation, and dispersion (Figure 1) [19].

During the adhesion stage, *S. aureus* planktonic cells use a range of different factors and relevant regulatory mechanisms, such as the expression of cell wall-anchored protein (CWP), adhesin, and eDNA, to promote the combination with the host [20]. The most characteristic of these regulations is the organization of microbial surface components recognizing adhesive matrix molecules (MSCRAMMs) that mainly include fibronectin-binding proteins (FnBPs, including FnbA and FnbB) [21], fibrinogen-binding protein (Fib), clumping factors (ClfA, ClfB) [22], and serine-aspartate repeat family proteins (SdrC, SdrD, and SdrE) [23]. These components mediate bacterial adhesion to natural tissues and biomaterial surfaces. In addition, bacterial appendages, such as flagella, cilia, and pili, allow for more permanent adherence of bacteria to surfaces [24].

After adhering to surfaces, the adherent bacterial cells begin to divide and accumulate in the presence of a sufficient nutrient source [19]. During the aggregation stage, bacteria regulate biofilm formation by sensing environmental signals that trigger regulatory networks and intracellular signal molecules. Then, bacteria continue to proliferate and thicken to form a biofilm [25]. The biofilm formed can provide resistance against the human immune system and antibiotics [26]. Bacterial cells proliferating in the matrix may lose direct contact with the graft surface and host protein and mainly depend on cell–cell and cell–EPS adhesion [27].

During the maturation stage, the structure of biofilm is highly structured and a compact three-dimensional mushroom or tower structure is formed [28]. A large number of pipes around the microcolony is constructed to promote the transport of nutrients to the deep layer of the biofilm [29]. Mature biofilms have diverse and unique metabolic structures that enable them to resist harmful environmental factors and stress drivers [30].

EPS can be attached by many single bacterial cells to form microcolonies, which are the basic units of biofilm structures [31]. Once microcolonies are formed, the bacterial biofilm continues to thicken and disperse the biofilm under the influence of specific genetic regulations or external factors. The process of dispersion may involve multiple steps, including the production of exoenzymes and surfactants to degrade EPS [32] and physiological changes that prepare cells for conditions outside the biofilm [33]. The dispersed bacteria form planktonic bacteria, which in turn can colonize other sites and form biofilm under certain conditions, thus forming a cyclic process.

The dispersion step is the final stage of the biofilm life cycle and the beginning of another life cycle [34]. During the growth and development of biofilm, surfactant phenol-soluble modulins (PSMs) are a key effector molecule for the dispersion and transmission of *S. aureus* biofilms [35]. PSMs are characterized by amphiphilicity α-helical secondary structure, which gives them surfactant-like properties [36]. These properties can destroy the non-covalent force in the biofilm matrix, forming the necessary channels to transport nutrients to the deeper layer of the biofilm. It also provides the necessary destructive force to spread the biofilm masses to the distal position [37]. PSMs of *S. aureus* not only exist in soluble form, but also aggregate into amyloid fibers to eliminate the biofilm-degrading activity of monomeric PSMs peptides and stabilize the biofilm structure [38]. In *S. aureus*, α-PSM1 and α-PSM4 peptides are the main amyloid proteins involved in the α-PSMs fibril production [39].

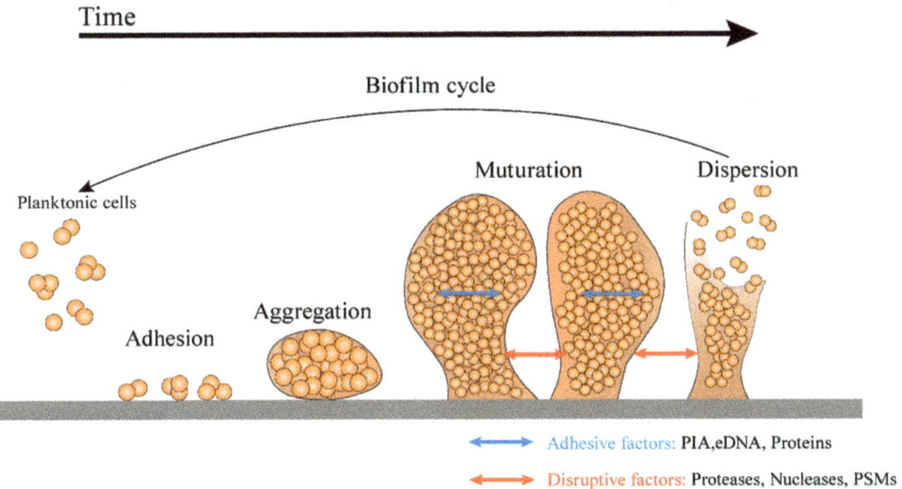

Figure 1. A model of the growth cycle of biofilm. In the first step of biofilm formation, planktonic cells attach to the surface via surface-associated proteins. After attachment, the cells gradually aggregate and begin to produce ECM, thus forming microcolonies. With cell division, a mature biofilm is gradually formed. Finally, in the separation stage, enzymes such as protease, nuclease, and a quorum sensing system promote the dispersion of biofilm, allowing the bacterial cells to detach from the biofilm and return to a planktonic state to colonize new ecological niches [40].

3. Biofilm Formation Mechanism

3.1. Polysaccharide Intercellular Adhesion(PIA)-Dependent Mechanism

Among the polymeric molecules involved in ECM, polysaccharide intercellular adhesion (PIA) (also known as poly-N-acetylglucosamine; PNAG) is an important factor in *S. aureus* biofilm formation [41]. PIA has a cationic property (Figure 2) and plays an important role in the adhesion and aggregation stage [42]. In mutual strains lacking PIA, the ability of bacterial cells adhering to each other is decreased significantly. In *S. aureus*, the mechanism of biofilm formation is controlled by the production of PIA through proteins encoded by *icaADBC* operon in the *ica* locus (Figure 3) [43]. In this mechanism, *icaA* and *icaD* genes are essential in the regulation of biofilm formation. The product of the *icaA* gene is N-acetylamino-glucosamine transferase which is a transmembrane protein [44]. The product of the *icaD* gene is the chaperone protein of *icaA*. It maintains the correct folding of *icaA* and increases the specificity of *icaA* to polymers [45]. The product of the *icaC* gene is a transmembrane protein that mediates the transfer of the newly synthesized PIA to the cell surface [46]. The product of the *icaB* gene is a deacetylase responsible for the deacetylation of mature PIA. This deacetylation gives the polymer a net positive charge, which is required for adhesion onto the cell surface and intercellular adhesion [47]. The maximum length of poly N-acetylglucosamine oligomer produced by *icaAD* is 20 residues. Longer oligomer chains are synthesized only when *icaAD* is co-expressed with *icaC*. PIA also increases biofilm retention and its resistance to antimicrobial peptides (AMPs) through deacetylation [48]. Non-deacetylated polyglucosamine in the homogenous *icaB* mutant cannot adhere to the surface of bacterial cells or mediate the biofilm formation [49]. The *icaADBC*-mediated polysaccharide production is an important mechanism for biofilm formation and contributes to the early growth of bacteria. Additionally, it is believed that the *ica* operon is under phase variation, which has a role in slipped strand mispairing and leads to an on/off switch for expressing the products [50].

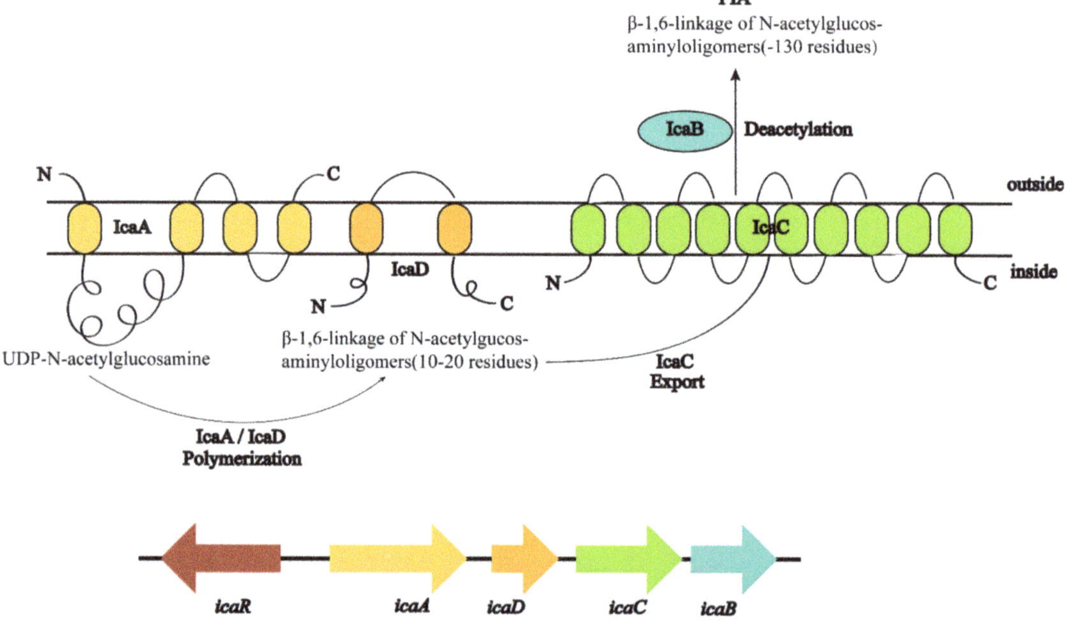

Figure 2. The structure of polysaccharide intercellular adhesin (PIA). PIA is a beta-linear homoglycan composed of 1,6-linked N-acetylglucosamine residues, 15–20% of which are deacetylated and therefore positively charged [41].

Figure 3. Genetic encoding and biosynthesis of PIA. PIA is synthesized from the product of *icaADBC* operon. *IcaA* and *IcaD* are two membrane proteins that polymerize N-acetylglucosamine from the activated precursor UDP-N-acetylglucosamine monomers. This chain may be exported by the membrane protein *IcaC*. *IcaB* is an acetylase attached to the outer surface of bacteria. By deacetylation of residues, *IcaB* introduces a positive charge into the originally neutral PIA molecule [41].

3.2. Protein-Dependent Mechanisms

Some studies reveal that the strains without *ica* operon can also form biofilm [51,52]. Among the strains forming biofilm, the mutation of the *ica* gene does not affect the formation of biofilm. When treated with proteases, the biofilm of these strains can be depolymerized [51]. This indicates that there are proteins involved in biofilm formation through other mechanisms independent of the *ica* operon.

The *bap* gene of *S. aureus* encodes a surface protein Bap (biofilm-associated protein) containing 2276 amino acids. Bap was identified as the main determinant of successful surface adhesion and intercellular adhesion during biofilm formation. It promotes the initial adhesion of bacteria to biological and abiotic surfaces and intercellular adhesion through an extracellular polysaccharides-independent mechanism [52]. All *S. aureus* strains carrying the *bap* gene have high adhesion and strong biofilm-forming ability, indicating that there is a strong correlation between the existence of this protein and the biofilm-forming ability onto an abiotic surface. The N-terminal region of Bap is released into ECM and assembled into amyloid fibers to help the construction of *S. aureus* biofilms [53]. The Bap core domain contains C repeats, which are predicted to fold into new structures and participate in cell adhesion. The Bap C-terminal contains a typical cell wall attachment region [54]. During infection, Bap promotes persistence in the mammary gland by enhancing adhesion to epithelial cells and binds to host receptors to prevent cellular internalization, thereby interfering with the FnBPs-mediated invasion pathway [53].

Fibronectin-binding proteins are multi-structural domain glycoproteins (440 kDa), which are found in almost all tissues and organs and biological fluids and play an important role in cell adhesion and migration [55]. The N-terminal A structure of fibronectin-binding proteins A (FnbA) is structurally and functionally related to cohesion factor ClfA and *Staphylococcus epidermidis* SdrG protein. FnbA binds to fibrinogen at the N2 and N3 ends of the A structural domain through changes in the DLL (dock, lock, and latch) mechanism to form a highly stable complex [56], promotes the accumulation phase and initial adhesion phase of the biofilm, and increases bacterial aggregation [27].

The attachment of clumping factors promotes colonization of *S. aureus* in the host, facilitates biofilm formation, and causes virulence by binding soluble fibrinogen for immune escape [57]. However, even in the absence of fibrinogen, the biofilm of some strains is dependent on increased ClfB activity in the absence of calcium. ClfB accumulates on the bacterial surface and mediates biofilm formation [58].

S. aureus surface protein G (SasG) causes intercellular adhesion and promotes biofilm formation through zinc-dependent dimerization [59]. The fibrillar nature of SasG can mask the binding of *S. aureus* MSCRAMM to their ligands and also promote biofilm formation [60]. *S. aureus* surface protein C (SasC) mediates cell cluster formation, intercellular adhesion, and biofilm formation, but SasC does not mediate the interaction with fibrinogen or fibronectin [61].

The carboxyl terminus of serine–aspartate repeat family proteins contains motifs required for cell-wall anchoring. SdrC mediates strong cellular interactions with hydrophobic surfaces, which may be related to the initial attachment of biological materials, the first stage of biofilm formation [62], while SdrC binds with low-affinity homophilic bonds and promotes cell adhesion as well as biofilm formation [62]. SdrD is an important key factor in the ability of *S. aureus* to survive and evade the blood's intrinsic immune system. SdrD promotes *S. aureus* adhesion to exfoliated nasal epithelial cells [63] and human keratin-forming cells in vitro [64]. It also promotes abscess formation in vivo [65]. SdrE traps the C-terminal tail of complement factor H (CFH) by a unique mechanism and isolates CFH on the surface of *S. aureus* to evade complement [66].

The collagen-binding adhesin (CAN) was originally reported to be necessary and sufficient for the binding of *S. aureus* to collagen-rich stromal cartilage [67]. CAN is a virulence factor in several animal models of infectious diseases. It also functions as an adhesin [68,69]. CAN is also a potential complement inhibitor that disrupts the molecular mechanism of complement activation and represents a potential immune evasion strategy that is associated with the development of multiple diseases [70].

3.3. Extracellular DNA (eDNA)-Dependent Mechanism

The mature *S. aureus* biofilm is sensitive to the external addition of DNase I, indicating that eDNA is a structural component of the biofilm matrix [71]. Due to the negative charge of DNA polymer, eDNA may participate in the early adhesion stage and mature stage of biofilms as an electrostatic polymer and play a basic structural function in the structural integrity of biofilms [72]. Its mechanism is to connect PIA and biofilm-related proteins and other biofilm components to stabilize the biofilm structure [73,74]. At the same time, eDNA also introduces favorable acid-base interactions to enhance adhesion and surface aggregation [75]. The accumulation of eDNA in the biofilm and infection site can acidify the local environment and promote antibiotic resistance phenotype [76]. It was found that eDNA also mediated horizontal gene transfer though conjugation of plasmids between cells in biofilms [77] and neutralized the important effector molecules of innate immunity such as AMPs by binding and isolating cations from the surrounding environment [78]. In *Staphylococcus epidermidis*, eDNA has also been found to be thermodynamically favorable interacting with positively charged vancomycin. It can reduce the potency of vancomycin, inhibit the transport of vancomycin in the biofilm, and thus protect the bacteria embedded in the biofilms [79]. eDNA is released through cell death and lysis and it is mainly regulated by the *cidA* gene. *cidA* encodes a responser of cell wall hydrolase activity and regulates cell death. *cidA* mediated cell lysis contributes to *S. aureus* biofilm formation in vivo and in vitro [80]. The inhibition of *cidA* activity by *lgrAB* operon can inhibit cell lysis and adhesion in the process of biofilm formation. Mutation of *lgrAB* operon leads to the formation of more adherent biofilms and higher eDNA contents [81]. Although these are the main components of the biofilm matrix, the exact composition of the biofilm matrix may vary and depends on matrix availability and physical factors.

4. Regulation Mechanism of Biofilm Formation

Biofilm formation is a social group behavior. Each of the steps from initial attachment to the dispersion and transmission of mature biofilm is strictly controlled by multiple regulatory systems or regulators [82].

4.1. Regulation of Quorum-Sensing System-Mediated Biofilm Formation

Intercellular signal transduction, commonly known as quorum sensing, is an internal communication system of bacteria. Bacteria detect the changes in the number of individual bacterial cells or other bacterial populations in the surrounding environment based on the changes in the concentration of a specific signal autoinducer. When the signal molecule reaches a certain threshold, the expression of relevant genes in bacteria is initiated to adapt to the changes in the environment [83]. The quorum sensing system usually involves signal transduction pathways that regulate biofilm formation, virulence, binding, antibiotic resistance, motility, and sporulation [84]. The quorum sensing system of *S. aureus* includes an accessory regulatory factor (Agr) system and LuxS/AI-2 system [85]. These two systems reduce biofilm formation in two different ways. The Agr system dissociates bacterial biofilm by upregulating the transcription of RNAIII, while the LuxS/AI-2 system reduces the expression of PIA.

The *agr* locus is a complex polygenic system (Figure 4). It responds to bacterial cell density and controls the expression of *S. aureus* adhesion and extracellular protein. The regulation of the Agr system on biofilm is multifaceted and is mainly involved in the adhesion, maturation, and dispersion stages [40]. The Agr system uses an autoinducing peptide (AIP) as the signal molecule of cell density [86]. The *agr* locus encodes a two-component quorum sensing system consisting of two relatively independent transcription units driven by P2 and P3 promoters [87]. The P2 promoter starts the transcription of RNAII. The RNAII transcript harbors four open reading frames including *agrA*, *agrB*, *agrC*, and *agrD*, which encode the proteins required for AIP biosynthesis, transport, signal sensing, and regulation of target genes [88]. AgrC is a signal transduction factor with sensor histidine protein kinase activity; AgrA is a response regulator; AgrD is a precursor of the AIP; and AgrB is a multifunctional endopeptidase and chaperone protein. On the one hand, AgrB participates in the processing of AgrD as a protease to make it a mature AIP, but on the other hand, it acts as an oligopeptide transporter that helps secrete mature AIP out of cells [89]. The P3 promoter initiates the transcription of RNAIII. When the concentration of AIP in the environment reaches a certain threshold, it binds to and activates the histidine kinase AgrC, which leads to autophosphorylation and initials the signal transduction process [90]. AgrC phosphorylates AgrA after activation, which in turn induces the P3 promoter to transcriptionally express RNAIII; the main effector molecule of quorum sensing system [88]. RNAIII positively regulates the expression of toxin genes, prevents the translation of the repressor of toxin (Rot) [91], and reduces the expression of several surface adhesins, which are negatively associated with biofilm formation [92]. At the same time, an increased level of the AIP can promote the depolymerization of *S. aureus* biofilms by increasing the secretion of extracellular protease [85,93]. Different from other targets, the production of PSMs is generated by the direct binding of AgrA to its promoter [94,95]. Upregulation of *agr* leads to an increase in PSMs and promotes maturation and spreading of biofilm. In addition, some nutrients can affect the biofilm formation through the Agr system. Studies have shown that glucose strongly inhibits the expression of the P3 promoter. In established biofilms, glucose consumption activates the Agr system and leads to biofilm diffusion [93]. In *S. aureus*, some regulatory systems are interconnected with the Agr system that regulates the response to changes in environmental conditions and the development of biofilms (Figure 5) [20].

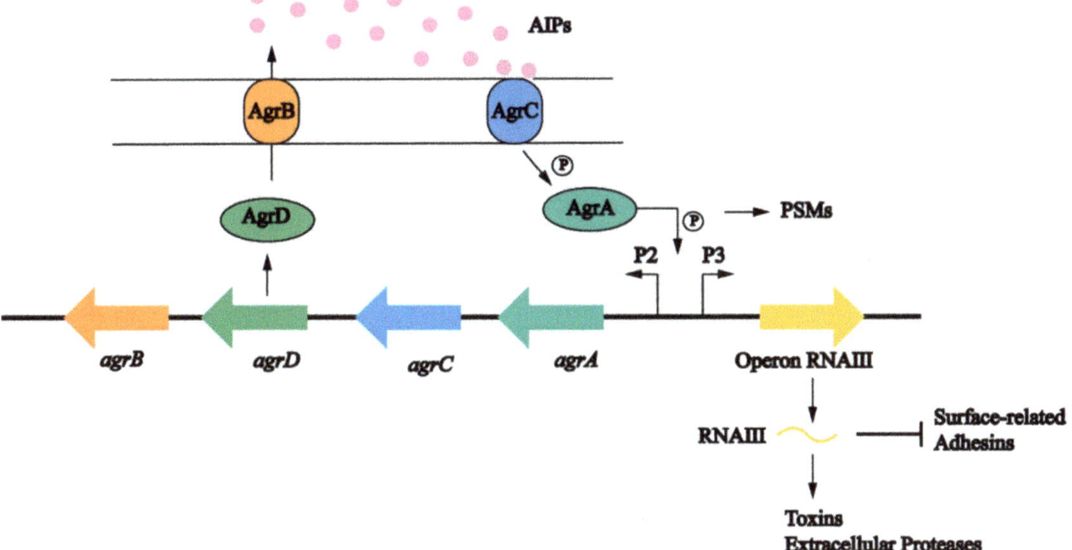

Figure 4. The role of the Agr quorum sensing system in biofilm formation in *S. aureus*. The Agr system is controlled by *agr* operon. AgrD is the precursor of self-induced peptides (AIPS), which is modified by AgrB and secreted into the extracellular matrix. AIPs are intracellular signal molecules that respond to cell density. When the bacterial density increases, AIP activates the transmembrane protein AgrC. Phosphorylated AgrC further activates AgrA and finally promotes the expression of target genes. AgrA acts on P2, which regulates the Agr protein, and P3, which can activate RNAIII expression. RNAIII is the effector molecule of *agr* locus. RNAIII induces the expression of virulence factors, such as protease and toxin. On the other hand, RNAIII also inhibits the expression of surface adhesion proteins [83].

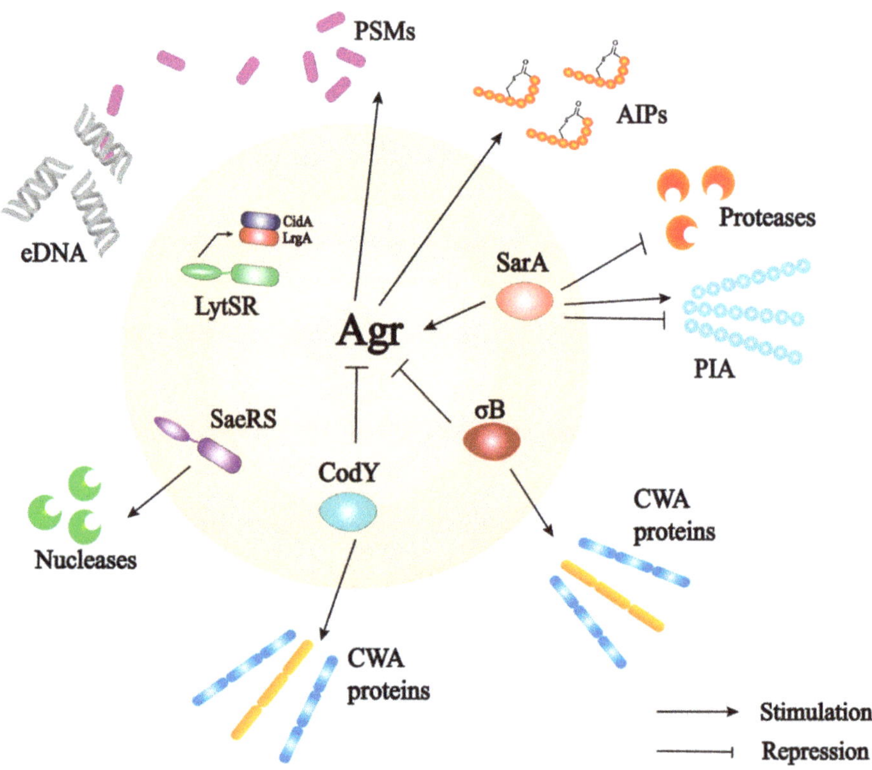

Figure 5. The interaction between Agr quorum sensing system and some important biofilms regulators (LytSR TCS, SigB, CodY and SarA) [20].

LuxS/AI-2 quorum sensing system is shared by both Gram-negative and Gram-positive bacteria. It is mediated by the signal molecule AI-2 (furanyl borate diester) synthesized by *luxS* gene (AI-2 synthase). It enables bacteria to make collective decisions about the expression of a specific set of genes. The precursor of AI-2 is 4,5-dihydroxy-2,3-pentanedione (DPD) [96]. LuxS/AI-2 quorum sensing system that exists in *S. aureus* plays a role in the regulation of biofilm formation [97]. Two early studies have shown that LuxS is a negative regulator of biofilm formation. Biofilm formation is significantly increased in *luxS* mutant strains compared with wild strains. A study reported that the transcript level of *icaR* was significantly reduced in *luxS* mutants, while *icaA* expression was significantly increased, suggesting that AI-2 represses *icaADBC* transcription through activation of *icaR* [98]. Moreover, when a low nanomolar concentration range of DPD was added, the biofilm formation was changed. On the contrary, when higher concentrations of DPD were added, no effect on biofilm formation was observed [98]. Another study found that Rbf was a positive regulator of biofilm. It promotes PIA-dependent biofilm formation in *S. aureus* by binding to the *sarX* promoter, upregulating *sarX* transcription, and indirectly downregulating *icaR* expression [99]. The transcription level of *rbf* was increased in *luxS* mutant strains. The transcription of the *rbf* gene could also be restored to normal when supplemented with *luxS*-containing plasmids or effective concentrations of exogenous AI-2. The results suggest that *luxS* may suppress *rbf* expression and reduce the transcription level of *icaA* through AI-2-mediated signaling, thereby reducing PIA-dependent biofilm formation [100]. In contrast to the above study, another study observed upregulation of the *icaADBC* by AI-2 in *S. epidermidis*. The expression of the *icaADBC* was also upregulated when micromolar concentrations of DPD were added [101]. This result suggests that the effect of signal molecule AI-2 may be very different for each species, even depending on the concentration.

4.2. The Global Response Regulator

Staphylococcal accessory regulator (SarA) is a DNA-binding protein encoded by *sarA* locus [102]. It is the main global regulator of many virulence determinants and directly regulates the expression of some virulence factors [103]. *sarA* locus is necessary for *ica* operon transcription, PIA/PNAG production, and biofilm formation of *S. aureus* [104]. SarA regulates biofilm formation through the *agr*-dependent pathway, binds to the *agr* promoter to stimulate transcription of RNAIII, and cascades and regulates downstream target genes [105,106]. SarA can control gene expression by directly interacting with target gene promoters through the *agr*-independent pathway [104]. In addition, SarA activates the P2 promoter and promotes transcription by bending the DNA region of the *agr* promoter, thereby enhancing *agr* dimer interactions to upregulate *agr* expression [107]. A sequence with 58% homology to the predicted recognition sequence of *sarA* exists at 70 nt upstream of *ica* start codon in *S. aureus*. The purified SarA binds to this sequence with high affinity to upregulate *ica* operon expression and promote biofilm formation [108]. SarA not only induces the transcription of *ica* operon but also of its suppressor *icaR*, indicating that SarA may prevent the excessive production of PIA [109]. Another role of SarA in biofilm formation is the inhibition of extracellular proteases production [110]. *sarA* mutants have a global impact on the abundance of many *S. aureus* transcripts, producing high levels of proteases, such as metalloproteinase Aur, serine protease SspA, cysteine protease SspB and ScpA, resulting in the inability to form biofilms [110]. Only by simultaneously eliminating the production of these extracellular proteases, the biofilm formation and virulence of *sarA* mutants can be fully restored [111]. In summary, SarA is an important regulator controlling the biofilm formation of *S. aureus* through a variety of mechanisms.

σ^B is a product of *sigB* operon and is the main regulator of *S. aureus* response to environmental stress. It plays an important role in the production of bacterial drug resistance, biofilm formation, and the expression regulation of virulence-related genes [112]. Under osmotic stress, the biofilm formation of *S. aureus* wild strain MA12 is significantly stimulated, but the *sigB* deletion mutation eliminates the biofilm formation. After supplying the *sigB*-containing plasmid, biofilm formation could be restored to normal under conventional conditions or stimulated by osmotic stress [113]. However, another study reported that the *sigB* deletion mutant still formed biofilm effectively, suggesting that *sigB* could regulate *S. aureus* biofilm in a strain-dependent manner [114]. σB also mediates an increase in SarA level and decreases the level of RNA III of the Agr system, leading to growth-stage-dependent differences in some virulence factors [115]. Moreover, it mediates the production of several cell surface proteins related to the early adhesion of biofilms, such as FnbA and ClfA. σ^B promotes the transcription of *fnbA* in early growth and significantly upregulates the transcription of *clfA* in late growth [116]. However, various exoprotein genes that are important for biofilm dispersal are repressed by σ^B [117].

In *S. aureus*, the global transcription factor CodY acts mainly as a repressor of metabolic and virulence genes, directly or indirectly regulating more than 200 genes [118]. Under adequate nutritional conditions, CodY interacts with its effector molecules, leading to conformational changes in CodY that enhance the affinity of CodY to its DNA binding sites [119]. CodY strongly inhibits the *agr* locus [120]. Some studies have shown that CodY may bind to a locus in *agrC* in vitro [121,122], but this binding does not affect the in vivo transcription of *agrA*, suggesting that CodY does not control *agr* gene expression through direct binding within the *agr* locus [122]. The identification of CodY regulatory Agr system targets reveals undoubtedly a new regulatory pathway affecting most virulence genes in *S. aureus*. CodY regulates biofilm formation by mediating *ica* expression and PIA production [117,123]. In the *codY* deletion mutant, the *icaA* was overexpressed; in the *icaR* deletion mutant, the transcript level of *icaA* gene was extremely low and the amount of biofilm formation was low; in the *icaR* and *codY* double mutant, the transcript level of *icaA* was similar to that of the *codY* mutant, but the amount of biofilm formation was three times more than that of the *codY* mutant, indicating that *codY* may override the *icaR*-mediated loss of the *ica* locus repression and that *codY* is epistatic to *icaR* [118]. In addition to regulating PIA-dependent biofilm, CodY also represses the exoprotease and thermonuclease (*nuc*), which are important regulators of biofilm formation [118,124]. *nuc* is directly controlled by the SaeRS system, and in *codY* null mutants, *nuc* overexpression requires SaeR, indicating that the *codY* overexpression phenotype is at least partially via the SaeRS system [118]. In addition, CodY regulates the *sae* locus in a Rot-dependent and Rot-independent manner [125].

4.3. *ica* Operon

The enzyme required for PIA synthesis is encoded by *icaADBC*. *icaR* are located upstream of *icaADBC* and belong to the TetR family encoding transcriptional repressors of the *ica* locus [126]. IcaR is a DNA-binding protein that negatively regulates *icaADBC* expression by binding to the upstream region of the *icaA* start codon [127]. In addition to IcaR, a second *icaADBC* direct repressor has been identified. This regulator, named TcaR (teicoplanin-associated locus regulator), belongs to the MarR family of transcriptional regulators [128]. The putative binding sequence of TcaR has been identified in the promoter region of the *icaADBC* operon and has been shown to function as a direct repressor by TcaR [129].

Existing studies have pointed out that *S. aureus* isolates carrying the deletion of the 5 bp motif "TATTT" between the *icaR* and *icaA* intergenic region exhibit a mucoid phenotype, resulting in the increased transcription of *ica* and inducing excessive production of PIA/PNAG in *S. aureus* [130]. Through the electrophoretic mobility shift assay and DNaseI footprint assay of recombinant IcaR, it was found that the recombinant IcaR protected a 42 bp region upstream of the *icaA* gene, but not with the region containing 5 bp motif. It shows that the transcriptional control of 5 bp at *ica* site is independent of *icaR* [131]. The effect of the 5 bp motif on *icaADBC* expression is considered to be controlled by other undetermined repressors. Whole-genome sequencing and microarray analysis revealed that another clinical isolate with a mucus phenotype contained a hypothetical transcriptional regulatory gene with a spontaneous mutation that was expressed at a significantly higher level than in a control strain without biofilm formation [132]. This gene was named "regulator of biofilm" (*rob*). Rob is a DNA-binding protein that shares homology with the TetR family. Rob was experimentally shown to recognize and bind a 25 bp region between the *icaR* and *icaA* intergenic regions (including the 5 bp motif). In the absence of 5 bp, Rob cannot bind to this region, resulting in excessive biofilm formation and is a novel repressor of the *ica* locus [132]. Rob was first discovered as the *gbaAB* (glucose-induced biofilm access gene) operon. GbaAB can participate in the regulation of the multicellular aggregation step of glucose responsive *S. aureus* biofilm formation through *ica* locus and PIA [133]. However, Rob affects biofilm formation in a glucose-independent manner [132]. Although the strains used in these two studies are different, the potential mechanisms need to be further explored.

4.4. *Two-Component Regulatory System*

The SrrAB TCS (*Staphylococcal* respiratory response regulator) is a major regulator of respiratory growth and virulence in *S. aureus*, which is critical for survival under environment conditions such as hypoxic and oxidative [134]. The oxygen utilization plays an important role in *S. aureus* virulence by regulating toxin production and biofilm formation [135]. SrrAB TCS sensor kinase SrrB responds to oxygen in the environment by autophosphorylation and the effector molecule SrrA activates or represses the regulation of target genes [136]. Purified phosphorylated SrrA binds to a 100 bp DNA sequence upstream of *icaA* in a concentration-dependent manner to induce transcription of *icaA* and PIA production under anaerobic conditions [137]. In particular, the introduction of insertional mutations in *srrA* resulted in increased PIA production but reduced biofilm formation [138]. Meanwhile, the cysteine residues within the SrrB Cache domain form redox-sensitive disulfide bonds, which are required for biofilm formation and virulence expression under anaerobic and microaerobic conditions [134]. Another study showed that SrrAB-dependent biofilms increased with decreased respiratory activity due to fermenting cells increasing eDNA and proteins in an AtlA murein hydrolase-dependent manner [139]. In *srrAB* mutants, biofilm formation decreased over time and cell death levels increased under static aeration conditions compared with wild strains [140], indicating that the SrrAB is important for long-term biofilm stability and survival. This difference could be caused by different ways of hypoxia production, different growth media and different culture time. All these factors emphasize the importance of biofilm growth models [141].

The *S. aureus* exoprotein expression (SaeRS) TCS is a key regulator of toxin and exoprotein, which has critical roles in evasion of innate immunity and pathogenesis [142]. SaeRS TCS is composed of sensor kinase SaeS and response regulator SaeR along with two auxiliary proteins SaeP and SaeQ [143]. SaeP and SaeQ form a protein complex with SaeS, which activates the phosphatase activity of the SaeS and returns to the pre-activation state via a negative feedback mechanism [144]. SaeRS regulates the expression of genes associated with biofilm adhesion proteins such as *fnbA, fnbB,* and *fib* [109,145,146]. The accumulation of biofilm-related proteins enhances biofilm formation [147]. SaeRS mutation limits the biofilm formation of *S. aureus*. However, in *S. aureus* Newman, a point mutation in SaeS (SaeSP) leads to overexpression of SaeRS activity, preventing this strain from forming a robust biofilm, and exhibiting a SaeRS polymorphism [148]. In addition to SrrAB, SaeRS also regulates fermentation biofilm formation, decreased respiration caused an increasing activity in SaeRS, leading to increased expression of AtlA and FnbA as well as biofilm formation [149]. Recently, overexpression of ScrA (*S. aureus* clumping regulator A) was found causing an increase in bacterial aggregation [150]. Through proteomics and transcriptomics, strain overexpressing ScrA was found to cause cell aggregation and biofilm formation through activation of SaeRS, resulting in upregulation of multiple adhesins and downregulation of secreted proteases [150].

The cell death and lysis of *S. aureus* are regulated by LytSR TCS through a bacteriophage-like holin/antiholin system [151]. Holin is encoded by *cidABC* operon, CidA oligomerizes and forms pores in the cytoplasmic membrane, leading to membrane depolarization, activation of murine protein hydrolase activity, and cell lysis [151,152]. The CidA mutant exhibits decreased lysis, resulting in reduced eDNA content and impaired biofilm formation [153]. Antiholin inhibits the activity of CidA, encoded by *lrgAB* operon [81], which is an inhibitor of these processes [154], counteracts CidA activity by interfering with the ability of CidA to depolarize membranes and cause subsequent death and lysis [77]. The *lrgAB* operon, together with the *cidABC* operon, have been shown to be the regulators of cell death and lysis during biofilm development [152]. In *S. aureus*, LytSR positively regulates *lrgAB* operon and CidR enhances the expression of *cidABC* and *lrgAB* [155]. Agr and SarA, like LytSR, are positively regulating *lrgAB* expression. Mutations in the *agr* locus reduce *lrgAB* expression by approximately sixfold, whereas mutations in the *sarA* reduce *lrgAB* expression to undetectable levels [156]. *lytS* mutation results in more eDNA produced in ECM, leading to a thicker and more adherent biofilm [157]. Some studies have shown that the activity of the *lrgAB* promoter is mainly expressed in the tower structure of *S. aureus* biofilms [158], while a small part of the tower structure formed by *lytS* mutant strain still shows significant *lrgAB* expression. The results suggest that there is a LytS-independent pathway of LytR activation [159]. Since mutations in *srrAB* lead to increased cell death during biofilm development, it is reported that SrrAB inhibits cell death by directly suppressing the expression of the *cidABC* under conditions of glucose overload [160]. The cell death could be due to the effect of *cidABC* expression leading to increased reactive oxygen species (ROS) accumulation [160].

ArlRS TCS was originally identified as a regulator of autolysis and biofilm formation [161]. MgrA is a global transcriptional regulator downstream of ArlRS that forms a regulatory cascade with ArlRS [162]. ArlRS activates the expression of MgrA and regulates a variety of genes through MgrA, including seven CWPs and virulence [162]. Among the 15 mutants of non-essential TCSs, ArlRS TCS, SrrAB TCS, and Agr are required for biofilm formation, with ArlRS playing a major role in the process [163]. ArlRS enhances the expression of *icaADBC* by suppressing the expression of *icaR*, which activates PNAG production, and in *arl* mutants, the synthesis of PNAG is lost [163]. However, overexpression of MgrA could not restore PNAG expression in ArlRS mutants [164]. In contrast, MgrA can act as a negative regulator of *psm* expression, negatively regulating biofilm formation and dispersion by directly binding to the *psm* promoter region to repress transcription of the *psm* operon in cultures and biofilms [165]. These results suggest that ArlRS is a key TCS for biofilm formation.

4.5. The Second Messenger

Cyclic dinucleotides are highly versatile signaling molecules in prokaryotes involved in the control of various important biological processes [166]. These intracellular signal nucleotides coordinate diverse aspects of bacterial colony behavior, including motility, biofilm formation, and virulence gene expression [167]. The second messenger, cyclic diguanylic acid (c-di-GMP), is synthesized by two molecules of GTP by the diguanylate cyclases (DGCs) and degraded by the c-di-GMP phosphodiesterases (PDEs) [167], which is the main regulator of the conversion between free movement and biofilm of Gram-negative bacteria [168]. A high c-di-GMP level reduces the expression and/or activity of flagella and stimulates the synthesis of adhesins and biofilm-related extracellular polysaccharide [169], promoting biofilm formation [170]. In *S. aureus*, the second messenger, c-di-AMP, is synthesized by two molecules of ATP by the diadenylyl cyclase (DAC) enzyme DacA and degraded by the c-di-AMP phosphodiesterase GdpP-containing GGDEF domain [171]. Similar to c-di-GMP, in the *S. aureus* SEJ1 strain with *gdpP* mutation, the level of c-di-AMP and the amount of biofilm are significantly increased [171]. However, the *S. aureus* USA300 LAC strain could not form a robust biofilm in this condition [171], indicating that the effect of c-di-AMP on biofilm might be strain dependent. The *gdpP* mutation also leads to reduced eDNA levels, indicating that c-di-AMP is also involved in the release of eDNA [172]. In addition, The c-di-AMP also interacts with the purine biosynthesis pathway. In purine biosynthesis mutant methicillin-sensitive *S. aureus* (MRSA) strains, ADP, ATP, c-di-AMP, and eDNA levels were lower, and biofilm formation was less; vancomycin binding and its induced cleavage were increased [173]. In *S. aureus*, GdpS, a protein containing the GGDEF domain, inhibits early biofilm formation and is independent of autolysis by reducing *lrgAB* and *cidABC*-mediated release of eDNA [174]. These results reveal the influence of c-di-AMP on biofilms, which may play an important role in the persistence of *S. aureus* biofilm infection.

4.6. sRNAs

In addition to regulating transcriptional proteins, RNA molecules are now recognized as key players in gene regulation in all organisms [175]. Small RNAs (sRNAs) are usually non-coding and function at the level of transcription, translation, or RNA degradation [176], playing a key role in biofilm formation through base pairing with target mRNAs or interactions with regulatory proteins, resulting in both positive and negative regulatory mechanisms [177,178]. sRNA RsaA inhibits the synthesis of MgrA and enhances biofilm formation by binding to two regions of the mRNA of mgrA [179]. In addition, the synthesis of RsaA is controlled by SigB. Thus, RsaA functionally connects the global regulators SigB and MgrA [179]. The sRNA RsaF binds the hyaluronate lyase HsyA and serine protease-like protein SplD; the disruption of RsaF leads to a decreasing activity in HysA, which in turn increases biofilm formation [180]. sRNA RsaI (or RsaOG) depress the formation of biofilms at high glucose concentrations by binding to the 3′-UTR of *icaR*. Inhibition of IcaR synthesis by stabilization of mRNA recycles and/or by counteracts binding to activators of IcaR mRNA translation, but the exact molecular mechanism has not been determined.

The unwinding of biofilm formation occurs by binding to the 3′-UTR of icaR. This result corroborates with previous reports of increased biofilm formation observed in hysA mutants [181]. The 5′-untranslated region of *sarA* contains two sRNAs, named teg49 and teg48, which are detectable in the P3-P1 and P1 regions of the *sarA* promoter, respectively [182]. sRNA teg58 plays an important role in regulating arginine biosynthesis and biofilm formation in *S. aureus*. teg58 inhibits arginine synthesis; the arginine biosynthesis gene (*argGH*) is upregulated in teg58 mutants. Biofilm formation is reduced in parental strains after supplementation with exogenous arginine or endogenous *argGH* [183]. However, teg49 does not affect biofilm formation. Biofilm-associated genes (such as *ica* locus) are not affected in the case of teg49 inactivation or overexpression [184]. Transcriptomic analysis suggests that teg49 may post-transcriptionally affect the SaeRS and LytRS TCS, but the exact mechanism is unclear [184]. The sRNA SprX is encoded in the pathogenicity island of the MRSA Newman strain. SprX1 is one of three copies of SprX. SprX1 interacts directly with mRNAs encoding ClfB and Hld. Cells overexpressing SprX1 exhibited increased intercellular aggregation and biofilm formation [185]. In addition, SprX may modulate its effect on biofilms by increasing the stability of RNAIII [185]. In general, the mechanism of sRNA in the regulatory pathway of biofilm formation is not very clear at present. Further exploration is still needed.

5. Conclusions and Future Perspectives

The presence of *S. aureus* and most bacteria in the form of a biofilm is a considerable challenge for the medical community. Biofilms are a survival strategy for bacteria, making them extremely difficult to treat due to their inherent immune response and antibiotic resistance. Therefore, it is necessary to further study the formation and regulation mechanism of *S. aureus* biofilm for research and development of anti-biofilm drugs that inhibit biofilm formation. The process of biofilm formation is complex and involves the co-expression of multiple genes. *S. aureus* relies on a broad network of regulatory systems that coordinate biofilm formation in a complementary or opposing way. At present, although progress has been made in the regulatory mechanism of the formation of *S. aureus* biofilm, researchers have not yet understood the synergy between different regulatory networks. Moreover, the regulation mechanism of biofilm is dynamic. For example, different growth times and different environments may have different regulatory genes. How to realize these signals thus needs further investigation. At the same time, little is known about whether the regulation mechanism of biofilm in vivo is the same as in vitro. Human organoid models may be one of the most exciting tools to understand the regulation mechanism of *S. aureus* biofilm. On the other hand, the experimental results of reference strains may not be similar to the results of clinical isolates. The results obtained from reference strains may have some differences from that of clinical isolates. Therefore, different clinical isolates can be used for further research to provide better clinical significance. In addition, in clinical infection, *S. aureus* can also form mixed species biofilm with other bacteria and/or fungi. It is also important to include mixed species biofilm in the study.

Author Contributions: Conceptualization, Q.P.; writing—original draft preparation, Q.P., W.D. and X.T.; writing—review and editing, W.Y. and N.S. All authors have read and agreed to the published version of the manuscript.

Funding: This work was supported by the National Natural Science Foundation of China (grant No. 81703333), Natural Science Foundation of Guangdong Province (grant No. 2020A1515011326 and 2021A1515011360), Guangzhou Science and Technology Project (grant No. 202102080469), Guangzhou Health Science and Technology Project (grant No. 2020A010015), and Medical Scientific Research Foundation of Guangdong Province (grant No. C2021077).

Institutional Review Board Statement: Not applicable.

Informed Consent Statement: Not applicable.

Data Availability Statement: Not applicable.

Conflicts of Interest: The authors declare no conflict of interest.

References

1. Karygianni, L.; Ren, Z.; Koo, H.; Thurnheer, T. Biofilm Matrixome: Extracellular Components in Structured Microbial Communities. *Trends Microbiol.* **2020**, *28*, 668–681. [CrossRef] [PubMed]
2. Di Martino, P. Extracellular polymeric substances, a key element in understanding biofilm phenotype. *AIMS Microbiol.* **2018**, *4*, 274–288. [CrossRef] [PubMed]
3. Guilhen, C.; Forestier, C.; Balestrino, D. Biofilm dispersal: Multiple elaborate strategies for dissemination of bacteria with unique properties. *Mol. Microbiol.* **2017**, *105*, 188–210. [CrossRef] [PubMed]
4. Jamal, M.; Ahmad, W.; Andleeb, S.; Jalil, F.; Imran, M.; Nawaz, M.A.; Hussain, T.; Ali, M.; Rafiq, M.; Kamil, M.A. Bacterial biofilm and associated infections. *J. Chin. Med. Assoc.* **2018**, *81*, 7–11. [CrossRef] [PubMed]
5. Vestby, L.K.; Grønseth, T.; Simm, R.; Nesse, L.L. Bacterial Biofilm and its Role in the Pathogenesis of Disease. *Antibiotics* **2020**, *9*, 59. [CrossRef]
6. Khatoon, Z.; McTiernan, C.D.; Suuronen, E.J.; Mah, T.-F.; Alarcon, E.I. Bacterial biofilm formation on implantable devices and approaches to its treatment and prevention. *Heliyon* **2018**, *4*, e01067. [CrossRef]
7. Lowy, F.D. *Staphylococcus aureus* Infections. *N. Engl. J. Med.* **1998**, *339*, 520–532. [CrossRef]
8. Zheng, Y.; He, L.; Asiamah, T.K.; Otto, M. Colonization of medical devices by staphylococci. *Environ. Microbiol.* **2018**, *20*, 3141–3153. [CrossRef]
9. Lee, J.; Zilm, P.S.; Kidd, S.P. Novel Research Models for *Staphylococcus aureus* Small Colony Variants (SCV) Development: Co-pathogenesis and Growth Rate. *Front. Microbiol.* **2020**, *11*, 321. [CrossRef]
10. Hall-Stoodley, L.; Costerton, J.W.; Stoodley, P. Bacterial biofilms: From the Natural environment to infectious diseases. *Nat. Rev. Genet.* **2004**, *2*, 95–108. [CrossRef]
11. Guo, H.; Tong, Y.; Cheng, J.; Abbas, Z.; Li, Z.; Wang, J.; Zhou, Y.; Si, D.; Zhang, R. Biofilm and Small Colony Variants—An Update on *Staphylococcus aureus* Strategies toward Drug Resistance. *Int. J. Mol. Sci.* **2022**, *23*, 1241. [CrossRef] [PubMed]
12. Stewart, P.S. Mechanisms of antibiotic resistance in bacterial biofilms. *Int. J. Med. Microbiol.* **2002**, *292*, 107–113. [CrossRef] [PubMed]
13. Kumar, P.; Lee, J.-H.; Beyenal, H.; Lee, J. Fatty Acids as Antibiofilm and Antivirulence Agents. *Trends Microbiol.* **2020**, *28*, 753–768. [CrossRef] [PubMed]
14. Beloin, C.; Renard, S.; Ghigo, J.-M.; Lebeaux, D. Novel approaches to combat bacterial biofilms. *Curr. Opin. Pharmacol.* **2014**, *18*, 61–68. [CrossRef] [PubMed]
15. Girish, V.M.; Liang, H.; Aguilan, J.T.; Nosanchuk, J.D.; Friedman, J.M.; Nacharaju, P. Anti-biofilm activity of garlic extract loaded nanoparticles. *Nanomedicine* **2019**, *20*, 102009. [CrossRef] [PubMed]
16. Donne, J.; Dewilde, S. The Challenging World of Biofilm Physiology. In *Recent Advances in Microbial Oxygen-Binding Proteins*; Poole, R.K., Ed.; Advances in Microbial Physiology; Academic Press: London, UK, 2015; Volume 67, pp. 235–292.
17. Ribeiro, S.M.; Felício, M.R.; Boas, E.V.; Gonçalves, S.; Costa, F.F.; Samy, R.P.; Santos, N.C.; Franco, O.L. New frontiers for anti-biofilm drug development. *Pharmacol. Ther.* **2016**, *160*, 133–144. [CrossRef]
18. Idrees, M.; Sawant, S.; Karodia, N.; Rahman, A. *Staphylococcus aureus* Biofilm: Morphology, Genetics, Pathogenesis and Treatment Strategies. *Int. J. Environ. Res. Public Health* **2021**, *18*, 7602. [CrossRef]
19. Moormeier, D.E.; Bayles, K.W. *Staphylococcus aureus* biofilm: A complex developmental organism. *Mol. Microbiol.* **2017**, *104*, 365–376. [CrossRef]
20. Schilcher, K.; Horswill, A.R. Staphylococcal Biofilm Development: Structure, Regulation, and Treatment Strategies. *Microbiol. Mol. Biol. Rev.* **2020**, *84*, e00026-19. [CrossRef]
21. O'Neill, E.; Pozzi, C.; Houston, P.; Humphreys, H.; Robinson, D.A.; Loughman, A.; Foster, T.J.; O'Gara, J.P. A Novel *Staphylococcus aureus* Biofilm Phenotype Mediated by the Fibronectin-Binding Proteins, FnBPA and FnBPB. *J. Bacteriol.* **2008**, *190*, 3835–3850. [CrossRef]
22. Wolz, C.; Goerke, C.; Landmann, R.; Zimmerli, W.; Fluckiger, U. Transcription of Clumping Factor A in Attached and Unattached *Staphylococcus aureus* In Vitro and during Device-Related Infection. *Infect. Immun.* **2002**, *70*, 2758–2762. [CrossRef] [PubMed]
23. Sabat, A.; Melles, D.C.; Martirosian, G.; Grundmann, H.; van Belkum, A.; Hryniewicz, W. Distribution of the Serine-Aspartate Repeat Protein-Encoding *sdr* Genes among Nasal-Carriage and Invasive *Staphylococcus aureus* Strains. *J. Clin. Microbiol.* **2006**, *44*, 1135–1138. [CrossRef] [PubMed]
24. Beaussart, A.; Feuillie, C.; El-Kirat-Chatel, S. The microbial adhesive arsenal deciphered by atomic force microscopy. *Nanoscale* **2020**, *12*, 23885–23896. [CrossRef] [PubMed]
25. Mangwani, N.; Kumari, S.; Das, S. Bacterial biofilms and quorum sensing: Fidelity in bioremediation technology. In *Biotechnology and Genetic Engineering Reviews*; Mayes, S., Harding, S.E., Eds.; Biotechnology & Genetic Engineering Reviews; Taylor & Francis Ltd.: Abingdon, UK, 2016; Volume 32, pp. 43–73.
26. Otto, M. Staphylococcal biofilms. In *Bacterial Biofilms*; Romeo, T., Ed.; Current Topics in Microbiology and Immunology; Springer: Berlin/Heidelberg, Germany, 2008; Volume 322, p. 207.
27. Herman-Bausier, P.; El-Kirat-Chatel, S.; Foster, T.J.; Geoghegan, J.A.; Dufrêne, Y.F. *Staphylococcus aureus* Fibronectin-Binding Protein A Mediates Cell-Cell Adhesion through Low-Affinity Homophilic Bonds. *mBio* **2015**, *6*, e00413-15. [CrossRef] [PubMed]

28. Rasamiravaka, T.; Labtani, Q.; Duez, P.; El Jaziri, M. The Formation of Biofilms by *Pseudomonas aeruginosa*: A Review of the Natural and Synthetic Compounds Interfering with Control Mechanisms. *BioMed Res. Int.* **2015**, *2015*, 759348. [CrossRef]

29. O'Toole, G.; Kaplan, H.B.; Kolter, R. Biofilm Formation as Microbial Development. *Annu. Rev. Microbiol.* **2000**, *54*, 49–79. [CrossRef]

30. Gupta, P.; Sarkar, S.; Das, B.; Bhattacharjee, S.; Tribedi, P. Biofilm, pathogenesis and prevention—A journey to break the wall: A review. *Arch. Microbiol.* **2016**, *198*, 1–15. [CrossRef]

31. Costerton, J. Introduction to biofilm. *Int. J. Antimicrob. Agents* **1999**, *11*, 217–221; discussion 237–239. [CrossRef]

32. Lister, J.L.; Horswill, A.R. *Staphylococcus aureus* biofilms: Recent developments in biofilm dispersal. *Front. Cell. Infect. Microbiol.* **2014**, *4*, 178. [CrossRef]

33. Boles, B.R.; Horswill, A.R. Staphylococcal biofilm disassembly. *Trends Microbiol.* **2011**, *19*, 449–455. [CrossRef]

34. Vandana; Das, S. Genetic regulation, biosynthesis and applications of extracellular polysaccharides of the biofilm matrix of bacteria. *Carbohydr. Polym.* **2022**, *291*, 119536. [CrossRef] [PubMed]

35. Periasamy, S.; Joo, H.-S.; Duong, A.C.; Bach, T.-H.L.; Tan, V.Y.; Chatterjee, S.S.; Cheung, G.Y.C.; Otto, M. How *Staphylococcus aureus* biofilms develop their characteristic structure. *Proc. Natl. Acad. Sci. USA* **2012**, *109*, 1281–1286. [CrossRef] [PubMed]

36. Cheung, G.Y.; Joo, H.-S.; Chatterjee, S.S.; Otto, M. Phenol-soluble modulins—Critical determinants of staphylococcal virulence. *FEMS Microbiol. Rev.* **2014**, *38*, 698–719. [CrossRef]

37. Le, K.Y.; Dastgheyb, S.; Ho, T.V.; Otto, M. Molecular determinants of staphylococcal biofilm dispersal and structuring. *Front. Cell. Infect. Microbiol.* **2014**, *4*, 167. [CrossRef] [PubMed]

38. Schwartz, K.; Syed, A.K.; Stephenson, R.E.; Rickard, A.H.; Boles, B.R. Functional Amyloids Composed of Phenol Soluble Modulins Stabilize *Staphylococcus aureus* Biofilms. *PLoS Pathog.* **2012**, *8*, e1002744. [CrossRef] [PubMed]

39. Marinelli, P.; Pallares, I.; Navarro, S.; Ventura, S. Dissecting the contribution of *Staphylococcus aureus* α-phenol-soluble modulins to biofilm amyloid structure. *Sci. Rep.* **2016**, *6*, 34552. [CrossRef]

40. Otto, M. Staphylococcal Infections: Mechanisms of Biofilm Maturation and Detachment as Critical Determinants of Pathogenicity. *Annu. Rev. Med.* **2013**, *64*, 175–188. [CrossRef] [PubMed]

41. Nguyen, H.T.; Nguyen, T.H.; Otto, M. The staphylococcal exopolysaccharide PIA—Biosynthesis and role in biofilm formation, colonization, and infection. *Comput. Struct. Biotechnol. J.* **2020**, *18*, 3324–3334. [CrossRef]

42. Mack, D.; Fischer, W.; Krokotsch, A.; Leopold, K.; Hartmann, R.; Egge, H.; Laufs, R. The intercellular adhesin involved in biofilm accumulation of *Staphylococcus epidermidis* is a linear beta-1,6-linked glucosaminoglycan: Purification and structural analysis. *J. Bacteriol.* **1996**, *178*, 175–183. [CrossRef]

43. Cramton, S.E.; Gerke, C.; Schnell, N.F.; Nichols, W.W.; Götz, F. The Intercellular Adhesion (*ica*) Locus Is Present in *Staphylococcus aureus* and Is Required for Biofilm Formation. *Infect. Immun.* **1999**, *67*, 5427–5433. [CrossRef] [PubMed]

44. O'Gara, J.P. *ica* and beyond: Biofilm mechanisms and regulation in *Staphylococcus epidermidis* and *Staphylococcus aureus*. *FEMS Microbiol. Lett.* **2007**, *270*, 179–188. [CrossRef]

45. Gerke, C.; Kraft, A.; Süßmuth, R.; Schweitzer, O.; Götz, F. Characterization of the N-Acetylglucosaminyltransferase Activity Involved in the Biosynthesis of the *Staphylococcus epidermidis* Polysaccharide Intercellular Adhesin. *J. Biol. Chem.* **1998**, *273*, 18586–18593. [CrossRef] [PubMed]

46. Atkin, K.E.; MacDonald, S.J.; Brentnall, A.S.; Potts, J.R.; Thomas, G.H. A different path: Revealing the function of staphylococcal proteins in biofilm formation. *FEBS Lett.* **2014**, *588*, 1869–1872. [CrossRef]

47. Vuong, C.; Kocianova, S.; Voyich, J.M.; Yao, Y.; Fischer, E.R.; DeLeo, F.R.; Otto, M. A Crucial Role for Exopolysaccharide Modification in Bacterial Biofilm Formation, Immune Evasion, and Virulence. *J. Biol. Chem.* **2004**, *279*, 54881–54886. [CrossRef] [PubMed]

48. Sedarat, Z.; Taylor-Robinson, A.W. Biofilm Formation by Pathogenic Bacteria: Applying a *Staphylococcus aureus* Model to Appraise Potential Targets for Therapeutic Intervention. *Pathogens* **2022**, *11*, 388. [CrossRef] [PubMed]

49. Marques, V.F.; Santos, H.A.; Santos, T.H.; Melo, D.A.; Coelho, S.M.; Coelho, I.S.; Souza, M.M. Expression of *icaA* and *icaD* genes in biofilm formation in *Staphylococcus aureus* isolates from bovine subclinical mastitis. *Pesquisa Veterinária Brasileira* **2021**, *41*, e06645. [CrossRef]

50. Brooks, J.L.; Jefferson, K.K. Phase Variation of Poly-N-Acetylglucosamine Expression in *Staphylococcus aureus*. *PLoS Pathog.* **2014**, *10*, e1004292. [CrossRef]

51. O'Neill, E.; Pozzi, C.; Houston, P.; Smyth, D.; Humphreys, H.; Robinson, D.A.; O'Gara, J.P. Association between Methicillin Susceptibility and Biofilm Regulation in *Staphylococcus aureus* Isolates from Device-Related Infections. *J. Clin. Microbiol.* **2007**, *45*, 1379–1388. [CrossRef] [PubMed]

52. Cucarella, C.; Tormo, M.A.; Ubeda, C.; Trotonda, M.P.; Monzón, M.; Peris, C.; Amorena, B.; Lasa, I.; Penadés, J.R. Role of Biofilm-Associated Protein Bap in the Pathogenesis of Bovine *Staphylococcus aureus*. *Infect. Immun.* **2004**, *72*, 2177–2185. [CrossRef]

53. Taglialegna, A.; Navarro, S.; Ventura, S.; Garnett, J.A.; Matthews, S.; Penades, J.R.; Lasa, I.; Valle, J. Staphylococcal Bap Proteins Build Amyloid Scaffold Biofilm Matrices in Response to Environmental Signals. *PLoS Pathog.* **2016**, *12*, e1005711. [CrossRef]

54. Arrizubieta, M.J.; Toledo-Arana, A.; Amorena, B.; Penadés, J.R.; Lasa, I. Calcium Inhibits Bap-Dependent Multicellular Behavior in *Staphylococcus aureus*. *J. Bacteriol.* **2004**, *186*, 7490–7498. [CrossRef] [PubMed]

55. Pietrocola, G.; Campoccia, D.; Motta, C.; Montanaro, L.; Arciola, C.R.; Speziale, P. Colonization and Infection of Indwelling Medical Devices by *Staphylococcus aureus* with an Emphasis on Orthopedic Implants. *Int. J. Mol. Sci.* **2022**, *23*, 5958. [CrossRef] [PubMed]

56. McCourt, J.; O'Halloran, D.P.; McCarthy, H.; O'Gara, J.P.; Geoghegan, J.A. Fibronectin-binding proteins are required for biofilm formation by community-associated methicillin-resistant *Staphylococcus aureus* strain LAC. *FEMS Microbiol. Lett.* **2014**, *353*, 157–164. [CrossRef]

57. Herman-Bausier, P.; Labate, C.; Towell, A.M.; Derclaye, S.; Geoghegan, J.A.; Dufrêne, Y.F. *Staphylococcus aureus* clumping factor A is a force-sensitive molecular switch that activates bacterial adhesion. *Proc. Natl. Acad. Sci. USA* **2018**, *115*, 5564–5569. [CrossRef]

58. Abraham, N.M.; Jefferson, K.K. *Staphylococcus aureus* clumping factor B mediates biofilm formation in the absence of calcium. *Microbiology* **2012**, *158*, 1504–1512. [CrossRef] [PubMed]

59. Formosa-Dague, C.; Speziale, P.; Foster, T.J.; Geoghegan, J.A.; Dufrêne, Y.F. Zinc-dependent mechanical properties of *Staphylococcus aureus* biofilm-forming surface protein SasG. *Proc. Natl. Acad. Sci. USA* **2016**, *113*, 410–415. [CrossRef] [PubMed]

60. Corrigan, R.M.; Rigby, D.; Handley, P.; Foster, T.J. The role of *Staphylococcus aureus* surface protein SasG in adherence and biofilm formation. *Microbiology* **2007**, *153*, 2435–2446. [CrossRef]

61. Schroeder, K.; Jularic, M.; Horsburgh, S.M.; Hirschhausen, N.; Neumann, C.; Bertling, A.; Schulte, A.; Foster, S.; Kehrel, B.E.; Peters, G.; et al. Molecular Characterization of a Novel *Staphylococcus Aureus* Surface Protein (SasC) Involved in Cell Aggregation and Biofilm Accumulation. *PLoS ONE* **2009**, *4*, e7567. [CrossRef]

62. Feuillie, C.; Formosa-Dague, C.; Hays, L.M.C.; Vervaeck, O.; Derclaye, S.; Brennan, M.P.; Foster, T.J.; Geoghegan, J.A.; Dufrêne, Y.F. Molecular interactions and inhibition of the staphylococcal biofilm-forming protein SdrC. *Proc. Natl. Acad. Sci. USA* **2017**, *114*, 3738–3743. [CrossRef]

63. Corrigan, R.M.; Miajlovic, H.; Foster, T.J. Surface proteins that promote adherence of *Staphylococcus aureus*to human desquamated nasal epithelial cells. *BMC Microbiol.* **2009**, *9*, 22. [CrossRef]

64. Askarian, F.; Ajayi, C.; Hanssen, A.-M.; van Sorge, N.; Pettersen, I.; Diep, D.B.; Sollid, J.U.E.; Johannessen, M. The interaction between *Staphylococcus aureus* SdrD and desmoglein 1 is important for adhesion to host cells. *Sci. Rep.* **2016**, *6*, 22134. [CrossRef] [PubMed]

65. Askarian, F.; Uchiyama, S.; Valderrama, J.A.; Ajayi, C.; Sollid, J.U.E.; van Sorge, N.M.; Nizet, V.; van Strijp, J.A.G.; Johannessen, M. Serine-Aspartate Repeat Protein D Increases *Staphylococcus aureus* Virulence and Survival in Blood. *Infect. Immun.* **2017**, *85*, e00559-16. [CrossRef]

66. Zhang, Y.; Wu, M.; Hang, T.; Wang, C.; Yang, Y.; Pan, W.; Zang, J.; Zhang, M.; Zhang, X. *Staphylococcus aureus* SdrE captures complement factor H's C-terminus via a novel 'close, dock, lock and latch' mechanism for complement evasion. *Biochem. J.* **2017**, *474*, 1619–1631. [CrossRef]

67. Patti, J.M.; Boles, J.O.; Hook, M. Identification and biochemical characterization of the ligand binding domain of the collagen adhesin from *Staphylococcus aureus*. *Biochemistry* **1993**, *32*, 11428–11435. [CrossRef] [PubMed]

68. Singh, K.V.; Nallapareddy, S.R.; Sillanpää, J.; Murray, B.E. Importance of the Collagen Adhesin Ace in Pathogenesis and Protection against *Enterococcus faecalis* Experimental Endocarditis. *PLoS Pathog.* **2010**, *6*, e1000716. [CrossRef]

69. Nakano, K.; Hokamura, K.; Taniguchi, N.; Wada, K.; Kudo, C.; Nomura, R.; Kojima, A.; Naka, S.; Muranaka, Y.; Thura, M.; et al. The collagen-binding protein of *Streptococcus mutans* is involved in haemorrhagic stroke. *Nat. Commun.* **2011**, *2*, 485. [CrossRef] [PubMed]

70. Kang, M.; Ko, Y.-P.; Liang, X.; Ross, C.L.; Liu, Q.; Murray, B.E.; Höök, M. Collagen-binding Microbial Surface Components Recognizing Adhesive Matrix Molecule (MSCRAMM) of Gram-positive Bacteria Inhibit Complement Activation via the Classical Pathway. *J. Biol. Chem.* **2013**, *288*, 20520–20531. [CrossRef]

71. Mann, E.E.; Rice, K.C.; Boles, B.R.; Endres, J.L.; Ranjit, D.; Chandramohan, L.; Tsang, L.H.; Smeltzer, M.S.; Horswill, A.R.; Bayles, K.W. Modulation of eDNA Release and Degradation Affects *Staphylococcus aureus* Biofilm Maturation. *PLoS ONE* **2009**, *4*, e5822. [CrossRef]

72. Campoccia, D.; Montanaro, L.; Arciola, C.R. Extracellular DNA (eDNA). A Major Ubiquitous Element of the Bacterial Biofilm Architecture. *Int. J. Mol. Sci.* **2021**, *22*, 9100. [CrossRef] [PubMed]

73. Blakeman, J.T.; Morales-García, A.L.; Mukherjee, J.; Gori, K.; Hayward, A.S.; Lant, N.J.; Geoghegan, M. Extracellular DNA Provides Structural Integrity to a *Micrococcus luteus* Biofilm. *Langmuir* **2019**, *35*, 6468–6475. [CrossRef] [PubMed]

74. Payne, D.E.; Boles, B.R. Emerging interactions between matrix components during biofilm development. *Curr. Genet.* **2016**, *62*, 137–141. [CrossRef] [PubMed]

75. Das, T.; Sharma, P.; Busscher, H.J.; van der Mei, H.C.; Krom, B.P. Role of Extracellular DNA in Initial Bacterial Adhesion and Surface Aggregation. *Appl. Environ. Microbiol.* **2010**, *76*, 3405–3408. [CrossRef]

76. Wilton, M.; Charron-Mazenod, L.; Moore, R.; Lewenza, S. Extracellular DNA Acidifies Biofilms and Induces Aminoglycoside Resistance in *Pseudomonas aeruginosa*. *Antimicrob. Agents Chemother.* **2016**, *60*, 544–553. [CrossRef]

77. Hall, C.W.; Mah, T.-F. Molecular mechanisms of biofilm-based antibiotic resistance and tolerance in pathogenic bacteria. *FEMS Microbiol. Rev.* **2017**, *41*, 276–301. [CrossRef]

78. Batoni, G.; Maisetta, G.; Esin, S. Antimicrobial peptides and their interaction with biofilms of medically relevant bacteria. *Biochim. Biophys. Acta* **2016**, *1858*, 1044–1060. [CrossRef]

79. Doroshenko, N.; Tseng, B.S.; Howlin, R.P.; Deacon, J.; Wharton, J.A.; Thurner, P.J.; Gilmore, B.F.; Parsek, M.R.; Stoodley, P. Extracellular DNA Impedes the Transport of Vancomycin in *Staphylococcus epidermidis* Biofilms Preexposed to Subinhibitory Concentrations of Vancomycin. *Antimicrob. Agents Chemother.* **2014**, *58*, 7273–7282. [CrossRef] [PubMed]
80. Rice, K.C.; Firek, B.A.; Nelson, J.B.; Yang, S.-J.; Patton, T.G.; Bayles, K.W. The *Staphylococcus aureus cidAB* Operon: Evaluation of Its Role in Regulation of Murein Hydrolase Activity and Penicillin Tolerance. *J. Bacteriol.* **2003**, *185*, 2635–2643. [CrossRef]
81. Groicher, K.H.; Firek, B.A.; Fujimoto, D.F.; Bayles, K.W. The *Staphylococcus aureus lrgAB* Operon Modulates Murein Hydrolase Activity and Penicillin Tolerance. *J. Bacteriol.* **2000**, *182*, 1794–1801. [CrossRef] [PubMed]
82. Parsek, M.R.; Greenberg, E. Sociomicrobiology: The connections between quorum sensing and biofilms. *Trends Microbiol.* **2005**, *13*, 27–33. [CrossRef] [PubMed]
83. Lu, L.; Hu, W.; Tian, Z.; Yuan, D.; Yi, G.; Zhou, Y.; Cheng, Q.; Zhu, J.; Li, M. Developing natural products as potential anti-biofilm agents. *Chin. Med.* **2019**, *14*, 11. [CrossRef] [PubMed]
84. Sionov, R.V.; Steinberg, D. Targeting the Holy Triangle of Quorum Sensing, Biofilm Formation, and Antibiotic Resistance in Pathogenic Bacteria. *Microorganisms* **2022**, *10*, 1239. [CrossRef]
85. Le, K.Y.; Otto, M. Quorum-sensing regulation in staphylococci—An overview. *Front. Microbiol.* **2015**, *6*, 1174. [CrossRef]
86. Dunman, P.M.; Murphy, E.; Haney, S.; Palacios, D.; Tucker-Kellogg, G.; Wu, S.; Brown, E.L.; Zagursky, R.J.; Shlaes, D.; Projan, S.J. Transcription Profiling-Based Identification of *Staphylococcus aureus* Genes Regulated by the *agr* and/or *sarA* Loci. *J. Bacteriol.* **2001**, *183*, 7341–7353. [CrossRef]
87. Harraghy, N.; Kerdudou, S.; Herrmann, M. Quorum-sensing systems in staphylococci as therapeutic targets. *Anal. Bioanal. Chem.* **2006**, *387*, 437–444. [CrossRef] [PubMed]
88. Ji, G.; Beavis, R.; Novick, R.P. Bacterial Interference Caused by Autoinducing Peptide Variants. *Science* **1997**, *276*, 2027–2030. [CrossRef] [PubMed]
89. Gordon, C.P.; Olson, S.D.; Lister, J.L.; Kavanaugh, J.S.; Horswill, A.R. Truncated Autoinducing Peptides as Antagonists of *Staphylococcus lugdunensis* Quorum Sensing. *J. Med. Chem.* **2016**, *59*, 8879–8888. [CrossRef]
90. Novick, R.P.; Geisinger, E. Quorum Sensing in Staphylococci. *Annu. Rev. Genet.* **2008**, *42*, 541–564. [CrossRef] [PubMed]
91. Mootz, J.M.; Benson, M.A.; Heim, C.E.; Crosby, H.A.; Kavanaugh, J.S.; Dunman, P.M.; Kielian, T.; Torres, V.J.; Horswill, A.R. Rot is a key regulator of *Staphylococcus aureus* biofilm formation. *Mol. Microbiol.* **2015**, *96*, 388–404. [CrossRef] [PubMed]
92. Yarwood, J.M.; Bartels, D.J.; Volper, E.M.; Greenberg, E.P. Quorum Sensing in *Staphylococcus aureus* Biofilms. *J. Bacteriol.* **2004**, *186*, 1838–1850. [CrossRef] [PubMed]
93. Boles, B.R.; Horswill, A.R. agr-Mediated Dispersal of *Staphylococcus aureus* Biofilms. *PLoS Pathog.* **2008**, *4*, e1000052. [CrossRef] [PubMed]
94. Queck, S.Y.; Jameson-Lee, M.; Villaruz, A.E.; Bach, T.-H.L.; Khan, B.A.; Sturdevant, D.E.; Ricklefs, S.M.; Li, M.; Otto, M. RNAIII-Independent Target Gene Control by the agr Quorum-Sensing System: Insight into the Evolution of Virulence Regulation in *Staphylococcus aureus*. *Mol. Cell* **2008**, *32*, 150–158. [CrossRef] [PubMed]
95. Le, K.Y.; Villaruz, A.E.; Zheng, Y.; He, L.; Fisher, E.L.; Nguyen, T.H.; Ho, T.V.; Yeh, A.J.; Joo, H.-S.; Cheung, G.Y.; et al. Role of Phenol-Soluble Modulins in *Staphylococcus epidermidis* Biofilm Formation and Infection of Indwelling Medical Devices. *J. Mol. Biol.* **2019**, *431*, 3015–3027. [CrossRef] [PubMed]
96. Schauder, S.; Shokat, K.; Surette, M.G.; Bassler, B.L. The LuxS family of bacterial autoinducers: Biosynthesis of a novel quorum-sensing signal molecule. *Mol. Microbiol.* **2001**, *41*, 463–476. [CrossRef] [PubMed]
97. Xu, L.; Li, H.; Vuong, C.; Vadyvaloo, V.; Wang, J.; Yao, Y.; Otto, M.; Gao, Q. Role of the *luxS* Quorum-Sensing System in Biofilm Formation and Virulence of *Staphylococcus epidermidis*. *Infect. Immun.* **2006**, *74*, 488–496. [CrossRef] [PubMed]
98. Yu, D.; Zhao, L.; Xue, T.; Sun, B. *Staphylococcus aureus* autoinducer-2 quorum sensing decreases biofilm formation in an *icaR*-dependent manner. *BMC Microbiol.* **2012**, *12*, 288. [CrossRef] [PubMed]
99. Cue, D.; Lei, M.G.; Lee, C.Y. Activation of *sarX* by Rbf Is Required for Biofilm Formation and *icaADBC* Expression in *Staphylococcus aureus*. *J. Bacteriol.* **2013**, *195*, 1515–1524. [CrossRef] [PubMed]
100. Ma, R.; Qiu, S.; Jiang, Q.; Sun, H.; Xue, T.; Cai, G.; Sun, B. AI-2 quorum sensing negatively regulates rbf expression and biofilm formation in *Staphylococcus aureus*. *Int. J. Med. Microbiol.* **2017**, *307*, 257–267. [CrossRef] [PubMed]
101. Xue, T.; Ni, J.; Shang, F.; Chen, X.; Zhang, M. Autoinducer-2 increases biofilm formation via an ica- and bhp-dependent manner in *Staphylococcus epidermidis* RP62A. *Microbes Infect.* **2015**, *17*, 345–352. [CrossRef]
102. Selvaraj, A.; Jayasree, T.; Valliammai, A.; Pandian, S.K. Myrtenol Attenuates MRSA Biofilm and Virulence by Suppressing *sarA* Expression Dynamism. *Front. Microbiol.* **2019**, *10*, 2027. [CrossRef]
103. Balamurugan, P.; Krishna, V.P.; Bharath, D.; Lavanya, R.; Vairaprakash, P.; Princy, S.A. *Staphylococcus aureus* Quorum Regulator SarA Targeted Compound, 2-[(Methylamino)methyl]phenol Inhibits Biofilm and Down-Regulates Virulence Genes. *Front. Microbiol.* **2017**, *8*, 1290. [CrossRef]
104. Valle, J.; Toledo-Arana, A.; Berasain, C.; Ghigo, J.-M.; Amorena, B.; Penadés, J.R.; Lasa, I. SarA and not σB is essential for biofilm development by *Staphylococcus aureus*. *Mol. Microbiol.* **2003**, *48*, 1075–1087. [CrossRef] [PubMed]
105. Chien, Y.-T.; Cheung, A.L. Molecular Interactions between Two Global Regulators, sar and agr, in *Staphylococcus aureus*. *J. Biol. Chem.* **1998**, *273*, 2645–2652. [CrossRef]
106. Cheung, A.L.; Nishina, K.; Manna, A.C. SarA of *Staphylococcus aureus* Binds to the *sarA* Promoter To Regulate Gene Expression. *J. Bacteriol.* **2008**, *190*, 2239–2243. [CrossRef]

107. Reyes, D.; Andrey, D.O.; Monod, A.; Kelley, W.L.; Zhang, G.; Cheung, A.L. Coordinated Regulation by AgrA, SarA, and SarR To Control *agr* Expression in *Staphylococcus aureus*. *J. Bacteriol.* **2011**, *193*, 6020–6031. [CrossRef]

108. Chien, Y.-T.; Manna, A.C.; Projan, S.J.; Cheung, A.L. SarA, a Global Regulator of Virulence Determinants in *Staphylococcus aureus*, Binds to a Conserved Motif Essential for sar-dependent Gene Regulation. *J. Biol. Chem.* **1999**, *274*, 37169–37176. [CrossRef]

109. Xiong, Y.Q.; Willard, J.; Yeaman, M.R.; Cheung, A.L.; Bayer, A.S. Regulation of *Staphylococcus aureus* α-Toxin Gene *(hla)* Expression by *agr*, *sarA*, and *sae* In Vitro and in Experimental Infective Endocarditis. *J. Infect. Dis.* **2006**, *194*, 1267–1275. [CrossRef]

110. Zielinska, A.K.; Beenken, K.E.; Mrak, L.N.; Spencer, H.J.; Post, G.R.; Skinner, R.A.; Tackett, A.J.; Horswill, A.R.; Smeltzer, M.S. *sarA*-mediated repression of protease production plays a key role in the pathogenesis of *Staphylococcus aureus* USA300 isolates. *Mol. Microbiol.* **2012**, *86*, 1183–1196. [CrossRef] [PubMed]

111. Loughran, A.J.; Atwood, D.N.; Anthony, A.C.; Harik, N.S.; Spencer, H.J.; Beenken, K.E.; Smeltzer, M.S. Impact of individual extracellular proteases on *Staphylococcus aureus* biofilm formation in diverse clinical isolates and their isogenic *sarA* mutants. *MicrobiologyOpen* **2014**, *3*, 897–909. [CrossRef]

112. Kullik, I.; Giachino, P.; Fuchs, T. Deletion of the Alternative Sigma Factor ς^B in *Staphylococcus aureus* Reveals Its Function as a Global Regulator of Virulence Genes. *J. Bacteriol.* **1998**, *180*, 4814–4820. [CrossRef]

113. Rachid, S.; Ohlsen, K.; Wallner, U.; Hacker, J.; Hecker, M.; Ziebuhr, W. Alternative Transcription Factor ς^B Is Involved in Regulation of Biofilm Expression in a *Staphylococcus aureus* Mucosal Isolate. *J. Bacteriol.* **2000**, *182*, 6824–6826. [CrossRef] [PubMed]

114. Valle, J.; Echeverz, M.; Lasa, I. σ^B Inhibits Poly-*N*-Acetylglucosamine Exopolysaccharide Synthesis and Biofilm Formation in *Staphylococcus aureus*. *J. Bacteriol.* **2019**, *201*, e00098-19. [CrossRef]

115. Bischoff, M.; Entenza, J.M.; Giachino, P. Influence of a Functional *sigB* Operon on the Global Regulators *sar* and *agr* in *Staphylococcus aureus*. *J. Bacteriol.* **2001**, *183*, 5171–5179. [CrossRef]

116. Entenza, J.-M.; Moreillon, P.; Senn, M.M.; Kormanec, J.; Dunman, P.M.; Berger-Bächi, B.; Projan, S.; Bischoff, M.; Aguilar-Be, I.; Zardo, R.D.S.; et al. Role of σ^B in the Expression of *Staphylococcus aureus* Cell Wall Adhesins ClfA and FnbA and Contribution to Infectivity in a Rat Model of Experimental Endocarditis. *Infect. Immun.* **2005**, *73*, 990–998. [CrossRef]

117. Atwood, D.N.; Loughran, A.J.; Courtney, A.P.; Anthony, A.C.; Meeker, D.G.; Spencer, H.J.; Gupta, R.K.; Lee, C.Y.; Beenken, K.E.; Smeltzer, M.S. Comparative impact of diverse regulatory loci on *Staphylococcus aureus* biofilm formation. *MicrobiologyOpen* **2015**, *4*, 436–451. [CrossRef]

118. Waters, N.R.; Samuels, D.J.; Behera, R.K.; Livny, J.; Rhee, K.Y.; Sadykov, M.R.; Brinsmade, S.R. A spectrum of CodY activities drives metabolic reorganization and virulence gene expression in *Staphylococcus aureus*. *Mol. Microbiol.* **2016**, *101*, 495–514. [CrossRef] [PubMed]

119. Stenz, L.; Francois, P.; Whiteson, K.; Wolz, C.; Linder, P.; Schrenzel, J. The CodY pleiotropic repressor controls virulence in gram-positive pathogens. *FEMS Immunol. Med. Microbiol.* **2011**, *62*, 123–139. [CrossRef] [PubMed]

120. Majerczyk, C.D.; Sadykov, M.R.; Luong, T.T.; Lee, C.; Somerville, G.A.; Sonenshein, A.L. *Staphylococcus aureus* CodY Negatively Regulates Virulence Gene Expression. *J. Bacteriol.* **2008**, *190*, 2257–2265. [CrossRef] [PubMed]

121. Majerczyk, C.D.; Dunman, P.M.; Luong, T.T.; Lee, C.Y.; Sadykov, M.R.; Somerville, G.A.; Bodi, K.; Sonenshein, A.L. Direct Targets of CodY in *Staphylococcus aureus*. *J. Bacteriol.* **2010**, *192*, 2861–2877. [CrossRef]

122. Roux, A.; Todd, D.A.; Velázquez, J.V.; Cech, N.B.; Sonenshein, A.L. CodY-Mediated Regulation of the *Staphylococcus aureus* Agr System Integrates Nutritional and Population Density Signals. *J. Bacteriol.* **2014**, *196*, 1184–1196. [CrossRef]

123. Mlynek, K.D.; Bulock, L.L.; Stone, C.J.; Curran, L.J.; Sadykov, M.R.; Bayles, K.W.; Brinsmade, S.R. Genetic and Biochemical Analysis of CodY-Mediated Cell Aggregation in *Staphylococcus aureus* Reveals an Interaction between Extracellular DNA and Polysaccharide in the Extracellular Matrix. *J. Bacteriol.* **2020**, *202*, e00593-19. [CrossRef]

124. Rom, J.S.; Atwood, D.N.; Beenken, K.E.; Meeker, D.G.; Loughran, A.J.; Spencer, H.J.; Lantz, T.L.; Smeltzer, M.S. Impact of *Staphylococcus aureus* regulatory mutations that modulate biofilm formation in the USA300 strain LAC on virulence in a murine bacteremia model. *Virulence* **2017**, *8*, 1776–1790. [CrossRef] [PubMed]

125. Mlynek, K.D.; Sause, W.E.; Moormeier, D.E.; Sadykov, M.R.; Hill, K.R.; Torres, V.J.; Bayles, K.W.; Brinsmade, S.R. Nutritional Regulation of the Sae Two-Component System by CodY in *Staphylococcus aureus*. *J. Bacteriol.* **2018**, *200*, e00012-18. [CrossRef] [PubMed]

126. Conlon, K.M.; Humphreys, H.; O'Gara, J.P. *icaR* Encodes a Transcriptional Repressor Involved in Environmental Regulation of *ica* Operon Expression and Biofilm Formation in *Staphylococcus epidermidis*. *J. Bacteriol.* **2002**, *184*, 4400–4408. [CrossRef] [PubMed]

127. Hoang, T.-M.; Zhou, C.; Lindgren, J.K.; Galac, M.R.; Corey, B.; Endres, J.E.; Olson, M.E.; Fey, P.D. Transcriptional Regulation of *icaADBC* by both IcaR and TcaR in *Staphylococcus epidermidis*. *J. Bacteriol.* **2019**, *201*, e00524-18. [CrossRef] [PubMed]

128. Jefferson, K.K.; Pier, D.B.; Goldmann, D.A.; Pier, G.B. The Teicoplanin-Associated Locus Regulator (TcaR) and the Intercellular Adhesin Locus Regulator (IcaR) Are Transcriptional Inhibitors of the *ica* Locus in *Staphylococcus aureus*. *J. Bacteriol.* **2004**, *186*, 2449–2456. [CrossRef] [PubMed]

129. Chang, Y.-M.; Jeng, W.-Y.; Ko, T.-P.; Yeh, Y.-J.; Chen, C.K.-M.; Wang, A.H.-J. Structural study of TcaR and its complexes with multiple antibiotics from *Staphylococcus epidermidis*. *Proc. Natl. Acad. Sci. USA* **2010**, *107*, 8617–8622. [CrossRef]

130. Schwartbeck, B.; Birtel, J.; Treffon, J.; Langhanki, L.; Mellmann, A.; Kale, D.; Kahl, J.; Hirschhausen, N.; Neumann, C.; Lee, J.C.; et al. Dynamic in vivo mutations within the ica operon during persistence of *Staphylococcus aureus* in the airways of cystic fibrosis patients. *PLoS Pathog.* **2016**, *12*, e1006024. [CrossRef]

131. Jefferson, K.K.; Cramton, S.E.; Götz, F.; Pier, G. Identification of a 5-nucleotide sequence that controls expression of the ica locus in *Staphylococcus aureus* and characterization of the DNA-binding properties of IcaR. *Mol. Microbiol.* **2003**, *48*, 889–899. [CrossRef]

132. Yu, L.; Hisatsune, J.; Hayashi, I.; Tatsukawa, N.; Sato'O, Y.; Mizumachi, E.; Kato, F.; Hirakawa, H.; Pier, G.B.; Sugai, M. A Novel Repressor of the *ica* Locus Discovered in Clinically Isolated Super-Biofilm-Elaborating *Staphylococcus aureus*. *mBio* **2017**, *8*, e02282-16. [CrossRef]

133. You, Y.; Xue, T.; Cao, L.; Zhao, L.; Sun, H.; Sun, B. *Staphylococcus aureus* glucose-induced biofilm accessory proteins, GbaAB, influence biofilm formation in a PIA-dependent manner. *Int. J. Med. Microbiol.* **2014**, *304*, 603–612. [CrossRef]

134. Tiwari, N.; López-Redondo, M.; Miguel-Romero, L.; Kulhankova, K.; Cahill, M.P.; Tran, P.M.; Kinney, K.J.; Kilgore, S.H.; Al-Tameemi, H.; Herfst, C.A.; et al. The SrrAB two-component system regulates *Staphylococcus aureus* pathogenicity through redox sensitive cysteines. *Proc. Natl. Acad. Sci. USA* **2020**, *117*, 10989–10999. [CrossRef] [PubMed]

135. Park, M.K.; Myers, R.A.M.; Marzella, L. Oxygen Tensions and Infections: Modulation of Microbial Growth, Activity of Antimicrobial Agents, and Immunologic Responses. *Clin. Infect. Dis.* **1992**, *14*, 720–740. [CrossRef] [PubMed]

136. Pragman, A.A.; Yarwood, J.M.; Tripp, T.J.; Schlievert, P.M. Characterization of Virulence Factor Regulation by SrrAB, a Two-Component System in *Staphylococcus aureus*. *J. Bacteriol.* **2004**, *186*, 2430–2438. [CrossRef] [PubMed]

137. Ulrich, M.; Bastian, M.; Cramton, S.E.; Ziegler, K.; Pragman, A.A.; Bragonzi, A.; Memmi, G.; Wolz, C.; Schlievert, P.; Cheung, A.; et al. The staphylococcal respiratory response regulator SrrAB induces *ica* gene transcription and polysaccharide intercellular adhesin expression, protecting *Staphylococcus aureus* from neutrophil killing under anaerobic growth conditions. *Mol. Microbiol.* **2007**, *65*, 1276–1287. [CrossRef] [PubMed]

138. Quoc, P.H.T.; Genevaux, P.; Pajunen, M.; Savilahti, H.; Georgopoulos, C.; Schrenzel, J.; Kelley, W.L. Isolation and Characterization of Biofilm Formation-Defective Mutants of *Staphylococcus aureus*. *Infect. Immun.* **2007**, *75*, 1079–1088. [CrossRef] [PubMed]

139. Mashruwala, A.A.; Van De Guchte, A.; Boyd, J.M. Impaired respiration elicits SrrAB-dependent programmed cell lysis and biofilm formation in *Staphylococcus aureus*. *eLife* **2017**, *6*, e23845. [CrossRef] [PubMed]

140. Kinkel, T.L.; Roux, C.M.; Dunman, P.M.; Fang, F.C. The *Staphylococcus aureus* SrrAB Two-Component System Promotes Resistance to Nitrosative Stress and Hypoxia. *mBio* **2013**, *4*, e00696-13. [CrossRef] [PubMed]

141. Lamret, F.; Varin-Simon, J.; Velard, F.; Terryn, C.; Mongaret, C.; Colin, M.; Gangloff, S.C.; Reffuveille, F. *Staphylococcus aureus* Strain-Dependent Biofilm Formation in Bone-Like Environment. *Front. Microbiol.* **2021**, *12*, 714994. [CrossRef] [PubMed]

142. Cho, H.; Jeong, D.-W.; Liu, Q.; Yeo, W.-S.; Vogl, T.; Skaar, E.P.; Chazin, W.J.; Bae, T. Calprotectin Increases the Activity of the SaeRS Two Component System and Murine Mortality during *Staphylococcus aureus* Infections. *PLoS Pathog.* **2015**, *11*, e1005026. [CrossRef]

143. Liu, Q.; Yeo, W.; Bae, T. The SaeRS Two-Component System of *Staphylococcus aureus*. *Genes* **2016**, *7*, 81. [CrossRef]

144. Jeong, D.-W.; Cho, H.; Jones, M.B.; Shatzkes, K.; Sun, F.; Ji, Q.; Liu, Q.; Peterson, S.N.; He, C.; Bae, T. The auxiliary protein complex SaePQ activates the phosphatase activity of sensor kinase SaeS in the SaeRS two-component system of *Staphylococcus aureus*. *Mol. Microbiol.* **2012**, *86*, 331–348. [CrossRef] [PubMed]

145. Liang, X.; Yu, C.; Sun, J.; Liu, H.; Landwehr, C.; Holmes, D.; Ji, Y. Inactivation of a Two-Component Signal Transduction System, SaeRS, Eliminates Adherence and Attenuates Virulence of *Staphylococcus aureus*. *Infect. Immun.* **2006**, *74*, 4655–4665. [CrossRef] [PubMed]

146. Olson, M.E.; Nygaard, T.K.; Ackermann, L.; Watkins, R.L.; Zurek, O.W.; Pallister, K.B.; Griffith, S.; Kiedrowski, M.R.; Flack, C.E.; Kavanaugh, J.S.; et al. *Staphylococcus aureus* Nuclease Is an SaeRS-Dependent Virulence Factor. *Infect. Immun.* **2013**, *81*, 1316–1324. [CrossRef] [PubMed]

147. Mrak, L.N.; Zielinska, A.K.; Beenken, K.E.; Mrak, I.N.; Atwood, D.N.; Griffin, L.M.; Lee, C.Y.; Smeltzer, M.S. *saeRS* and *sarA* Act Synergistically to Repress Protease Production and Promote Biofilm Formation in *Staphylococcus aureus*. *PLoS ONE* **2012**, *7*, e38453. [CrossRef] [PubMed]

148. Cue, D.; Junecko, J.M.; Lei, M.G.; Blevins, J.S.; Smeltzer, M.; Lee, C.Y. SaeRS-Dependent Inhibition of Biofilm Formation in *Staphylococcus aureus* Newman. *PLoS ONE* **2015**, *10*, e0123027. [CrossRef] [PubMed]

149. Mashruwala, A.A.; Gries, C.M.; Scherr, T.D.; Kielian, T.; Boyd, J.M. SaeRS Is Responsive to Cellular Respiratory Status and Regulates Fermentative Biofilm Formation in *Staphylococcus aureus*. *Infect. Immun.* **2017**, *85*, e00157-17. [CrossRef]

150. Wittekind, M.A.; Frey, A.; Bonsall, A.E.; Briaud, P.; Keogh, R.A.; Wiemels, R.E.; Shaw, L.N.; Carroll, R.K. The novel protein ScrA acts through the SaeRS two-component system to regulate virulence gene expression in *Staphylococcus aureus*. *Mol. Microbiol.* **2022**, *117*, 1196–1212. [CrossRef]

151. Ranjit, D.K.; Endres, J.L.; Bayles, K.W. *Staphylococcus aureus* CidA and LrgA Proteins Exhibit Holin-Like Properties. *J. Bacteriol.* **2011**, *193*, 2468–2476. [CrossRef] [PubMed]

152. Rice, K.C.; Bayles, K.W. Molecular Control of Bacterial Death and Lysis. *Microbiol. Mol. Biol. Rev.* **2008**, *72*, 85–109. [CrossRef]

153. Rice, K.C.; Mann, E.E.; Endres, J.L.; Weiss, E.C.; Cassat, J.E.; Smeltzer, M.S.; Bayles, K.W. The *cidA* murein hydrolase regulator contributes to DNA release and biofilm development in *Staphylococcus aureus*. *Proc. Natl. Acad. Sci. USA* **2007**, *104*, 8113–8118. [CrossRef] [PubMed]

154. Leroy, S.; Lebert, I.; Andant, C.; Micheau, P.; Talon, R. Investigating Extracellular DNA Release in *Staphylococcus xylosus* Biofilm In Vitro. *Microorganisms* **2021**, *9*, 2192. [CrossRef] [PubMed]

155. Yang, S.-J.; Dunman, P.M.; Projan, S.J.; Bayles, K.W. Characterization of the *Staphylococcus aureus* CidR regulon: Elucidation of a novel role for acetoin metabolism in cell death and lysis. *Mol. Microbiol.* **2006**, *60*, 458–468. [CrossRef] [PubMed]

156. Fujimoto, D.F.; Brunskill, E.W.; Bayles, K.W. Analysis of Genetic Elements Controlling *Staphylococcus aureus* lrgAB Expression: Potential Role of DNA Topology in SarA Regulation. *J. Bacteriol.* **2000**, *182*, 4822–4828. [CrossRef] [PubMed]
157. Sharma-Kuinkel, B.K.; Mann, E.E.; Ahn, J.-S.; Kuechenmeister, L.J.; Dunman, P.M.; Bayles, K.W. The *Staphylococcus aureus* LytSR Two-Component Regulatory System Affects Biofilm Formation. *J. Bacteriol.* **2009**, *191*, 4767–4775. [CrossRef] [PubMed]
158. Moormeier, D.E.; Endres, J.L.; Mann, E.E.; Sadykov, M.R.; Horswill, A.R.; Rice, K.C.; Fey, P.D.; Bayles, K.W. Use of Microfluidic Technology To Analyze Gene Expression during *Staphylococcus aureus* Biofilm Formation Reveals Distinct Physiological Niches. *Appl. Environ. Microbiol.* **2013**, *79*, 3413–3424. [CrossRef]
159. Lehman, M.K.; Bose, J.L.; Sharma-Kuinkel, B.K.; Moormeier, D.E.; Endres, J.L.; Sadykov, M.R.; Biswas, I.; Bayles, K.W. Identification of the amino acids essential for LytSR-mediated signal transduction in *Staphylococcus aureus* and their roles in biofilm-specific gene expression. *Mol. Microbiol.* **2015**, *95*, 723–737. [CrossRef] [PubMed]
160. Windham, I.H.; Chaudhari, S.S.; Bose, J.L.; Thomas, V.C.; Bayles, K.W. SrrAB Modulates *Staphylococcus aureus* Cell Death through Regulation of *cidABC* Transcription. *J. Bacteriol.* **2016**, *198*, 1114–1122. [CrossRef] [PubMed]
161. Fournier, B.; Hooper, D.C. A New Two-Component Regulatory System Involved in Adhesion, Autolysis, and Extracellular Proteolytic Activity of *Staphylococcus aureus*. *J. Bacteriol.* **2000**, *182*, 3955–3964. [CrossRef] [PubMed]
162. Crosby, H.A.; Schlievert, P.M.; Merriman, J.A.; King, J.M.; Salgado-Pabón, W.; Horswill, A.R. The *Staphylococcus aureus* Global Regulator MgrA Modulates Clumping and Virulence by Controlling Surface Protein Expression. *PLoS Pathog.* **2016**, *12*, e1005604. [CrossRef]
163. Burgui, S.; Gil, C.; Solano, C.; Lasa, I.; Valle, J. A Systematic Evaluation of the Two-Component Systems Network Reveals That ArlRS Is a Key Regulator of Catheter Colonization by *Staphylococcus aureus*. *Front. Microbiol.* **2018**, *9*, 342. [CrossRef] [PubMed]
164. Trotonda, M.P.; Tamber, S.; Memmi, G.; Cheung, A.L. MgrA Represses Biofilm Formation in *Staphylococcus aureus*. *Infect. Immun.* **2008**, *76*, 5645–5654. [CrossRef] [PubMed]
165. Jiang, Q.; Jin, Z.; Sun, B. MgrA Negatively Regulates Biofilm Formation and Detachment by Repressing the Expression of *psm* Operons in *Staphylococcus aureus*. *Appl. Environ. Microbiol.* **2018**, *84*, e01008-18. [CrossRef]
166. Jenal, U.; Reinders, A.; Lori, C. Cyclic di-GMP: Second messenger extraordinaire. *Nat. Rev. Genet.* **2017**, *15*, 271–284. [CrossRef]
167. Ha, D.-G.; O'Toole, G.A. c-di-GMP and its Effects on Biofilm Formation and Dispersion: A *Pseudomonas Aeruginosa* Review. *Microbiol. Spectr.* **2015**, *3*, MB-0003-2014. [CrossRef] [PubMed]
168. Kaplan, J. Biofilm Dispersal: Mechanisms, Clinical Implications, and Potential Therapeutic Uses. *J. Dent. Res.* **2010**, *89*, 205–218. [CrossRef] [PubMed]
169. Xiong, Z.-Q.; Fan, Y.-Z.; Song, X.; Liu, X.-X.; Xia, Y.-J.; Ai, L.-Z. The second messenger c-di-AMP mediates bacterial exopolysaccharide biosynthesis: A review. *Mol. Biol. Rep.* **2020**, *47*, 9149–9157. [CrossRef] [PubMed]
170. Hengge, R. Principles of c-di-GMP signalling in bacteria. *Nat. Rev. Microbiol.* **2009**, *7*, 263–273. [CrossRef] [PubMed]
171. Corrigan, R.M.; Abbott, J.C.; Burhenne, H.; Kaever, V.; Gründling, A. c-di-AMP Is a New Second Messenger in *Staphylococcus aureus* with a Role in Controlling Cell Size and Envelope Stress. *PLoS Pathog.* **2011**, *7*, e1002217. [CrossRef]
172. DeFrancesco, A.S.; Masloboeva, N.; Syed, A.K.; DeLoughery, A.; Bradshaw, N.; Li, G.-W.; Gilmore, M.S.; Walker, S.; Losick, R. Genome-wide screen for genes involved in eDNA release during biofilm formation by *Staphylococcus aureus*. *Proc. Natl. Acad. Sci. USA* **2017**, *114*, E5969–E5978. [CrossRef] [PubMed]
173. Li, L.; Li, Y.; Zhu, F.; Cheung, A.L.; Wang, G.; Bai, G.; Proctor, R.A.; Yeaman, M.R.; Bayer, A.S.; Xiong, Y.Q. New Mechanistic Insights into Purine Biosynthesis with Second Messenger c-di-AMP in Relation to Biofilm-Related Persistent Methicillin-Resistant *Staphylococcus aureus* Infections. *mBio* **2021**, *12*, e0208121. [CrossRef] [PubMed]
174. Fischer, A.; Kambara, K.; Meyer, H.; Stenz, L.; Bonetti, E.-J.; Girard, M.; Lalk, M.; Francois, P.; Schrenzel, J. GdpS contributes to *Staphylococcus aureus* biofilm formation by regulation of eDNA release. *Int. J. Med. Microbiol.* **2014**, *304*, 284–299. [CrossRef]
175. Fechter, P.; Caldelari, I.; Lioliou, E.; Romby, P. Novel aspects of RNA regulation in *Staphylococcus aureus*. *FEBS Lett.* **2014**, *588*, 2523–2529. [CrossRef] [PubMed]
176. Storz, G.; Vogel, J.; Wassarman, K.M. Regulation by Small RNAs in Bacteria: Expanding Frontiers. *Mol. Cell* **2011**, *43*, 880–891. [CrossRef] [PubMed]
177. Desgranges, E.; Marzi, S.; Moreau, K.; Romby, P.; Caldelari, I. Noncoding RNA. *Microbiol. Spectr.* **2019**, *7*, 7-2. [CrossRef] [PubMed]
178. Ghaz-Jahanian, M.A.; Khodaparastan, F.; Berenjian, A.; Jafarizadeh-Malmiri, H. Influence of Small RNAs on Biofilm Formation Process in Bacteria. *Mol. Biotechnol.* **2013**, *55*, 288–297. [CrossRef] [PubMed]
179. Romilly, C.; Lays, C.; Tomasini, A.; Caldelari, I.; Benito, Y.; Hammann, P.; Geissmann, T.; Boisset, S.; Romby, P.; Vandenesch, F. A Non-Coding RNA Promotes Bacterial Persistence and Decreases Virulence by Regulating a Regulator in *Staphylococcus aureus*. *PLoS Pathog.* **2014**, *10*, e1003979. [CrossRef] [PubMed]
180. Patel, N.; Nair, M. The small RNA RsaF regulates the expression of secreted virulence factors in *Staphylococcus aureus* Newman. *J. Microbiol.* **2021**, *59*, 920–930. [CrossRef]
181. Ibberson, C.B.; Parlet, C.P.; Kwiecinski, J.; Crosby, H.A.; Meyerholz, D.K.; Horswill, A.R. Hyaluronan Modulation Impacts *Staphylococcus aureus* Biofilm Infection. *Infect. Immun.* **2016**, *84*, 1917–1929. [CrossRef]
182. Kim, S.; Reyes, D.; Beaume, M.; Francois, P.; Cheung, A. Contribution of teg49 Small RNA in the 5′ Upstream Transcriptional Region of *sarA* to Virulence in *Staphylococcus aureus*. *Infect. Immun.* **2014**, *82*, 4369–4379. [CrossRef]
183. Manna, A.C.; Leo, S.; Girel, S.; González-Ruiz, V.; Rudaz, S.; Francois, P.; Cheung, A.L. Teg58, a small regulatory RNA, is involved in regulating arginine biosynthesis and biofilm formation in *Staphylococcus aureus*. *Sci. Rep.* **2022**, *12*, 14963. [CrossRef]

184. Manna, A.C.; Kim, S.; Cengher, L.; Corvaglia, A.; Leo, S.; Francois, P.; Cheung, A.L. Small RNA teg49 Is Derived from a *sarA* Transcript and Regulates Virulence Genes Independent of SarA in *Staphylococcus aureus*. *Infect. Immun.* **2018**, *86*, e00635-17. [CrossRef] [PubMed]
185. Kathirvel, M.; Buchad, H.; Nair, M. Enhancement of the pathogenicity of *Staphylococcus aureus* strain Newman by a small noncoding RNA SprX1. *Med. Microbiol. Immunol.* **2016**, *205*, 563–574. [CrossRef] [PubMed]

 *antibiotics*

Article

Eradication of *Staphylococcus aureus* Biofilm Infection by Persister Drug Combination

Rebecca Yee [1,†], Yuting Yuan [1,‡], Andreina Tarff [2], Cory Brayton [3], Naina Gour [4], Jie Feng [5] and Ying Zhang [6,*]

[1] Department of Molecular Microbiology and Immunology, Bloomberg School of Public Health, Johns Hopkins University, Baltimore, MD 21205, USA
[2] Department of Graduate Medical Education, Louis A. Weiss Memorial Hospital, Chicago, IL 60640, USA
[3] Department of Comparative Medicine, Johns Hopkins University School of Medicine, Baltimore, MD 21205, USA
[4] The Solomon H. Snyder Department of Neuroscience, Johns Hopkins University School of Medicine, Baltimore, MD 21205, USA
[5] School of Basic Medical Sciences, Lanzhou University, Lanzhou 730000, China
[6] State Key Laboratory for Diagnosis and Treatment of Infectious Diseases, National Clinical Research Center for Infectious Diseases, National Medical Center for Infectious Diseases, Collaborative Innovation Center for Diagnosis and Treatment of Infectious Diseases, The First Affiliated Hospital, Zhejiang University School of Medicine, Hangzhou 310003, China
* Correspondence: yzhang207@zju.edu.cn; Tel.: +86-0571-87232219
† Current address: Department of Pathology, The George Washington University School of Medicine and Health Sciences, Washington, DC 20037, USA.
‡ Current address: Department of Pharmacology and Molecular Sciences, Johns Hopkins University School of Medicine, Baltimore, MD 21205, USA.

Citation: Yee, R.; Yuan, Y.; Tarff, A.; Brayton, C.; Gour, N.; Feng, J.; Zhang, Y. Eradication of *Staphylococcus aureus* Biofilm Infection by Persister Drug Combination. *Antibiotics* **2022**, *11*, 1278. https://doi.org/10.3390/antibiotics11101278

Academic Editors: Ding-Qiang Chen, Yulong Tan, Ren-You Gan, Guanggang Qu, Zhenbo Xu, Junyan Liu and Marc Maresca

Received: 9 August 2022
Accepted: 16 September 2022
Published: 20 September 2022

Publisher's Note: MDPI stays neutral with regard to jurisdictional claims in published maps and institutional affiliations.

Abstract: *Staphylococcus aureus* can cause a variety of infections, including persistent biofilm infections, which are difficult to eradicate with current antibiotic treatments. Here, we demonstrate that combining drugs that have robust anti-persister activity, such as clinafloxacin or oritavancin, in combination with drugs that have high activity against growing bacteria, such as vancomycin or meropenem, could completely eradicate *S. aureus* biofilm bacteria in vitro. In contrast, single or two drugs, including the current treatment doxycycline plus rifampin for persistent *S. aureus* infection, failed to kill all biofilm bacteria in vitro. In a chronic persistent skin infection mouse model, we showed that the drug combination clinafloxacin + meropenem + daptomycin which killed all biofilm bacteria in vitro completely eradicated *S. aureus* biofilm infection in mice while the current treatments failed to do so. The complete eradication of biofilm bacteria is attributed to the unique high anti-persister activity of clinafloxacin, which could not be replaced by other fluoroquinolones including moxifloxacin, levofloxacin, or ciprofloxacin. We also compared our persister drug combination with the current approaches for treating persistent infections, including gentamicin + fructose and ADEP4 + rifampin in the *S. aureus* biofilm infection mouse model, and found neither treatment could eradicate the biofilm infection. Our study demonstrates an important treatment principle, the Yin–Yang model, for persistent infections by targeting both growing and non-growing heterogeneous bacterial populations, utilizing persister drugs for the more effective eradication of persistent and biofilm infections. Our findings have implications for the improved treatment of other persistent and biofilm infections in general.

Keywords: *Staphylococcus aureus*; persisters; biofilm; antimicrobial activity; drug combination

1. Introduction

As a virulent opportunistic pathogen, *Staphylococcus aureus* is the most common cause of skin infections and can also cause chronic persistent infections such as endocarditis and osteomyelitis [1–3]. For conditions, such as endocarditis and osteomyelitis, and prosthetic joint infections, treatments with vancomycin as a monotherapy or drug combination for at least six weeks are recommended. Drug combinations, such as doxycycline + rifampin for up to 10 days, vancomycin + gentamicin + rifampin for at least six weeks, are recommended to treat chronic infections, such as recurrent tissue infections and endocarditis on prosthetic valves, respectively (The Johns Hopkins Antibiotics Guide). In particular, indwelling devices are conducive to biofilm formation, complicating treatment and leading to prolonged infections. Up to 80% of human infections are biofilm infections, and globally, chronic persistent and biofilm infections represent a huge burden to public health as they increase the length of hospital stay, cause relapse, cost of treatment, and risk of death by at least three-fold [4]. Bacteria in biofilms are more tolerant to antibiotics compared to planktonic cells [5]. Studies have shown that antibiotics can penetrate the biofilm but they do not always kill the bacteria, suggesting that tolerance to treatment is not due to impaired antibiotic penetration or genetic resistance [6,7], but due to dormant non-growing or slow growing persister bacteria. Bacteria inside the biofilm are quite heterogeneous as some cells grow slowly, which are representative of stationary phase bacteria, while others form dormant persister cells due to the high cell density, nutrient, and oxygen limiting environment inside the biofilm matrix [8].

Persisters were first described in 1942 by Hobby et al., who found that while 99% of *S. aureus* cells were killed by penicillin, about 1% of residual metabolically quiescent or dormant cells called persisters were not killed [9]. The persisters were not resistant to penicillin, and hence did not undergo genetic changes, but were phenotypic variants that became tolerant to antibiotics [10]. Similarly, a clinical observation was also made as penicillin failed to clear chronic infections due to the presence of persister cells found in patients [10]. While the mechanisms of *S. aureus* persistence were largely unknown for a long time, recent studies have shown that pathways involved in quorum sensing, pigmentation production, and metabolic processes, such as oxidative phosphorylation, glycolysis, amino acid, and energy metabolism [11–15], are involved. Despite the observation of persister bacteria from 1940s and their implications in causing prolonged treatment and post-treatment relapse, the importance of persister bacteria and drugs that target persister bacteria in clinical settings has been ignored largely because no persister drugs have been found that can cure or shorten treatment duration or reduce relapse in clinically relevant persistent infections. The importance of persister drugs to more effectively cure persistent infections is only recognized in the case of tuberculosis persister drug pyrazinamide (PZA), which shortens the treatment from 9–12 months to six months after its inclusion in a drug combination setting [16]. PZA's activity in killing *M. tuberculosis* persisters, unlike the other drugs used to treat tuberculosis, is crucial in developing a shorter treatment due to its unique mechanisms of action by inhibiting persister targets, including energy metabolism (CoA synthesis via PanD) and protein degradation pathways (RpsA and ClpC1) essential for persister survival [16–19]. The drug PZA validates an important principle of using a persister drug in combination with other drugs targeting both non-growing persisters and growing bacteria in formulating an effective therapy for chronic persistent infections [20,21]. In support of this idea, more recently, a similar approach was used to identify the effective drug combination using persister drug daptomycin in combination with doxycycline and cefuroxime, which completely eradicated biofilm-like structures of *Borrelia burgdorferi* in vitro [22] and in mice [23].

Using this approach, in a recent study aimed at identifying drugs targeting non-growing persisters, we used a stationary phase culture of *S. aureus* as a drug screen model and identified several drugs, such as clinafloxacin and tosufloxacin, with high activity against *S. aureus* persisters [24]. However, their activities alone and in drug combinations in killing biofilms have not been evaluated in vitro or in related *S. aureus* persistent infections in vivo. In this study, we developed persister drug combinations utilizing persister drug clinafloxacin that can more effectively eradicate *S. aureus* biofilms by formulating drug combinations that have high activities against growing bacteria and non-growing persisters in a biofilm model in vitro initially. Then, we established a persistent skin infection mouse model for *S. aureus* using biofilm "persister seeding" [21,25] and evaluated drug combinations in clearing the biofilm infection in this persistent infection model. Here, we show that combining meropenem and daptomycin targeting growing bacteria, with persister drug clinafloxacin targeting biofilm persister bacteria led to the complete eradication of *S. aureus* biofilms not only in vitro, but more importantly also in vivo in a murine model of persistent skin infection, whereas other approaches for treating persistent infections and the currently recommended drug combination treatment without persister drugs failed to do so.

2. Materials and Methods

2.1. Culture Media, Antibiotics, and Chemicals

Staphylococcus aureus strains Newman, USA300 (a biofilm-proficient common circulating strain of community acquired-MRSA, CA-MRSA) and clinical strains CA-409, CA-127, and GA-656 were obtained from American Type Tissue Collections (Manassas, VA, USA) and cultivated in tryptic soy broth (TSB) and tryptic soy agar (TSA) (Becton Dickinson, Franklin Lakes, NJ, USA) at 37 °C. Vancomycin, gentamicin, rifampicin, levofloxacin, ciprofloxacin, moxifloxacin, and oritavancin were obtained from Sigma-Aldrich Co. (St. Louis, MO, USA). Daptomycin, meropenem, tosufloxacin, and clinafloxacin were obtained from AK Scientific, Inc. (Union City, CA, USA). Stock solutions were prepared in the laboratory, filter-sterilized, and used at indicated concentrations.

2.2. Microtiter Plate Biofilm Drug Exposure Assay and the SYBR Green I/PI Assay

S. aureus strains grown overnight in TSB were diluted 1:100 and 100 µL aliquots of each diluted culture, placed into a 96-well flat-bottom microtiter plate, and statically incubated for 24 h at 37 °C [26]. Planktonic cells were removed and discarded from the microtiter plates. Drugs at the indicated Cmax concentrations were then added to the biofilms attached to the bottom of the microtiter plate in a total volume of 100 µL in TSB medium and incubated at 37 °C without shaking for 4 days. To determine the cell and biofilm density, the supernatant was removed from the well and the biofilms were washed twice with PBS (1×). To enumerate bacterial cell counts, the biofilms in the wells were resuspended in TSB and scraped with a pipette tip before serial dilution and plating on TSA plates for CFU counts.

SYBR Green I/PI assay was used to assess biofilm cell viability using the ratio of green:red fluorescence to determine the ratio of live:dead cells, respectively, as described [27]. Briefly, the staining dyes were prepared by mixing SYBR Green I/PI (10,000× stock, Invitrogen, Carlsbad, CA, USA) with propidium iodide (20 mM, Sigma-Aldrich) in distilled water at a ratio 1:3 in 100 µL distilled H_2O. The SYBR Green I/PI dye mix (10 µL) was added to each 100 µL of sample and incubated at room temperature in the dark for 20 min. With excitation wavelengths of 485 nm and 538 nm and 612 nm for green and red emission, respectively, the green and red fluorescence intensity was determined for each sample using a Synergy H1 microplate reader by BioTek Instruments (Winooski, VT, USA).

2.3. Mouse Skin Infection Model

Female Swiss-Webster mice of 6 weeks of age were obtained from Charles River. They were housed 3–5 per cage under BSL-2 housing conditions. All animal procedures were approved by the Johns Hopkins University Animal Care and Use Committee (protocol code MO17H167 and 23 May 2017 approval). *S. aureus* strain USA300 and strain Newman were used in the mouse infection experiments. The details of the biofilm mouse infection model were described in our previous study [25]. Briefly, mice were anesthetized and then shaved to remove a patch of fur of approximately 3 cm by 2 cm. For the preparation of biofilm inoculum for infection, biofilms were first grown in microtiter plates as we described previously [25], and then resuspended and scraped up with a pipette tip. Quantification of all inoculum was performed by serial dilution and plating. Bacteria of indicated inoculum size (10^8 CFU/mL) were subcutaneously injected into the mice. Treatment was started after 1 week of infection with different drugs and drug combinations for 1 week. For details on drugs, drug dosage, and route of administration, please refer to Table 1. Skin lesion sizes were measured at the indicated time points using a caliper. The area of the lesion size was calculated by measuring the lesion as an oval shape [area = $\pi \times$ (length/2) $\times$ (width/2)]. Mice were euthanized after 1 week post-treatment, and skin tissues were removed, homogenized, and serially diluted for bacterial counting on TSA plates.

Table 1. Drug dosage, scheduling, and administration.

Drug	Dosage	Route	Times Treated	Cmax Tested (Clinically Achievable Concentrations) *
Vancomycin	110 mg/kg	Intraperitoneally	Twice/daily	20 µg/mL
Daptomycin	50 mg/kg	Intraperitoneally	Once/daily	80 µg/mL
Meropenem	50 mg/kg	Intraperitoneally	Once/daily	20 µg/mL
Clinafloxacin	50 mg/kg	Intraperitoneally	Once/daily	2 µg/mL
Doxycycline	100 mg/kg	Oral	Twice/daily	5 µg/mL
Rifampin	10 mg/kg	Oral	Twice/daily	5 µg/mL
Moxifloxacin	100 mg/kg	Oral	Once/daily	4 µg/mL
ADEP4	25 mg/kg and 35 mg/kg	Intraperitoneally	Twice/daily	
Rifampin (for ADEP4 combination)	30 mg/kg	Intraperitoneally	Once/daily	
Gentamicin	20 mg/kg	Intraperitoneally	Once/daily	
Fructose	1.5 g/kg	Intraperitoneally	Once/daily	
Oritavancin	—	—	—	5 µg/mL
Ciprofloxacin	—	—	—	10 µg/mL
Levofloxacin	—	—	—	10 µg/mL

* Cmax concentrations refer to maximum blood drug concentrations in humans and were derived from Johns Hopkins Antibiotics Guides (https://www.hopkinsguides.com/hopkins/index/Johns_Hopkins_ABX_Guide/Antibiotics, accessed on 20 June 2019).

2.4. Histopathology

Skin tissues were dissected, laid flat, and fixed for 24 hrs with neutral buffered formalin. Tissues were embedded in paraffin, cut into 5-µm sections, and mounted on glass slides. Tissue sections were stained with hematoxylin and eosin for histopathological scoring. Tissue sections were evaluated for lesion crust formation, ulcer formation, hyperplasia, inflammation, gross size, and bacterial count and were assigned a score on a 0–3 scale (0 = none, 1 = mild, 2 = moderate, and 3 = severe). The cumulative pathology score represented the sum of each individual pathology parameter. Scoring was performed by an observer in consultation with a veterinary pathologist. Representative images were taken using a Keyence BZ-X710 Microscope (Itasca, IL, USA).

2.5. Statistical Analyses

Statistical analyses were performed using ANOVA or Student's *t*-test where appropriate. Mean differences were considered statistically significant if the *p* value was <0.05. All experiments were performed in triplicates. Analyses were performed using GraphPad Prism version 8.0.2 and Microsoft Office Excel 2016.

3. Results

3.1. Commonly Used Treatments for MRSA Have Poor Activity against Biofilms In Vitro

We first evaluated the activity of the above drugs in killing biofilm bacteria in vitro using traditional bacterial CFU counts (Figure 1A) and viability assessment by SYBR Green I/PI staining that has been developed to screen for drugs targeting Borrelia persister bacteria [27] (Figure 1B). We found that such clinically used combinations were not completely effective against biofilms, as shown by different assays, i.e., CFU assay (Figure 1A), SYBR Green/PI assay (Figure 1B). After four days of treatment, biofilm bacteria were not completely eradicated by any of the current treatments with vancomycin alone, or doxycycline + rifampin or vancomycin + gentamicin + rifampin as shown by significant numbers of bacteria remaining

(Figure 1). However, of note, vancomycin + gentamicin + rifampin was more active than vancomycin alone or doxycycline + rifampin in killing biofilm bacteria (Figure 1A).

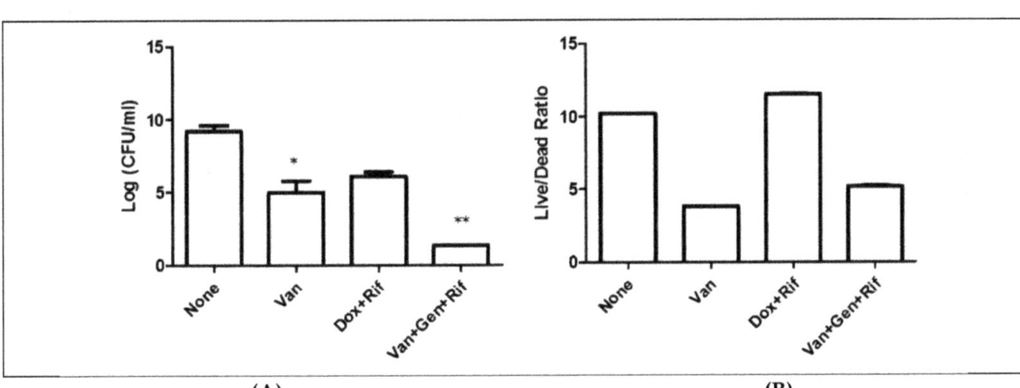

Figure 1. Clinically recommended treatments for chronic *S. aureus* infections only partially killed *S. aureus* USA300 biofilm bacteria in vitro. Treatments (all 50 μM) of vancomycin alone, doxycycline + rifampin, and the combination of vancomycin + gentamicin + rifampin for 4 days were evaluated for biofilm killing by CFU enumeration (**A**) and viability staining using SYBR Green I/PI (**B**). See Methods section for details. Vancomycin, Van; Doxycycline, Dox; Rifampin, Rif; Gentamicin, Gen. Student's *t*-test, * $p < 0.05$, ** $p < 0.005$. Error bars indicate standard deviation.

3.2. Identification of Drug Combinations with Strong Anti-Biofilm Activity

To address the clinical unmet need for better treatments for persistent biofilm infections, we hypothesize that a drug combination that includes drugs that act on growing bacteria, such as cell wall (e.g., vancomycin, meropenem) or cell membrane inhibitors (e.g., daptomycin), plus a drug that acts on persister bacteria will be more potent in eradicating biofilm bacteria. Previous studies from our lab identified tosufloxacin and clinafloxacin as having strong anti-persister activity against *S. aureus* [24]. In order to identify a potent combination, we tested various drug combinations that include drugs against both growing bacteria and non-growing persisters in a biofilm model with *S. aureus* USA300, a common biofilm-proficient clinical MRSA strain. While we previously showed that tosufloxacin had robust activity against *S. aureus* persister cells [24], the drug combination of vancomycin/meropenem + daptomycin + tosufloxacin achieved only partial eradication, with 10^5 CFU/mL biofilm bacteria remaining after treatment (Figure 2A,B). In contrast, the combination of vancomycin/meropenem + daptomycin + clinafloxacin showed complete eradication of biofilms after 4-day treatment, as shown by 0 CFU and a live/dead ratio below the limit of detection in SYBR Green/PI assay (Figure 2A,B). Although we used the same molar concentration of each individual drug at 50 μM for comparison of relative drug activity, to evaluate the activity of the drug combination in a more clinically relevant manner, we treated the biofilms with the drugs at their Cmax concentrations (Table 1). Since the drugs used in the anti-biofilm studies are mostly clinically used drugs, they would not be expected to have significant cytotoxicity at their respective Cmax (maximum human blood) drug concentrations. Our findings with Cmax drug concentrations were confirmatory as the drug combination of vancomycin/meropenem + daptomycin + clinafloxacin still achieved complete eradication, while drug-free control and the clinically used combination of doxycycline + rifampin could not. The clearance of biofilms was confirmed by both CFU counts and the SYBR Green I/PI viability stain (Figure 2C,D).

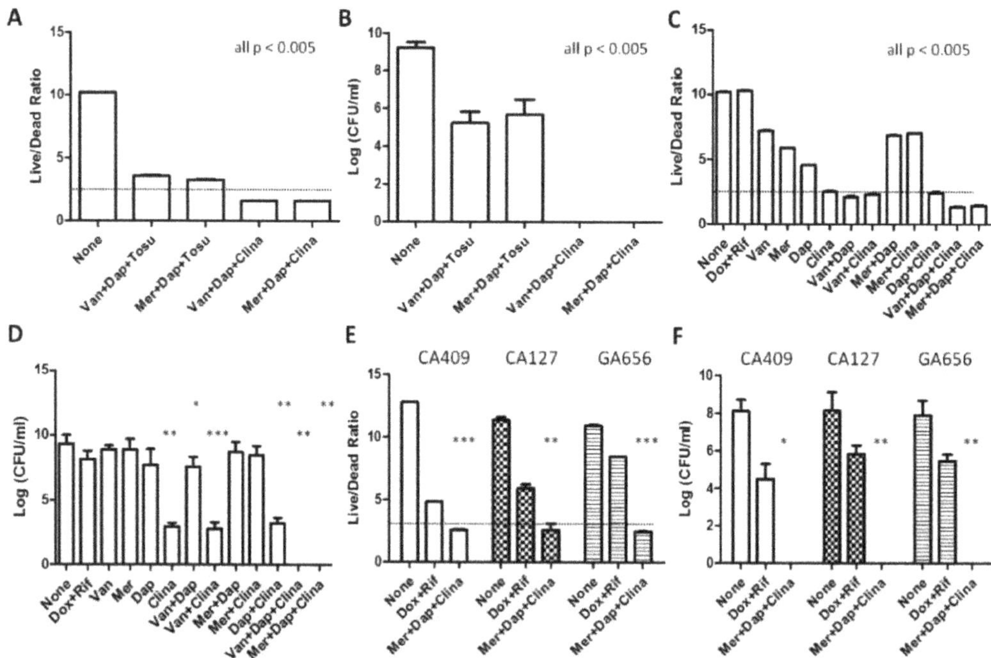

Figure 2. Identification of drug combinations that kill MRSA biofilms. Various drug combinations consisting of drugs (at 50 μM) highly active against the growing phase and persister cells were incubated with *S. aureus* USA300 biofilm bacteria for 4 days followed by assessing the anti-biofilm activity using SYBR Green I/PI viability staining (**A**) and CFU enumeration (**B**). Drug combinations with sterilizing activity against USA300 biofilms were tested at clinically achievable Cmax concentrations using SYBR Green I/PI viability staining (**C**) and CFU enumeration (**D**). Validation of meropenem + daptomycin + clinafloxacin in killing biofilms of various MRSA clinical isolates (**E,F**). Vancomycin, Van; Meropenem, Mer; Daptomycin, Dap; Tosufloxacin, Tosu; Clinafloxacin, Clina; Doxycycline, Dox; Rifampin, Rif. One-way ANOVA and post hoc Tukey's test was used for multiple group comparisons. * $p < 0.05$, ** $p < 0.005$, *** $p < 0.0005$. Error bars indicate standard deviation.

We then tested the potential of the drug combination of meropenem + daptomycin + clinafloxacin to eradicate biofilm bacteria from different MRSA *S. aureus* strains, including CA-MRSA clinical isolates CA-409, CA-127, and hospital-acquired MRSA strain GA-656. Complete eradication (0 CFU/mL) and undetectable levels of live cells (under the limit of detection) were found for all of the MRSA strains tested after 4 days of treatment with the combination meropenem + daptomycin + clinafloxacin (Figure 2E,F).

3.3. Unique Anti-Persister Activity of Clinafloxacin Could Not Be Replaced by Other Fluoroquinolone Drugs

Clinafloxacin is a member of the fluoroquinolone antibiotics which inhibits DNA replication by binding to DNA gyrase. As our results suggest (Figure 2D), clinafloxacin is a powerful anti-persister drug. Hence, we wanted to rank the anti-biofilm activity of different fluoroquinolones to determine whether the robust anti-biofilm activity of clinafloxacin is unique or can be replaced by other fluoroquinolones by using CFU assay and SYBR Green/PI viability assay. We used the *S. aureus* Newman strain due to its susceptibility to many fluoroquinolones to avoid any confounding factors due to inherent drug resistance. While other fluoroquinolones such as ciprofloxacin, levofloxacin, and moxifloxacin had certain anti-biofilm activity when used in combination with meropenem and daptomycin after four days of treatment, the drug combination with clinafloxacin was indeed the most active and was the only combination that achieved complete sterilization as seen by CFU count (Figure 3A). By contrast, biofilms treated with other quinolone (ciprofloxacin, moxifloxacin, levofloxacin) combinations still had 10^4–10^8 CFU/mL bacteria remaining (Figure 3A). The results with the SYBR Green/PI viability assay, where lower live/dead ratios would indicate lower number of viabile bacteria, were consistent with the CFU assay. However, the SYBR Green/PI assay, while being quick and showing general agreement with the CFU assay, was not as discriminative or accurate as the CFU assay (Figure 3B) as it could not distinguish between Mer+Dap+Clina, Mer+Dap+Cipro, and Mer+Dap+Moxi as the assay reached the limit of detection near live/dead ratios of 1–2 (Figure 3B). When used in combination, the activity of the quinolones from strongest to weakest as ranked by both viability assessment and viable cell counts is as follows: clinafloxacin, ciprofloxacin, moxifloxacin, and levofloxacin (Figure 3). Hence, clinafloxacin has unique potent activity against biofilm bacteria compared to other fluoroquinolone counterparts.

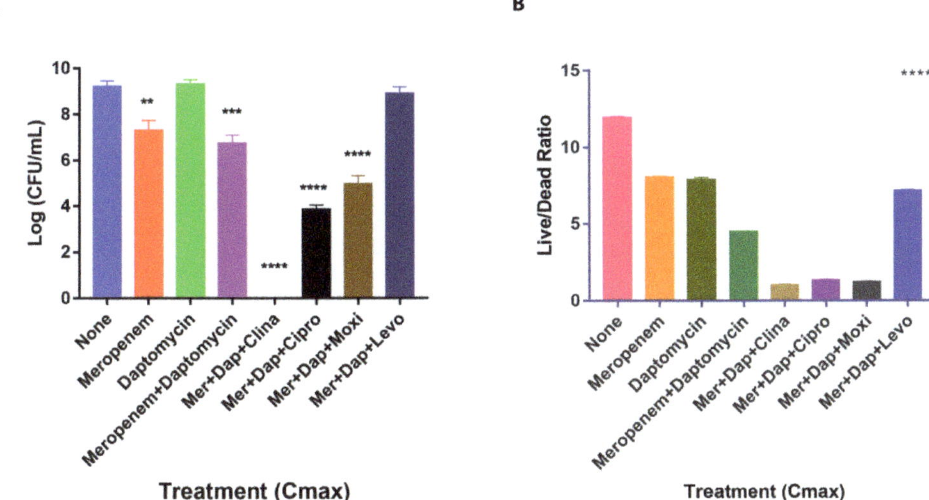

Figure 3. Ranking of fluoroquinolones in drug combinations at Cmax concentrations for activity against *S. aureus* USA300 biofilm bacteria. Commonly used fluoroquinolones ciprofloxacin, moxifloxacin, levofloxacin were evaluated together with clinafloxacin in combination with meropenem+daptomycin in drug exposure tests for 4 days with *S. aureus* Newman biofilm bacteria, when the effect of the treatment was assessed by (**A**) CFU count and (**B**) SYBR Green/PI viability assay as described in Methods. Meropenem, Mer; Daptomycin, Dap; Clinafloxacin, Clina; Ciprofloxacin, Cipro; Moxifloxacin, Moxi; Levofloxacin, Levo. One-way ANOVA and post hoc Tukey's test was used for multiple group comparisons. ** $p < 0.05$, *** $p < 0.005$, **** $p < 0.0001$. Error bars indicate standard deviation.

3.4. Anti-Biofilm Activity of Oritavancin and Dalbavancin

Thus far, our data suggest that the inclusion of a drug with great anti-biofilm activity can be beneficial in killing biofilm bacteria (Figures 2 and 3). To identify other potential anti-biofilm drug candidates, we turned to the new generation of lipoglycopeptides such as oritavancin and dalbavancin. These drugs have multiple mechanisms of action: inhibition of transglycosylation, transpeptidation, and cell membrane disruption, a property of persister drugs [17]. We first tested the activity of oritavancin and dalbavancin in killing *S. aureus* biofilm bacteria in comparison with their parent compound vancomycin, and the results revealed oritavancin was the best in killing biofilm among the three drugs

(Figure 4A). After six days of drug exposure, oritavancin killed 10^6 CFU/mL bacteria as compared with dalbavancin or vancomycin, which killed only about 10^2 CFU/mL.

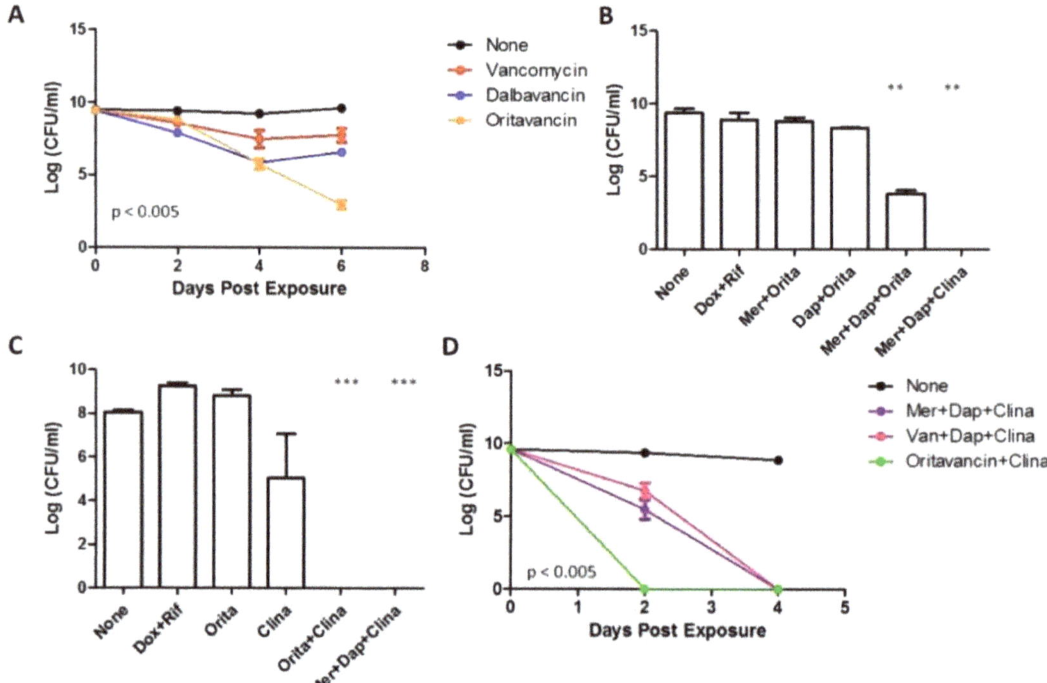

Figure 4. Evaluation of oritavancin in killing *S. aureus* USA300 biofilms as a single drug or in combinations. (**A**) Comparison of novel lipoglycopeptides oritavancin and dalbavancin with vancomycin (at 50 µM) in their activity to kill biofilm bacteria. (**B**) Evaluating oritavancin in killing persisters in combination with meropenem + daptomycin, and (**C**) in killing growing phase bacteria in combination with clinafloxacin (at Cmax concentrations). (**D**) Time-kill curve of biofilms comparing the top drug combination candidates (at Cmax concentrations). Meropenem, Mer; Daptomycin, Dap; Oritavancin, Orita; Clinafloxacin, Clina; Doxycycline, Dox; Rifampin, Rif. One- or two-way ANOVA and post hoc Tukey's test was used for multiple group comparisons., ** $p < 0.005$, *** $p < 0.0005$. Error bars indicate standard deviation.

Since oritavancin showed strong anti-biofilm activity, we next evaluated oritavancin's activity in drug combinations. After replacing clinafloxacin with oritavancin, we observed that the combination of meropenem + daptomycin + oritavancin exhibited partial activity against biofilms, causing a decrease of 10^5 CFU/mL, which is much better than the activity achieved by single drugs or two-drug combinations, but still was inferior to the clinafloxacin combination (Figure 4B).

Due to oritavancin's strong activity against growing *S. aureus* (MIC = 0.03 mg/L) [28,29] and its dual mechanism of action that mimics cell wall + cell membrane inhibitor, we tested oritavancin in place of meropenem and daptomycin. Surprisingly, the combination of oritavancin + clinafloxacin was also able to achieve complete eradication of biofilms suggesting that oritavancin can replace the drugs against growing bacteria (Figure 4C). It is also important to note that oritavancin alone could not kill biofilms (no change in CFU after 4-day treatment), which further validates the importance of drug combinations for biofilm bacteria.

To compare the activity of the three combinations tested thus far with clinafloxacin, we performed a time-kill experiment which revealed that oritavancin + clinafloxacin could kill all biofilms by day 2 of treatment whereas it took four days for meropenem/vancomycin + daptomycin + clinafloxacin to eradicate the biofilm bacteria (Figure 4D). Overall, our data suggest that the inclusion of an anti-biofilm drug in a drug combination to treat biofilms is paramount and these combinations possess much better activity than current clinically used regimens based on our in vitro studies (Figure 4B,C).

3.5. The Drug Combination Meropenem + Daptomycin +Clinafloxacin Eradicated Biofilm Infections in the Mouse Skin Persistent Infection Model

Given the robust activity of our drug combinations in eradicating biofilms in vitro, we were interested to know if the drug combination could also eradicate persistent infections in vivo. To do so, we chose to infect mice with biofilm bacteria from *S. aureus* strain USA300, a clinical MRSA strain most representative to cause persistent infections in humans. We allowed the infection to develop for seven days, followed by seven days of treatment with different regimens (Figure 5A). Previously, we have shown that mice infected with biofilm bacteria developed more chronic skin lesions [25]. Administration of the combination of doxycycline + rifampin (a control group as a clinically used treatment) or drug combination vancomycin + daptomycin + clinafloxacin decreased the bacteria load (about 1-log of bacteria) but did not clear the infection (Figure 5B). Other reported treatments which supposedly eradicate chronic *S. aureus* infections, such as ADEP4 + rifampin [30] or gentamicin + fructose [31], did not show sterilizing activity in our biofilm infection model, but instead caused increased lesion size and inflammation (Figure 5C,H). Remarkably, the combination of meropenem + daptomycin + clinafloxacin cleared the infection completely, decreased the size of lesions, reduced histopathology scores, and healed the lesions completely (Figure 5B,C,G), while other treatments, such as the control treatment with doxycycline + rifampin still had considerable lesions (Figure 5B,C,F). Intriguingly, in gross lesion examination, consistent with the lesion size data in Figure 5C, ADEP4 + rifampin [30] or gentamicin + fructose [31] produced worsened lesions (Figure 5H).

To verify the above observation, we performed a separate study with a methicillin-susceptible *S. aureus* (MSSA) Newman strain and compared clinafloxacin's activity with other quinolones. We were able to confirm the above findings with the MRSA strain and found that despite moxifloxacin and clinafloxacin having the same MIC for the Newman strain, the combination of meropenem + daptomycin + moxifloxacin was not effective in clearing the biofilm infection. In contrast, the meropenem + daptomycin + clinafloxacin combination indeed eradicated the infection completely (Figure 5I). This indicates clinafloxacin combination works for both MRSA and MSSA strains, and the unique sterilizing activity of clinafloxacin cannot be replaced by moxifloxacin. The above data support our hypothesis that a drug combination targeting both growing (e.g., meropenem, daptomycin) and persister bacteria (e.g., clinafloxacin) is essential in clearing chronic infections in vivo, such as the persistent skin infection caused by *S. aureus* biofilm.

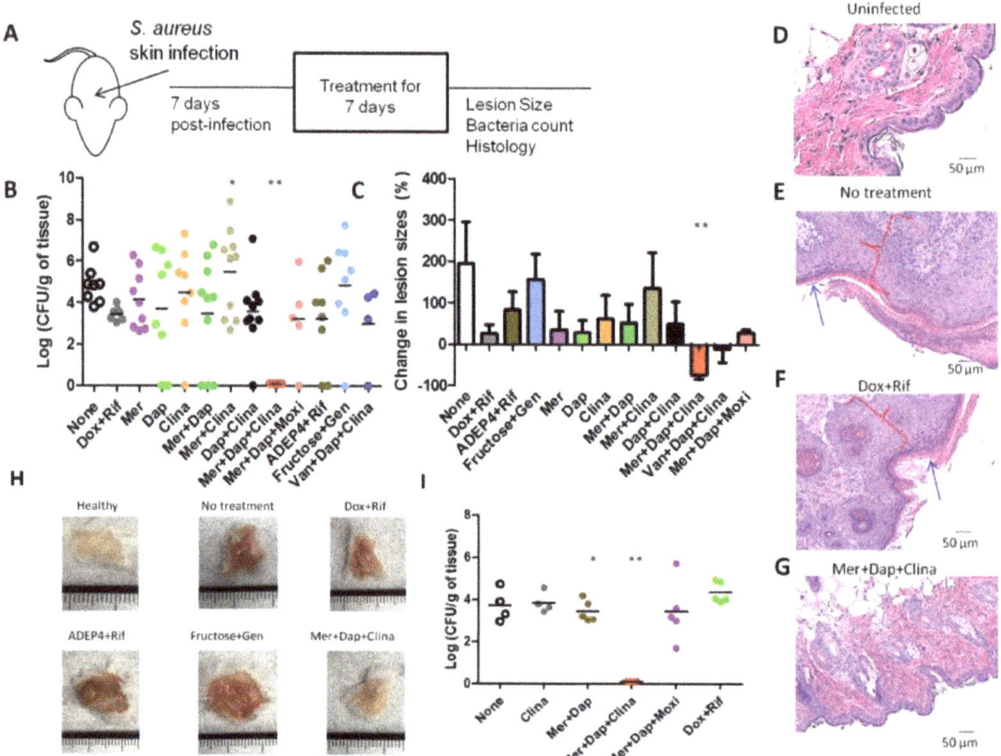

Figure 5. Validation of drug combinations in a chronic skin infection model established with *S. aureus* USA300 biofilm bacteria. (**A**) Study design of mouse treatment studies. (**B**) Bacterial load in the skin lesions. (**C**) Changes in lesion sizes and histology of skin tissues of mice infected with USA300 biofilm bacteria and treated 7-days with drug combinations and respective controls were measured. Histopathology of uninfected mice (**D**), infected mice receiving 7-days of no treatment (**E**), or treated with doxycycline + rifampin (**F**), or treated with meropenem + daptomycin + clinafloxacin (**G**) was analyzed. (**H**) Gross pathology lesions of skin tissues of mice treated 7-days with drug combinations and respective controls (in mm). (**I**) Bacterial loads in the skin tissues of mice infected with MSSA Newman strain and treated for 7-days with drug combinations or control treatments were enumerated. Images were taken at 200 X magnification. Meropenem, Mer; Daptomycin, Dap; Clinafloxacin, Clina; Doxycycline, Dox; Rifampin, Rif, Gentamicin, Gen. Blue arrows indicate crust formation, red brackets indicate hyperplasia and cellular infiltration. One-way ANOVA and post hoc Tukey's test was used for multiple group comparisons. * $p < 0.05$, ** $p < 0.005$. Error bars indicate standard deviation.

4. Discussion

Numerous studies have documented how resilient biofilms are to antibiotic treatments [32–35]. Since persister cells that are embedded in the biofilm are mostly responsible for the recalcitrance of biofilms to antibiotic treatments, many attempts have been made to identify novel effective treatments and synthetic compounds that kill bacterial persisters [36,37]. Some approaches include resuscitating or altering the metabolic status of persisters [38,39] or enhancing the activity of aminoglycoside antibiotics with sugars such as fructose [31] or activating protease by ADEP4 plus rifampin [30]. Although these new therapeutic approaches showed promising results in vitro and in some cases in animal models [30,36,40], not all treatments achieved complete sterilization and their utility in more persistent biofilm infections remains to be demonstrated. The animal models used either rely on immunosuppressant agents or short-term infections [30,31,36], which do not reflect true persistent infections clinically. Here, we used a recently developed more relevant persistent infection mouse model using biofilm inocula that mimic human infections without the use of immunosuppressant agents [25] for evaluation of drug combinations. Previous studies have mostly used log phase bacteria as inocula for infection in animal models [25]. However, we showed that *S. aureus* biofilm inocula produced a more severe lesion and more persistent infection than the log phase bacteria and the current treatments failed to eradicate the biofilm infection [25]. This biofilm-inocula model could serve as a useful model for evaluating treatment regimens against biofilm infections in vivo in general. Importantly, we demonstrate that the persister drug combination meropenem + daptomycin + clinafloxacin completely eradicated the biofilm infection in the mouse model, while the currently recommended treatment doxycycline+rifampin or daptomycin+meropenem and the well-known experimental persister treatment regimens, such as ADEP4+rifampin [30] or gentamicin+fructose [31], failed to do so (Figure 5B). To our knowledge, this is the first time a biofilm infection is completely eradicated by a specific persister drug combination but not by the current clinically used standard antibiotic treatments or other known experimental treatments, such as ADEP4+rifampin [30] or gentamicin+fructose [31], for persistent infections.

The mechanisms by which combination treatment effectively eradicates biofilm-derived bacteria are worth commenting upon. The well-known experimental regimens for killing persisters and treating persistent biofilm infections include ADEP4+rifampin [30] and gentamicin+fructose [31]. ADEP4 is a ClpP protease activator that kills persisters by degrading numerous cellular proteins, forcing cells to self-digest, and its combination with rifampin was shown to completely eradicate *S. aureus* biofilms in vitro and in a mouse model [30]. The other experimental approach to kill persisters is to use fructose to cause increased generation of proton-motive force (PMF) which facilitates aminoglycoside uptake for enhanced drug activity [31]. However, the highly potent anti-biofilm agent clinafloxacin seems to work differently. We previously hypothesized, per the Yin–Yang model, that a drug combination approach using drugs targeting different bacterial populations, i.e., non-growing persisters (Yin) and growing bacteria (Yang), will be required to more effectively cure persistent and biofilm infections [21]. Thus, it is likely that the sterilizing activity of the clinafloxacin drug combination (meropenem + daptomycin + clinafloxacin) in eradicating the biofilm bacteria in vitro and persistent skin infections in mice is due to the combined action of the unique anti-persister activity of clinafloxacin targeting persisters in combination with meropenem + daptomycin targeting growing bacteria. The strong activity of meropenem (cell wall inhibitor) and daptomycin (cell membrane disruptor) against *S. aureus* growing bacteria [41,42] allows for the rapid killing of growing bacteria in the biofilm bacterial population. While meropenem and daptomycin are directed at killing growing bacteria, they also have certain activity against persisters. Meropenem used in combination with polymyxin B has been shown to eradicate persisters in *Acinetobacter baumannii* [43]. Similarly, daptomycin has activity against *S. aureus* biofilms [44]. Daptomycin in combination with doxycycline and cefoperazone or cefuroxime has been shown to kill biofilm-like microcolony persisters of B. burgdorferi in vitro and in mice [23,45]. Daptomycin could disrupt the bacterial membrane structure and cause rapid depolarization of the membrane [46], which may affect the viability of some persisters. Nevertheless, Daptomycin alone or with other drug combinations had limited activity against *S. aureus* biofilm bacteria and could not achieve sterilization (Figures 2 and 5). To kill non-growing biofilm persisters, clinafloxacin has been shown to be crucial in the combination as moxifloxacin or levofloxacin in place of clinafloxacin in the above combination failed to eradicate the biofilm bacteria in vitro and in mice (Figures 3A and 5B). This occurs despite the MICs for clinafloxacin and moxifloxacin for *S. aureus* being the same, yet they differ in their activity to kill persisters both singly and in drug combinations both in vitro and in mice (Figure 5B,I). As a quinolone, clinafloxacin inhibits bacterial DNA gyrase and topoisomerase and thereby interferes with DNA synthesis. However, not all quinolones have significant anti-persister activity [24] (Figure 3). The unique anti-persister and biofilm activity of clinafloxacin seems to be beyond its common mechanism of action on inhibition of DNA synthesis and may be due to its additional chemical groups that attack critical persister targets. Comparing the chemical structure of clinafloxacin to the other quinolones that have relatively weak anti-persister activity (ciprofloxacin, moxifloxacin, and levofloxacin), a chloride group attached to the benzene ring appears to be unique to clinafloxacin and may be responsible for its high anti-biofilm activity. Further studies are required to explore the mechanism of clinafloxacin's unique ability to kill persisters, including the possibility that clinafloxacin may bind to the target gyrase with higher affinity or it might have preferential activity against targets crucial for persister survival.

It is important to note that the chronic persistent infection status of the mice inoculated with biofilm bacteria is a key component of our disease model. While Conlon et al. used a 1:100 dilution of stationary phase culture (2×10^6) as inocula to infect their mice and caused a deep-seated infection, the infection was allowed to develop for only 24 hours before treatment which cannot be a persistent infection; and the mice were made neutropenic [30], a condition that may not apply to most patients suffering from chronic *S. aureus* infections. Such differences in the animal models may explain why ADEP4 + rifampin, which was claimed to have sterilizing activity and eradicated persistent infection [30], failed to eradicate the more persistent infection in our model established with biofilm inocula and allowed to develop for one week before treatment. In addition, although the combination of an aminoglycoside + sugars has been proposed as an approach for killing persisters and treating persistent infections and was shown to be effective in an E. coli urinary tract infection mouse model [31], this approach was not effective in our biofilm infection model either. Allison et al. showed that gentamicin + fructose reduced 1.5-fold *S. aureus* biofilms in vitro after four hours of treatment [31] but it was not tested in animals. In our study here, we showed that mice treated for seven days with gentamicin + fructose still harbored 10^5 CFU/mL in skin tissues (Figure 5B). Surprisingly, gentamicin + fructose and ADEP4 + rifampin caused an increase in lesion size despite the treatment (Figure 5C,H), which is hard to explain. In both the above cases, the discrepancy could be due to differences in the disease models, as ours is a more persistent biofilm skin infection model established with biofilm inocula and would be expected to be more difficult to cure than those in the other studies that did not use biofilm inocula for the infection. Although our persistent biofilm infection was established using USA300 strain, a biofilm-proficient clinical MRSA strain, we fully expect that this model can be reproduced with the more clinically essential clones such as ST93, ST80, ST30. We noted that the *S. aureus* Newman strain had a point mutation (Pro18Leu) in the SaeS of the two-component system SaeRS, causing a defect in biofilm formation [47]. However, our study was performed mostly on the biofilm-proficient clinical MRSA strain USA300 and only used the Newman strain for confirmation as a drug susceptible strain control (see Figure 5I). In both cases with USA300 and Newman, the clinafloxacin drug combination (meropenem+daptomycin+clinafloxacin) completely eradicated the biofilm infection (Figure 5B,I). Therefore, there should be no concern about the validity of the results on the eradication of biofilm bacteria by the clinafloxacin drug combination, because the main findings of the study were taken on the USA300 strain (Figures 1–5, except Figure 5I) but not on the biofilm defective Newman strain. It is worth noting that our focus is to identify more powerful drug combinations that eradicate biofilm infections without any residual surviving bacteria but not on host immune factors that clear the bacterial infection. The near normal or minimal tissue inflammation and pathology after treatment with the most effective regimen meropenem+daptomycin+clinafloxacin (Figure 5G) is meant to show the effectiveness of the treatment in comparison with the standard treatment with doxycycline+rifampin, which still had significant tissue pathology (Figure 5F). Future studies to examine the role of immune control of the persistent biofilm infection in this model would be of interest.

Clinafloxacin shows impressive anti-biofilm activity in this study and both oral and intravenous formulations have been developed [48,49]. The drugs used in the antibiofilm studies are mostly clinically used drugs, and as such they would not be expected to have significant cytotoxicity at their respective Cmax (maximum human blood drug) concentrations. In addition, clinafloxacin administration drastically improved the condition of a cystic fibrosis patient who had a chronic Burkholderia cenocepacia infection and was not responding to other antibiotic treatments [50]. A human trial with patients having native or prosthetic valve endocarditis also showed that clinafloxacin was an effective treatment [51]. Nevertheless, clinafloxacin is not FDA-approved due to photosensitivity and hypoglycemia. However, the topical use of clinafloxacin or its drug combinations for chronic skin infections would be of interest and could be explored in the future. Further studies with new analogs of clinafloxacin without significant side effects but maintaining high anti-persister activity are needed in the future. Our in vitro data also suggested that oritavancin used in combination with clinafloxacin had robust activity against biofilms, killing all the bacteria after a short treatment of two days. The administration of oritavancin is a single 1200-mg dose given in a slow, three-hour infusion, which may also be of interest for patients due to the ease of administration and long half-life. Hence, further preclinical studies in mice to test oritavancin's activity in chronic infection models are warranted.

Currently used regimens for treating persistent infections are lengthy and ineffective and the inability to clear the bacteria in a timely fashion may also increase the chance of developing antibiotic resistance. A drug combination approach that targets both growing bacteria and non-growing persister cells as proposed in the Yin–Yang model [21] has promising potential in developing a more effective therapy for treating chronic persistent infections including biofilm infections. This study provides further validation of the Yin–Yang model [21] for treating persistent infections [23,52] and emphasizes the importance of persister drugs such as clinafloxacin in eradicating a persistent infection. This treatment algorithm takes into account the heterogeneous population of bacterial cells that exists upon encountering stress. With this principle in mind, this study reports novel drug combinations that are effective in killing *S. aureus* biofilms and treating chronic infections. In further support of this Yin–Yang model of targeting both growing and non-growing bacteria for more effective treatment of persistent infections, we also demonstrated, in a separate study, the complete eradication of a different persistent pulmonary infection with Pseudomonas aeruginosa in a mouse model of cystic fibrosis [52].

In conclusion, we showed that commonly used treatments for *S. aureus* infections have poor activity against biofilms in vitro and in vivo, and importantly, we were able to identify promising persister drug clinafloxacin drug combinations, meropenem (or vancomycin) + daptomycin + clinafloxacin, or oritavancin + clinafloxacin, that achieved complete eradication of *S. aureus* biofilm bacteria in vitro. More importantly, we demonstrated that the persister drug combination meropenem + daptomycin + clinafloxacin completely eradicated the bacterial load and healed lesions promptly with reduced pathology and inflammation in the mouse model of persistent biofilm infection, while the current treatments failed to do so. Our approach of combining drugs targeting both growing and non-growing bacteria with persister drugs to completely eradicate biofilm infections may have implications for developing more effective treatments against other persistent infections caused by different bacterial pathogens, fungi, and even cancer.

Author Contributions: Conceptualization, Y.Z.; methodology, R.Y., Y.Y., A.T., C.B. and N.G.; validation, R.Y., Y.Y., A.T. and C.B.; formal analysis, R.Y. and J.F.; writing—original draft preparation, R.Y. and Y.Z.; writing—review & editing, Y.Z. and N.G.; supervision, Y.Z.; funding Acquisition, Y.Z. All authors have read and agreed to the published version of the manuscript.

Funding: This research received no external funding.

Institutional Review Board Statement: The study was conducted according to the guidelines of the Declaration of Helsinki, and approved by the Johns Hopkins University Animal Care and Use Committee (protocol code MO17H167 and 23 May 2017 approval).

Informed Consent Statement: Not applicable.

Data Availability Statement: Not applicable.

Conflicts of Interest: The authors declare no conflict of interest.

References

1. Chan, L.C.; Chaili, S.; Filler, S.G.; Miller, L.S.; Solis, N.V.; Wang, H.; Johnson, C.W.; Lee, H.K.; Diaz, L.F.; Yeaman, M.R. Innate Immune Memory Contributes to Host Defense against Recurrent Skin and Skin Structure Infections Caused by Methicillin-Resistant *Staphylococcus aureus*. *Infect. Immun.* **2017**, *85*, e00876-16. [CrossRef] [PubMed]
2. David, M.Z.; Daum, R.S. Community-associated methicillin-resistant *Staphylococcus aureus*: Epidemiology and clinical consequences of an emerging epidemic. *Clin. Microbiol. Rev.* **2010**, *23*, 616–687. [CrossRef] [PubMed]
3. Dryden, M.S. Complicated skin and soft tissue infection. *J. Antimicrob. Chemother.* **2010**, *65*, iii35–iii44. [CrossRef]
4. Romling, U.; Balsalobre, C. Biofilm infections, their resilience to therapy and innovative treatment strategies. *J. Intern. Med.* **2012**, *272*, 541–561. [CrossRef] [PubMed]
5. de Oliveira, A.; Cataneli, P.V.; Pinheiro, L.; Moraes Riboli, D.F.; Benini Martins, K.; de Souza da Cunha, M.L.R. Antimicrobial Resistance Profile of Planktonic and Biofilm Cells of *Staphylococcus aureus* and Coagulase-Negative Staphylococci. *Int. J. Mol. Sci.* **2016**, *17*, 1423. [CrossRef]
6. Stewart, P.S.; Davison, W.M.; Steenbergen, J.N. Daptomycin rapidly penetrates a Staphylococcus epidermidis biofilm. *Antimicrob. Agents Chemother.* **2009**, *53*, 3505–3507. [CrossRef]
7. Kirker, K.R.; Fisher, S.T.; James, G.A. Potency and penetration of telavancin in staphylococcal biofilms. *Int. J. Antimicrob. Agents* **2015**, *46*, 451–455. [CrossRef]
8. Kavanaugh, J.S.; Horswill, A.R. Impact of Environmental Cues on Staphylococcal Quorum Sensing and Biofilm Development. *J. Biol. Chem.* **2016**, *291*, 12556–12564. [CrossRef]
9. Hobby, G.L.; Meyer, K.; Chaffee, E. Observations on the mechanism of action of penicillin. *Proc. Soc. Exp. Biol. Med.* **1942**, *50*, 281–285. [CrossRef]
10. Bigger, J.W. Treatment of staphylococcal infections with penicillin by intermittent sterilisation. *Lancet* **1944**, *244*, 4. [CrossRef]
11. Wang, W.; Chen, J.; Chen, G.; Du, X.; Cui, P.; Wu, J.; Zhao, J.; Wu, N.; Zhang, W.; Li, M.; et al. Transposon Mutagenesis Identifies Novel Genes Associated with *Staphylococcus aureus* Persister Formation. *Front. Microbiol.* **2015**, *6*, 1437. [CrossRef] [PubMed]
12. Conlon, B.P.; Rowe, S.E.; Gandt, A.B.; Nuxoll, A.S.; Donegan, N.P.; Zalis, E.A.; Clair, G.; Adkins, J.N.; Cheung, A.L.; Lewis, K. Persister formation in *Staphylococcus aureus* is associated with ATP depletion. *Nat. Microbiol.* **2016**, *1*, 16051. [CrossRef] [PubMed]
13. Yee, R.; Cui, P.; Shi, W.; Feng, J.; Zhang, Y. Genetic Screen Reveals the Role of Purine Metabolism in *Staphylococcus aureus* Persistence to Rifampicin. *Antibiotics* **2015**, *4*, 627–642. [CrossRef] [PubMed]
14. Xu, T.; Wang, X.Y.; Cui, P.; Zhang, Y.M.; Zhang, W.H.; Zhang, Y. The Agr Quorum Sensing System Represses Persister Formation through Regulation of Phenol Soluble Modulins in *Staphylococcus aureus*. *Front. Microbiol.* **2017**, *8*, 2189. [CrossRef]
15. Sahukhal, G.S.; Pandey, S.; Elasri, M.O. msaABCR operon is involved in persister cell formation in *Staphylococcus aureus*. *BMC Microbiol.* **2017**, *17*, 218. [CrossRef]
16. Zhang, Y.; Mitchison, D. The curious characteristics of pyrazinamide: A review. *Int. J. Tuberc. Lung Dis.* **2003**, *7*, 6–21.
17. Zhang, Y.; Wade, M.M.; Scorpio, A.; Zhang, H.; Sun, Z. Mode of action of pyrazinamide: Disruption of *Mycobacterium tuberculosis* membrane transport and energetics by pyrazinoic acid. *J. Antimicrob. Chemother.* **2003**, *52*, 790–795. [CrossRef]
18. Shi, W.; Zhang, X.; Jiang, X.; Yuan, H.; Lee, J.S.; Barry, C.E., 3rd; Wang, H.; Zhang, W.; Zhang, Y. Pyrazinamide inhibits trans-translation in *Mycobacterium tuberculosis*. *Science* **2011**, *333*, 1630–1632. [CrossRef] [PubMed]
19. Zhang, S.; Chen, J.; Shi, W.; Liu, W.; Zhang, W.; Zhang, Y. Mutations in panD encoding aspartate decarboxylase are associated with pyrazinamide resistance in *Mycobacterium tuberculosis*. *Emerg. Microbes Infect.* **2013**, *2*, e34. [CrossRef]
20. Zhang, Y.; Shi, W.; Zhang, W.; Mitchison, D. Mechanisms of Pyrazinamide Action and Resistance. *Microbiol. Spectr.* **2013**, *2*, MGM2-0023-2013.
21. Zhang, Y. Persisters, Persistent Infections and the Yin-Yang Model. *Emerg. Microbes Infect.* **2014**, *3*, 10. [CrossRef] [PubMed]
22. Feng, J.; Weitner, M.; Shi, W.; Zhang, S.; Zhang, Y. Eradication of Biofilm-Like Microcolony Structures of Borrelia burgdorferi by Daunomycin and Daptomycin but not Mitomycin C in Combination with Doxycycline and Cefuroxime. *Front. Microbiol.* **2016**, *7*, 62. [CrossRef] [PubMed]
23. Feng, J.; Li, T.; Yee, R.; Yuan, Y.; Bai, C.; Cai, M.; Shi, W.; Embers, M.; Brayton, C.; Saeki, H.; et al. Stationary phase persister/biofilm microcolony of Borrelia burgdorferi causes more severe disease in a mouse model of Lyme arthritis: Implications for understanding persistence, Post-treatment Lyme Disease Syndrome (PTLDS), and treatment failure. *Discov. Med.* **2019**, *27*, 125–138.
24. Niu, H.; Cui, P.; Yee, R.; Shi, W.; Zhang, S.; Feng, J.; Sullivan, D.; Zhang, W.; Zhu, B.; Zhang, Y. A Clinical Drug Library Screen Identifies Tosufloxacin as Being Highly Active against *Staphylococcus aureus* Persisters. *Antibiotics* **2015**, *4*, 329–336. [CrossRef] [PubMed]
25. Yee, R.; Yuan, Y.; Shi, W.; Brayton, C.; Tarff, A.; Feng, J.; Wang, J.; Behrens, A.; Zhang, Y. Infection with persister forms of *Staphylococcus aureus* causes a persistent skin infection with more severe lesions in mice: Failure to clear the infection by the current standard of care treatment. *Discov. Med.* **2019**, *28*, 7–16. [PubMed]
26. O'Toole, G.A. Microtiter dish biofilm formation assay. *J. Vis. Exp.* **2011**, *30*, 2437. [CrossRef]
27. Feng, J.; Wang, T.; Zhang, S.; Shi, W.; Zhang, Y. An optimized SYBR Green I/PI assay for rapid viability assessment and antibiotic susceptibility testing for Borrelia burgdorferi. *PLoS ONE* **2014**, *9*, e111809. [CrossRef] [PubMed]

28. Pfaller, M.A.; Sader, H.S.; Flamm, R.K.; Castanheira, M.; Mendes, R.E. Oritavancin in vitro activity against gram-positive organisms from European and United States medical centers: Results from the SENTRY Antimicrobial Surveillance Program for 2010-2014. *Diagn. Microbiol. Infect. Dis.* **2018**, *91*, 199–204. [CrossRef]
29. Yan, Q.; Karau, M.J.; Patel, R. In vitro activity of oritavancin against biofilms of staphylococci isolated from prosthetic joint infection. *Diagn. Microbiol. Infect. Dis.* **2018**, *92*, 155–157. [CrossRef]
30. Conlon, B.P.; Nakayasu, E.S.; Fleck, L.E.; LaFleur, M.D.; Isabella, V.M.; Coleman, K.; Leonard, S.N.; Smith, R.D.; Adkins, J.N.; Lewis, K. Activated ClpP kills persisters and eradicates a chronic biofilm infection. *Nature* **2013**, *503*, 365–370. [CrossRef]
31. Allison, K.R.; Brynildsen, M.P.; Collins, J.J. Metabolite-enabled eradication of bacterial persisters by aminoglycosides. *Nature* **2011**, *473*, 216–220. [CrossRef] [PubMed]
32. Pinto, H.; Simões, M.; Borges, A. Prevalence and Impact of Biofilms on Bloodstream and Urinary Tract Infections: A Systematic Review and Meta-Analysis. *Antibiotics* **2021**, *10*, 825. [CrossRef] [PubMed]
33. Høiby, N.; Bjarnsholt, T.; Moser, C.; Bassi, G.L.; Coenye, T.; Donelli, G.; Hall-Stoodley, L.; Holá, V.; Imbert, C.; Kirketerp-Møller, K.; et al. ESCMID Study Group for Biofilms and Consulting External Expert Werner Zimmerli. ESCMID guideline for the diagnosis and treatment of biofilm infections 2014. *Clin. Microbiol. Infect.* **2015**, *21*, S1–S25. [CrossRef] [PubMed]
34. Hughes, G.; Webber, M.A. Novel approaches to the treatment of bacterial biofilm infections. *Br. J. Pharm.* **2017**, *174*, 2237–2246. [CrossRef] [PubMed]
35. Jacqueline, C.; Caillon, J. Impact of bacterial biofilm on the treatment of prosthetic joint infections. *J. Antimicrob Chemother.* **2014**, *69* Suppl. S1, i37–i40. [CrossRef]
36. Kim, W.; Zhu, W.; Hendricks, G.L.; Van Tyne, D.; Steele, A.D.; Keohane, C.E.; Fricke, N.; Conery, A.L.; Shen, S.; Pan, W.; et al. A new class of synthetic retinoid antibiotics effective against bacterial persisters. *Nature* **2018**, *556*, 103–107. [CrossRef]
37. Mohamed, M.F.; Abdelkhalek, A.; Seleem, M.N. Evaluation of short synthetic antimicrobial peptides for treatment of drug-resistant and intracellular *Staphylococcus aureus*. *Sci. Rep.* **2016**, *6*, 29707. [CrossRef]
38. Zhang, Y.; Yew, W.W.; Barer, M.R. Targeting persisters for tuberculosis control. *Antimicrob. Agents Chemother.* **2012**, *56*, 2223–2230. [CrossRef]
39. Joers, A.; Kaldalu, N.; Tenson, T. The frequency of persisters in Escherichia coli reflects the kinetics of awakening from dormancy. *J. Bacteriol.* **2010**, *192*, 3379–3384. [CrossRef]
40. Mandell, J.B.; Deslouches, B.; Montelaro, R.C.; Shanks, R.M.Q.; Doi, Y.; Urish, K.L. Elimination of Antibiotic Resistant Surgical Implant Biofilms Using an Engineered Cationic Amphipathic Peptide WLBU2. *Sci. Rep.* **2017**, *7*, 18098. [CrossRef]
41. Humphries, R.M.; Pollett, S.; Sakoulas, G. A current perspective on daptomycin for the clinical microbiologist. *Clin. Microbiol. Rev.* **2013**, *26*, 759–780. [CrossRef] [PubMed]
42. CLSI. *Performance Standards for Antimicrobial Performance Standards for Antimicrobial Susceptibility Testing-26th Edition: CLSI Supplement M100S*; NCCLS: Wayne, PA, USA, 2016.
43. Gallo, S.W.; Ferreira, C.A.S.; de Oliveira, S.D. Combination of polymyxin B and meropenem eradicates persister cells from *Acinetobacter baumannii* strains in exponential growth. *J. Med. Microbiol.* **2017**, *66*, 1257–1260. [CrossRef]
44. Carpenter, C.F.; Chambers, H.F. Daptomycin: Another novel agent for treating infections due to drug-resistant gram-positive pathogens. *Clin. Infect. Dis.* **2004**, *38*, 994–1000. [CrossRef] [PubMed]
45. Feng, J.; Auwaerter, P.G.; Zhang, Y. Drug Combinations against Borrelia burgdorferi Persisters In Vitro: Eradication Achieved by Using Daptomycin, Cefoperazone and Doxycycline. *PLoS ONE* **2015**, *10*, e0117207. [CrossRef] [PubMed]
46. Pogliano, J.; Pogliano, N.; Silverman, J.A. Daptomycin-mediated reorganization of membrane architecture causes mislocalization of essential cell division proteins. *J. Bacteriol.* **2012**, *194*, 4494–4504. [CrossRef]
47. Cue, D.; Junecko, J.M.; Lei, M.G.; Blevins, J.S.; Smeltzer, M.S.; Lee, C.Y. SaeRS-Dependent Inhibition of Biofilm Formation in *Staphylococcus aureus* Newman. *PLoS ONE* **2015**, *10*, e0123027. [CrossRef]
48. Barrett, M.S.; Jones, R.N.; Erwin, M.E.; Johnson, D.M.; Briggs, B.M. Antimicrobial activity evaluations of two new quinolones, PD127391 (CI-960 and AM-1091) and PD131628. *Diagn. Microbiol. Infect. Dis.* **1991**, *14*, 389–401. [CrossRef]
49. Cohen, M.A.; Yoder, S.L.; Huband, M.D.; Roland, G.E.; Courtney, C.L. In vitro and in vivo activities of clinafloxacin, CI-990 (PD 131112), and PD 138312 versus enterococci. *Antimicrob. Agents Chemother.* **1995**, *39*, 2123–2127. [CrossRef]
50. Balwan, A.; Nicolau, D.P.; Wungwattana, M.; Zuckerman, J.B.; Waters, V. Clinafloxacin for Treatment of Burkholderia cenocepacia Infection in a Cystic Fibrosis Patient. *Antimicrob. Agents Chemother.* **2016**, *60*, 1–5. [CrossRef]
51. Levine, D.P.; Holley, H.P.; Eiseman, I.; Willcox, P.; Tack, K. Clinafloxacin for the treatment of bacterial endocarditis. *Clin. Infect. Dis.* **2004**, *38*, 620–631. [CrossRef]
52. Yuan, R.Y.; Gour, N.; Dong, X.Z.; Jie, F.; Shi, W.L.; Zhang, Y. Ranking of Major Classes of Antibiotics for Activity against Stationary Phase *Pseudomonas aeruginosa* and Identification of Clinafloxacin + Cefuroxime + Gentamicin Drug Combination that Eradicates Persistent *P. aeruginosa* Infection in a Murine Cystic Fibrosis Model. *bioRxiv*, 2019. [CrossRef]

 *antibiotics*

Article

Evaluation of Catechin Synergistic and Antibacterial Efficacy on Biofilm Formation and *acrA* Gene Expression of Uropathogenic *E. coli* Clinical Isolates

Najwan Jubair [1,*], Mogana R. [1,*], Ayesha Fatima [2], Yasir K. Mahdi [1] and Nor Hayati Abdullah [3]

1   Faculty of Pharmaceutical Sciences, UCSI University, Kuala Lumpur 56000, Malaysia
2   Beykoz Institute of Life Sciences and Biotechnology, Bezmialem Vakif University, 34820 Istanbul, Turkey
3   Forest Research Institute Malaysia (FRIM), Kuala Lumpur 52109, Malaysia
*   Correspondence: najwanjubair@yahoo.com (N.J.); mogana@ucsiuniversity.edu.my (M.R.)

Citation: Jubair, N.; R., M.; Fatima, A.; Mahdi, Y.K.; Abdullah, N.H. Evaluation of Catechin Synergistic and Antibacterial Efficacy on Biofilm Formation and *acrA* Gene Expression of Uropathogenic *E. coli* Clinical Isolates. *Antibiotics* **2022**, *11*, 1223. https://doi.org/10.3390/antibiotics11091223

Academic Editors: Zhenbo Xu, Ding-Qiang Chen, Junyan Liu, Ren-You Gan, Guanggang Qu and Yulong Tan

Received: 19 July 2022
Accepted: 29 August 2022
Published: 9 September 2022

Publisher's Note: MDPI stays neutral with regard to jurisdictional claims in published maps and institutional affiliations.

Abstract: Uropathogenic *Escherichia coli* has a propensity to build biofilms to resist host defense and antimicrobials. Recurrent urinary tract infection (UTI) caused by multidrug-resistant, biofilm-forming *E. coli* is a significant public health problem. Consequently, searching for alternative medications has become essential. This study was undertaken to investigate the antibacterial, synergistic, and antibiofilm activities of catechin isolated from *Canarium patentinervium* Miq. against three *E. coli* ATCC reference strains (ATCC 25922, ATCC 8739, and ATCC 43895) and fifteen clinical isolates collected from UTI patients in Baghdad, Iraq. In addition, the expression of the biofilm-related gene, *acrA*, was evaluated with and without catechin treatment. Molecular docking was performed to evaluate the binding mode between catechin and the target protein using Autodock Vina 1.2.0 software. Catechin demonstrated significant bactericidal activity with a minimum inhibitory concentration (MIC) range of 1–2 mg/mL and a minimum bactericidal concentration (MBC) range of 2–4 mg/mL and strong synergy when combined with tetracycline at the MBC value. In addition, catechin substantially reduced *E. coli* biofilm by downregulating the *acrA* gene with a reduction percent $\geq$ 60%. In silico analysis revealed that catechin bound with high affinity ($\Delta G = -8.2$ kcal/mol) to AcrB protein (PDB-ID: 5ENT), one of the key AcrAB-TolC efflux pump proteins suggesting that catechin might inhibit the *acrA* gene indirectly by docking at the active site of AcrB protein.

Keywords: *Canarium patentinervium* Miq.; multidrug resistance; antibiofilm activity; AcrAB-TolC efflux pump; molecular docking; Autodock Vina

1. Introduction

Escherichia coli is a Gram-negative multifaceted bacterium that comprises commensal *E. coli*, which normally colonizes in the gastrointestinal tract of humans and animals a few hours after birth [1], and pathogenic *E. coli*, which is the most common cause of gastrointestinal infections such as diarrheal disease caused by enterotoxigenic *E. coli* (ETEC) and extraintestinal infections such as urinary tract infections (UTI) caused by uropathogenic *E. coli* (UPEC) [2]. Some *E. coli* clones became pathogenic when they gain a number of specialized virulence factors, such as different adhesins, toxins, siderophores, and iron acquisition systems that interfere with the way the host cell works [3].

UTIs are commonly classified as community-acquired or healthcare-associated UTIs, and as complicated or uncomplicated UTIs, depending on the severity of the infection [4]. UTI is one of the most common bacterial infections worldwide, with an estimated 150 million UTI cases occurring each year [5]. Due to the anatomy of the female urinary tract, it is estimated that 50–60% of women will develop an UTI at some point in their lives [6]. UTIs are treatable in most cases. However, the progression of multidrug-resistant strains results in recurrent infections, treatment failure, and complications associated with increased rates of mortality and morbidity [7].

E. coli resists antibiotics through several mechanisms, such as production of enzymes called "beta-lactamases", which is a group of more than 2800 compounds derived from environmental sources, including extended spectrum beta-lactamase (ESBL), AmpC beta-lactamase that acquires resistance to penicillin and cephalosporins, New Delhi metallo-beta-lactamase, and Carbapenem hydrolyzing oxacillinase-48 [8]. Furthermore, efflux pump activity such as the resistance-nodulation-division (RND) tripartite efflux pump, AcrAB-TolC, represents the major contributor to intrinsic multidrug resistance in *E. coli* [9]. In addition, *E. coli* possesses several virulence factors to ensure survival, such as hemolysis and biofilm formation [10].

Biofilm formation enables *E. coli* colonies to evade the immune system and antibiotics, making it difficult to eradicate and resulting in multiple antibiotic resistance [11]. Increased drug resistance and the emergence of bacterial infections with no treatment options yet available make research into new antimicrobial agents urgent. On the other hand, medicinal plants provide a promising source for drug discovery. Since ancient times, Indigenous peoples have utilized medicinal plants to treat various conditions [12]. Today, people continue to depend on herbal therapy, particularly in developing nations where 80% of the population uses traditional medicine to treat a variety of illnesses [13].

Canarium patentinervium Miq. is a rare tropical plant belonging to the family Burseraceae, genus *Canarium* L., native to Asiatic–Pacific region, Malaysia, and Brunei [14]. It has been used for wound healing by Indigenous peoples of Malaysia [14]. Our team previously reported the antibacterial activity of the crude extracts, fractions, and isolated compounds from the leaves and bark of this plant [15,16]. The current study aimed to comprehensively evaluate the antibacterial activity of catechin (Figure 1), one of the active compounds isolated previously from the leaves of *Canarium patentinervium* Miq., against biofilm-forming, uropathogenic *E. coli* isolates from UTI urine samples collected in Baghdad, Iraq during the period November–December 2021. Molecular docking was carried out to achieve a deep insight into the molecular mechanism and to analyze the binding mode between catechin and the target protein proposed in the study.

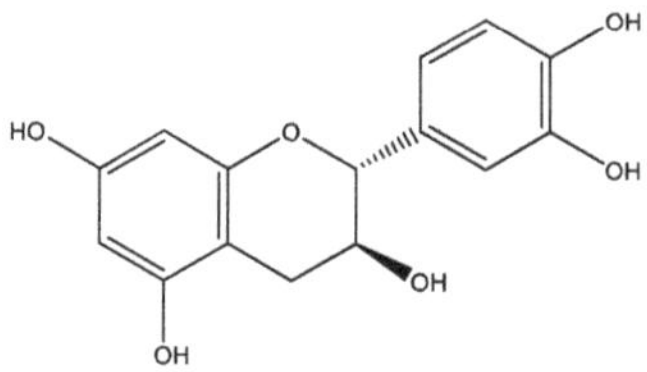

Figure 1. The chemical structure of catechin.

2. Results

2.1. Evaluation of the Antibacterial Activity

The study investigated the antibacterial activity of catechin isolated from the leaves of *Canarium patentinervium* Miq. against three *E. coli* ATCC strains (*E. coli* ATCC 25922, *E. coli* ATCC 8739, and *E. coli* 43895) and 15 multidrug-resistant *E. coli* clinical isolates. The results are displayed in Table 1. Among the tested antibiotics, tetracycline, erythromycin, clindamycin, and vancomycin showed insignificant inhibition zones in all *E. coli* ATCC strains. *E. coli* ATCC 8739 had intermediate sensitivity to rifampin (17.24 mm) and gentamicin (13.14 mm). In addition, gentamicin displayed moderate inhibition against *E. coli* ATCC 43895 (13.6 mm), although it showed a significant zone of inhibition against *E. coli* ATCC 25922 (31.32 mm). Nevertheless, *E. coli* clinical isolates displayed insignificant inhibition zones for most of the tested antibiotics. In regard to catechin, *E. coli* ATCC 8739 was resistant (9 mm), although catechin displayed remarkable inhibition against the remaining *E. coli* strains ATCC 25922 (10.47 mm), *E. coli* ATCC 43895 (11.8 mm) and most of the clinical isolates (zone of inhibition ranges from 9.93 mm to 13.92 mm). The synergistic effect of catechin in combination with the tested antibiotics was evaluated in this study. Results are displayed in Table 1. The strongest synergistic effect was observed between catechin and tetracycline, through which all the isolates (n = 15) showed synergism with no antagonism reported. In addition, combinations between catechin with azithromycin, gentamicin, erythromycin, and clindamycin displayed a high percentage of synergism with additive effects reported in three isolates, while the combination of catechin with rifampin had the lowest synergy with high antagonism reported.

Table 1. Zone of inhibition (mm) for the antimicrobial agents and catechin alone and in combination against *E. coli* isolates.

| Bacteria | Antimicrobials | Zone of Inhibition (ZI) (mm) | | | | |
		ZI of Antimicrobials	ZI Breakpoints According to CLSI *	ZI of Catechin	ZI of Catechin in Combination with Antimicrobials	Outcome
Escherichia coli ATCC 25922 [#]	Rifampin	7.42 ± 0.0	R		16 ± 0.0	Antagonism
	Tetracycline	-	R		10.4 ± 0.0	Additive
	Erythromycin	-	R		10.0 ± 0.0	Antagonism
	Clindamycin	-	R	10.47 ± 0.0	10.45 ± 0.0	Additive
	Azithromycin	10.95 ± 0.0	R		22.5	Synergy
	Vancomycin	-	R		10.4 ± 0.0	Additive
	Gentamicin	31.32 ± 0.0	S		35 ± 0.0	Antagonism
Escherichia coli ATCC 8739 [#]	Rifampin	17.24 ± 0.0	I		26 ± 0.0	Additive
	Tetracycline	-	R		9.1 ± 0.0	Additive
	Erythromycin	-	R		8.8 ± 0.0	Antagonism
	Clindamycin	-	R	9 ± 0.0	9.0 ± 0.0	Additive
	Azithromycin	15.15 ± 0.0	S		25 ± 0.0	Synergy
	Vancomycin	-	R		9.5 ± 0.0	Synergy
	Gentamicin	13.14 ± 0.0	I		22 ± 0.0	Additive
Escherichia coli ATCC 43895 [#]	Rifampin	-	R		11.5 ± 0.0	Antagonism
	Tetracycline	-	R		12.0 ± 0.0	Synergy
	Erythromycin	-	R		11.5 ± 0.0	Antagonism
	Clindamycin	-	R	11.8 ± 0.0	11.45 ± 0.0	Antagonism
	Azithromycin	6.03 ± 0.0	R		17 ± 0.0	Synergy
	Vancomycin	-	R		11.8 ± 0.0	Additive
	Gentamicin	13.6 ± 0.0	I		22	Antagonism
Escherichia coli (isolate 1)	Rifampin	2.45 ± 4.2	R		13.89 ± 3.6	Antagonism
	Tetracycline	13.85 ± 2.5	I		28.58 ± 1.0	Synergy
	Erythromycin	11.97 ± 7.9	I		25.04 ± 0.6	Synergy
	Clindamycin	7.71 ± 8.7	R	12.27 ± 1.3	22.19 ± 1.3	Synergy
	Azithromycin	11.58 ± 3.2	R		24.8 ± 0.5	Synergy
	Vancomycin	13.72 ± 7.2	I		26.43 ± 0.5	Synergy
	Gentamicin	10.42 ± 5.6	R		23.58 ± 0.4	Synergy
Escherichia coli (isolate 2)	Rifampin	-	R		14.0 ± 0.0	Additive
	Tetracycline	16.83 ± 1.0	S		31.5 ± 10.0	Synergy
	Erythromycin	18.97 ± 3.5	I		33.1 ± 0.0	Synergy
	Clindamycin	17.75 ± 0.4	I	13.92 ± 3.6	31.5 ± 4.5	Additive
	Azithromycin	12.60 ± 2.84	I		27.2 ± 0.5	Synergy
	Vancomycin	17.87 ± 2.2	S		31.6 ± 0.9	Additive
	Gentamicin	16.95 ± 0.3	S		30.7 ± 0.8	Additive
Escherichia coli (isolate 3)	Rifampin	-	R		9.0 ± 5.2	Antagonism
	Tetracycline	12.35 ± 0.4	I		26.12 ± 0.2	Synergy
	Erythromycin	16.19 ± 1.0	I		31.5 ± 2.1	Synergy
	Clindamycin	15.37 ± 3.8	I	12.9 ± 2.4	28.2 ± 3.1	Additive
	Azithromycin	12.43 ± 2.2	I		26.3 ± 0.3	Synergy
	Vancomycin	18.87 ± 0.0	S		32.0 ± 0.7	Synergy
	Gentamicin	16.24 ± 1.3	S		29.0 ± 0.1	Additive

Table 1. *Cont.*

Bacteria	Antimicrobials	Zone of Inhibition (ZI) (mm)				
		ZI of Antimicrobials	ZI Breakpoints According to CLSI *	ZI of Catechin	ZI of Catechin in Combination with Antimicrobials	Outcome
Escherichia coli (isolate 4)	Rifampin	10.87 ± 0.0	R	13.09 ± 0.0	20.21 ± 1.6	Antagonism
	Tetracycline	15.26 ± 0.5	S		30.0 ± 1.6	Synergy
	Erythromycin	20.61 ± 4.0	I		34.0 ± 0.0	Synergy
	Clindamycin	16.42 ± 1.1	I		29.5 ± 1.2	Additive
	Azithromycin	12.42 ± 0.7	I		26.1 ± 0.5	Synergy
	Vancomycin	20.03 ± 5.2	S		25.5 ± 0.1	Additive
	Gentamicin	-	R		13.1 ± 2.4	Additive
Escherichia coli (isolate 5)	Rifampin	-	R	10.68 ± 5.6	10.71 ± 1.1	Additive
	Tetracycline	16.82 ± 2.3	S		30.23 ± 0.0	Synergy
	Erythromycin	16.5 ± 0.0	I		28.0 ± 4.4	Synergy
	Clindamycin	-	R		12.1 ± 3.3	Synergy
	Azithromycin	12.12 ± 1.5	I		24.0 ± 0.8	Synergy
	Vancomycin	17.71 ± 3.4	S		29.0 ± 0.3	Synergy
	Gentamicin	-	R		11.0 ± 0.2	Synergy
Escherichia coli (isolate 6)	Rifampin	-	R	9.93 ± 3.6	9.0 ± 4.5	Additive
	Tetracycline	12.26 ± 1.0	I		25.0 ± 0.0	Synergy
	Erythromycin	14.1 ± 0.6	I		25.0 ± 0.0	Synergy
	Clindamycin	-	R		10.5 ± 0.0	Synergy
	Azithromycin	12.31 ± 2.6	I		23.2 ± 0.4	Synergy
	Vancomycin	16.82 ± 0.5	I		27.0 ± 0.5	Synergy
	Gentamicin	11.45 ± 0.0	R		21.8 ± 0.3	Synergy
Escherichia coli (isolate 7)	Rifampin	-	R	12.75 ± 5.6	14.74 ± 8.3	Synergy
	Tetracycline	12.93 ± 0.2	I		27.3 ± 0.9	Synergy
	Erythromycin	14.59 ± 1.4	I		28.1 ± 3.4	Synergy
	Clindamycin	-	R		13.22 ± 0.2	Synergy
	Azithromycin	12.94 ± 0.6	I		26.6 ± 0.9	Synergy
	Vancomycin	18.21 ± 1.5	S		30.89 ± 0.6	Additive
	Gentamicin	11.81 ± 0.0	R		25.6 ± 0.4	Synergy
Escherichia coli (isolate 8)	Rifampin	-	R	10.08 ± 4.3	14.69 ± 0.8	Synergy
	Tetracycline	12.12 ± 11.0	I		24.6 ± 5.3	Synergy
	Erythromycin	17.73 ± 8.76	I		28.4 ± 0.0	Synergy
	Clindamycin	-	R		11.54 ± 1.3	Synergy
	Azithromycin	12.04 ± 5.6	I		23.1 ± 0.2	Synergy
	Vancomycin	-	R		10.0 ± 3.4	Additive
	Gentamicin	11.02 ± 7.0	R		22.0 ± 1.1	Synergy
Escherichia coli (isolate 9)	Rifampin	9.46 ± 1.1	R	13.58 ± 2.3	23.1 ± 2.5	Additive
	Tetracycline	13.46 ± 2.3	I		27.89 ± 6.8	Synergy
	Erythromycin	19.09 ± 0.6	I		33.1 ± 0.5	Synergy
	Clindamycin	-	R		14.0 ± 0.0	Synergy
	Azithromycin	12.82 ± 0.0	I		27.0 ± 0.4	Synergy
	Vancomycin	-	R		13.5 ± 0.6	Additive
	Gentamicin	11.1 ± 0.0	R		26.4 ± 0.8	Synergy
Escherichia coli (isolate 10)	Rifampin	-	R	12.9 ± 1.7	12.6 ± 1.3	Additive
	Tetracycline	13.4 ± 0.5	I		27.0 ± 1.1	Synergy
	Erythromycin	14.34 ± 4.1	I		28.5 ± 8.85	Synergy
	Clindamycin	15.38 ± 3.2	I		30.0 ± 5.5	Synergy
	Azithromycin	12.55 ± 8.5	I		26.4 ± 0.5	Synergy
	Vancomycin	-	R		12.82 ± 0.9	Additive
	Gentamicin	13.52 ± 6.1	I		27.2 ± 0.4	Synergy

Table 1. *Cont.*

Bacteria	Antimicrobials	Zone of Inhibition (ZI) (mm)				
		ZI of Antimicrobials	ZI Breakpoints According to CLSI *	ZI of Catechin	ZI of Catechin in Combination with Antimicrobials	Outcome
Escherichia coli (isolate 11)	Rifampin	-	R	10.38 ± 0.6	10.4 ± 0.5	Additive
	Tetracycline	12.28 ± 0.0	I		24.5 ± 0.7	Synergy
	Erythromycin	-	R		10.5 ± 1.4	Additive
	Clindamycin	16.69 ± 1.1	I		28.1 ± 1.1	Synergy
	Azithromycin	12.21 ± 0.9	I		22.5 ± 0.5	Additive
	Vancomycin	16.59 ± 3.1	I		29.0 ± 0.7	Synergy
	Gentamicin	13.76 ± 2.5	I		25.4 ± 1.1	Synergy
Escherichia coli (isolate 12)	Rifampin	8.76 ± 5.3	R	13.17 ± 1.5	22.5 ± 6.4	Synergy
	Tetracycline	18.34 ± 7.1	I		31.9 ± 0.5	Synergy
	Erythromycin	-	R		13.4 ± 11.0	Additive
	Clindamycin	15.71 ± 2.6	I		30.5 ± 0.0	Synergy
	Azithromycin	12.77 ± 1.5	I		26.4 ± 0.2	Synergy
	Vancomycin	15.24 ± 1.1	I		28.5 ± 0.4	Additive
	Gentamicin	13.73 ± 0.6	I		27.5 ± 0.9	Synergy
Escherichia coli (isolate 13)	Rifampin	-	R	13.18 ± 8.7	13.0 ± 0.0	Additive
	Tetracycline	9.82 ± 4.3	R		25.4 ± 1.4	Synergy
	Erythromycin	-	R		14.21 ± 0.0	Synergy
	Clindamycin	18.27 ± 5.2	I		32.0 ± 0.5	Synergy
	Azithromycin	12.67 ± 9.7	I		27.1 ± 0.0	Synergy
	Vancomycin	16.37 ± 10.0	I		29.45 ± 0.5	Additive
	Gentamicin	13.14 ± 8.7	I		27.2 ± 0.3	Synergy
Escherichia coli (isolate 14)	Rifampin	7.67 ± 2.6	R	12.4 ± 1.3	20.0 ± 3.8	Additive
	Tetracycline	10.22 ± 11.0	R		23.4 ± 1.5	Synergy
	Erythromycin	-	R		12.42 ± 0.5	Additive
	Clindamycin	-	R		13.1 ± 0.6	Synergy
	Azithromycin	11.28 ± 5.4	I		23.5 ± 2.7	Additive
	Vancomycin	16.24 ± 8.6	I		29.0 ± 0.4	Synergy
	Gentamicin	11.81 ± 5.9	I		25.1 ± 0.2	Synergy
Escherichia coli (isolate 15)	Rifampin	-	R	12.62 ± 6.6	12.5 ± 7.8	Additive
	Tetracycline	15.89 ± 7.4	S		29.1 ± 0.0	Synergy
	Erythromycin	9.42 ± 11.0	R		24.01 ± 11.6	Synergy
	Clindamycin	-	R		14.0 ± 4.3	Synergy
	Azithromycin	-	R		12.5 ± 0.9	Additive
	Vancomycin	15.37 ± 8.3	I		28.0 ± 0.9	Additive
	Gentamicin	-	R		13.2 ± 0.6	Synergy

"#" means statistical data are unavailable, "-" means no activity, and "*" indicates that the results are categorized according to Clinical and Laboratory Standards Institute (CLSI) guidelines into R resistant; I intermediately resistant; and S sensitive. All the data were obtained from three independent experiments and is expressed as mean ± SD.

The minimum inhibitory concentration is the lowest antimicrobial concentration that inhibits the visible growth of a bacterium following overnight incubation [17]. Azithromycin showed potent antibacterial activity with a minimum inhibitory concentration (MIC) of 0.5 mg/mL against *E. coli* ATCC 8739 and 0.5–1 mg/mL against *E. coli* clinical isolates (Table 2). The MICs for the rest of the antibiotics ranged from 2 to 32 mg/mL, and they were only moderately effective against all of the strains tested. Catechin exhibited moderate activity (MIC = 1 mg/mL) against *E. coli* ATCC 25922, *E. coli* ATCC 43895, and *E. coli* clinical isolates, weak activity against *E. coli* ATCC 8739 (MIC = 2 mg/mL), and potent activity against two clinical isolates (isolate numbers 9 and 10) with a MIC of 0.5 mg/mL. According to the literature, strong antibacterial activity is considered with MIC values ranging from 0.05–0.5 mg/mL, moderate activity with MIC values ranging from 0.6–1.5 mg/mL, and weak antibacterial activity when the MIC values exceed 1.5 mg/mL [18].

Table 2. Minimum bactericidal concentration (MBC) (mg/mL), minimum inhibitory concentration (MIC) (mg/mL), and MBC/MIC ratio for the antimicrobial agents and catechin against *E. coli* isolates.

Bacteria	Effect	Rifampin	Tetracycline	Erythromycin	Clindamycin	Azithromycin	Vancomycin	Gentamycin	Catechin
Escherichia coli ATCC 25922 [#]	MBC	8	32	32	8	8	64	4	2
	MIC	4	16	8	4	4	16	2	1
	MBC/MIC	2(+)	2(+)	4(+)	2(+)	2(+)	4(+)	2(+)	2(+)
Escherichia coli ATCC 8739 [#]	MBC	4	32	32	16	1	32	4	4
	MIC	2	16	16	8	0.5	16	2	2
	MBC/MIC	2(+)	2(+)	2(+)	2(+)	2(+)	2(+)	2(+)	2(+)
Escherichia coli ATCC 43895 [#]	MBC	8	32	16	8	16	16	4	2
	MIC	4	16	8	4	8	8	2	1
	MBC/MIC	2(+)	2(+)	2(+)	2(+)	2(+)	2(+)	2(+)	2(+)
Escherichia coli (isolate 1)	MBC	32	32	32	32	2	32	8	4
	MIC	16	8	4	4	0.5	8	4	1
	MBC/MIC	2(+)	4(+)	8(−)	8(−)	4(+)	4(+)	2(+)	4(+)
Escherichia coli (isolate 2)	MBC	16	32	16	32	2	32	8	4
	MIC	8	4	4	4	1	8	4	1
	MBC/MIC	2(+)	8(−)	4(+)	8(−)	2(+)	4(+)	2(+)	4(+)
Escherichia coli (isolate 3)	MBC	16	32	32	32	1	16	8	4
	MIC	8	8	4	4	0.5	2	4	1
	MBC/MIC	2(+)	4(+)	8(−)	8(−)	2(+)	8(−)	2(+)	4(+)
Escherichia coli (isolate 4)	MBC	32	64	32	32	1	32	8	4
	MIC	16	4	8	8	0.5	8	4	1
	MBC/MIC	2(+)	16(−)	4(+)	4(+)	2(+)	4(+)	2(+)	4(+)
Escherichia coli (isolate 5)	MBC	16	64	32	32	1	32	8	4
	MIC	8	4	8	2	0.5	8	4	1
	MBC/MIC	2(+)	16(−)	4(+)	16(−)	2(+)	4(+)	2(+)	4(+)
Escherichia coli (isolate 6)	MBC	16	32	16	32	2	64	4	4
	MIC	8	4	4	8	1	16	1	1
	MBC/MIC	2(+)	8(−)	4(+)	4(+)	2(+)	4(+)	4(+)	4(+)
Escherichia coli (isolate 7)	MBC	32	32	64	32	2	64	16	4
	MIC	8	8	8	2	1	8	8	1
	MBC/MIC	4(+)	4(+)	8(−)	16(−)	2(+)	8(−)	2(+)	4(+)
Escherichia coli (isolate 8)	MBC	64	32	32	32	4	32	32	4
	MIC	8	8	4	4	1	16	16	1
	MBC/MIC	8(−)	4(+)	8(−)	8(−)	4(+)	2(+)	2(+)	4(+)
Escherichia coli (isolate 9)	MBC	64	32	32	32	2	32	4	2
	MIC	16	8	8	4	1	8	2	0.5
	MBC/MIC	4(+)	4(+)	4(−)	8(−)	2(+)	4(+)	2(+)	4(+)
Escherichia coli (isolate 10)	MBC	64	32	32	32	4	64	4	2
	MIC	32	8	4	4	1	16	2	0.5
	MBC/MIC	2(+)	4(+)	8(−)	8(−)	4(+)	4(+)	2(+)	4(+)
Escherichia coli (isolate 11)	MBC	64	32	16	32	2	16	8	4
	MIC	32	8	4	4	0.5	4	2	1
	MBC/MIC	2(+)	4(+)	4(+)	8(−)	4(+)	4(+)	4(+)	4(+)
Escherichia coli (isolate 12)	MBC	64	32	64	32	4	32	8	4
	MIC	32	8	8	4	1	8	2	1
	MBC/MIC	2(+)	4(+)	8(−)	8(−)	4(+)	4(+)	4(+)	4(+)
Escherichia coli (isolate 13)	MBC	64	32	16	16	2	32	8	4
	MIC	32	4	8	4	0.5	8	4	1
	MBC/MIC	2(+)	8(−)	2(+)	4(+)	4(+)	4(+)	2(+)	4(+)
Escherichia coli (isolate 14)	MBC	16	64	32	32	2	32	8	4
	MIC	8	32	4	4	1	4	4	1
	MBC/MIC	2(+)	2(+)	8(−)	8(−)	2(+)	8(−)	2(+)	4(+)
Escherichia coli (isolate 15)	MBC	32	64	32	32	1	32	8	4
	MIC	16	32	4	4	0.5	16	4	1
	MBC/MIC	2(+)	2(+)	8(−)	8(−)	2(+)	2(+)	2(+)	4(+)

"#" means statistical data are unavailable. All the data were obtained from three independent experiments. For the MBC/MIC ratio, (+) bactericidal; (−) bacteriostatic.

The minimum bactericidal concentration (MBC) is the lowest antimicrobial concentration that will stop an organism from growing after being subcultured on an antibiotic-free medium [17]. According to the ratio of MBC/MIC, the antibacterial effect is considered bactericidal if the MBC/MIC $\leq$ 4, and the effect is considered bacteriostatic if the MBC/MIC > 4 [19]. The MBC, MIC, and MBC/MIC ratio for the tested antibiotics and catechin are displayed in Table 2. All the tested antibiotics showed bactericidal activity against the tested strains except for tetracycline, erythromycin, and clindamycin, which showed bacteriostatic activity against some of the clinical isolates. Catechin exhibited bactericidal action against all the tested *E. coli* strains. Azithromycin was the most active antibiotic and *E. coli* ATCC 8739 was the most susceptible strain (MIC = 0.5 mg/mL, MBC = 1 mg/mL, MBC/MIC = 2). It is noteworthy that *E. coli* ATCC 8739 was resistant to catechin and all antibiotics except azithromycin. Interestingly, *E. coli* ATCC 43895 was the most sensitive strain to catechin (MIC = 1 mg/mL, MBC = 2 mg/ mL, MBC/MIC = 2), although it was resistant to all the tested antibiotics. Both catechin and azithromycin had significant antibacterial activity against *E. coli* clinical isolates with bactericidal effects, while the remaining antibiotics were inactive against the tested strains (MIC ranges of 2–32 mg/mL, MBC ranges of 4–64 mg/mL).

2.2. Biofilm Inhibition

The results for the biofilm inhibition assay of this study are shown in Figure 2. All the tested *E. coli* strains were biofilm producers. *E. coli* ATCC 43894 and the clinical isolates were strong biofilm formers (mean OD = 0.25), while *E. coli* ATCC 8739 and *E. coli* ATCC 25922 were moderate biofilm formers (mean OD = 0.18). According to the literature, mean OD values > 0.24 indicate that the tested bacterium can form strong biofilm, mean OD values 0.12–0.24 indicate that the test bacterium has moderate biofilm-forming ability, and mean OD values < 0.12 indicate weak or no biofilm formation [20]. Catechin at the MIC level distorted the biofilm formation of all the tested *E. coli*. The results indicated that catechin inhibited *E. coli* adhesion at a percentage of inhibition equal to 90.3% for *E. coli* ATCC 25922, 60% for *E. coli* ATCC 8739, 100% for *E. coli* ATCC 43895, and an average of 82% biofilm inhibition for *E. coli* clinical isolates. Inhibition of biofilm formation by gentamicin was modest against all of the examined *E. coli* strains.

2.3. In Silico Study

In a previous study, the relationship of the *acrA* gene with the biofilm formation ability of *E. coli* was established [11]. Thus, a molecular docking study was carried out to check catechin binding affinity to this protein. The AcrA protein is one of the AcrAB-TolC multidrug-resistant efflux pump proteins (Figure 3). AcrAB-TolC is a member of the resistance nodulation division family (RND) that is present in Gram-negative bacteria and has a crucial role in the *E. coli* resistance mechanism to broad spectrum antibiotics [21]. The pump consists of three major proteins; an outer membrane channel called TolC acts as an exit pathway of substrates; periplasmic protein AcrA is responsible for the stability of the connection between AcrB and TolC; and an inner membrane protein called AcrB, which is the target site for substrate binding [22–25]. In addition to the small residue that was identified recently, AcrZ affects substrate preference of AcrB [26].

It is noteworthy that the AcrA protein has no substrate binding site and AcrB is the substrate binding site in several antimicrobial agents such as Puromycin (PDB-ID: 5NC5), Erythromycin (PDB-ID: 3AOC), Rifampicin (PDB-ID: 3AOD), Levofloxacin (PDB-ID: 7B8T), Doxycycline (PDB-ID: 7B8R), Linezolid (PDB-ID: 4K7Q), Fusidic acid (PDB-ID: 6Q4P), and Minocycline (PDB-ID: 5ENT). In this regard, catechin binding affinity to AcrB was tested and the result was presented in Figure 2. Catechin showed high binding affinity to AcrB (PDB-ID: 5ENT) with $\Delta G = -8.2$ Kcal/mol compared to the control ligand (minocycline) with $\Delta G = -8.8$ Kcal/mol (Table 3). Nevertheless, further in vitro tests on the catechin effect on the AcrAB-TolC are required to confirm efflux pump inhibition of catechin.

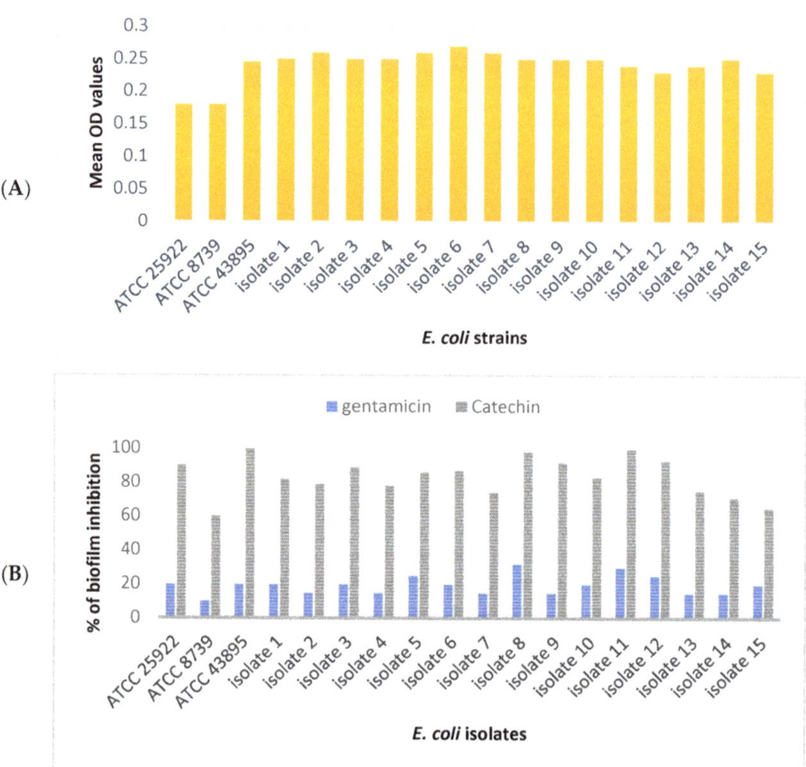

Figure 2. Biofilm formation ability of *E. coli* ATCC and clinical isolates (**A**), and biofilm inhibition effect of catechin (**B**). Gentamicin is the positive control. All the data were obtained from three independent experiments, and the mean values were presented.

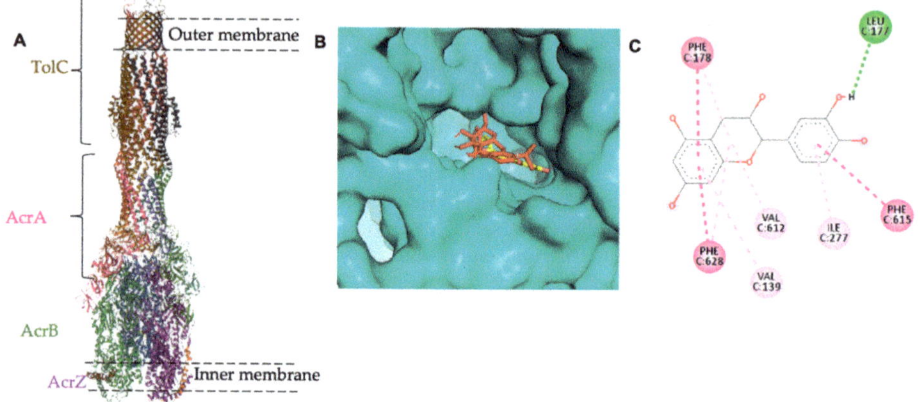

Figure 3. (**A**) Cryo-EM asymmetric structure of AcrBZ-Tolc pump of *E. coli* at 6.50 Å resolution (PDB-ID: 5NG5). (**B**) Binding affinity of catechin to AcrB protein of *E. coli* (docked catechin in yellow whereas docked minocycline is in red color). (**C**) the amino acids involved in the binding site of catechin to AcrB protein.

Table 3. Binding affinity of catechin to AcrB protein of *E. coli.*

Compound	Molecular Docking Binding Affinity ΔG (Kcal/mol)	Residue Involved in the Binding Site	Bonds Involved in the Binding Site
Minocycline (control)	−8.8	PHE-178 (chain C), ASN-274 (chain C) VAL-612 (chain C), ALA-279 (chain C) ILE-277 (chain C), PHE-615 (chain C)	Hydrogen bond Pi-Pi stacked Pi-Sigma
Catechin	−8.2	GLY-179 (chain C) LEU-177 (chain C) PHE-178 (chain C), VAL-612 (chain C), ILE-277 (chain C)	Hydrogen bond Carbon Hydrogen bond Pi-Pi stacked

2.4. Gene Expression

In the current study, fifteen clinical *E. coli* isolates were tested for the expression of the biofilm related gene, *acrA*, using the SYBR green assay for real-time PCR [27]. The results are summarized in Figure 4 and Table 4. A housekeeping gene, 16S rRNA, was used, which is commonly used in PCR analysis of bacteria [28]. The *acrA* gene was expressed in nine *E. coli* isolates (60% target gene expression). However, the expression was depressed in *E. coli* isolates treated with catechin at sub-MIC (0.5 mg/mL).

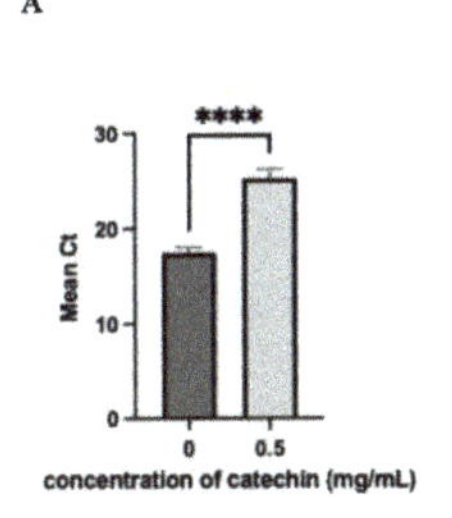

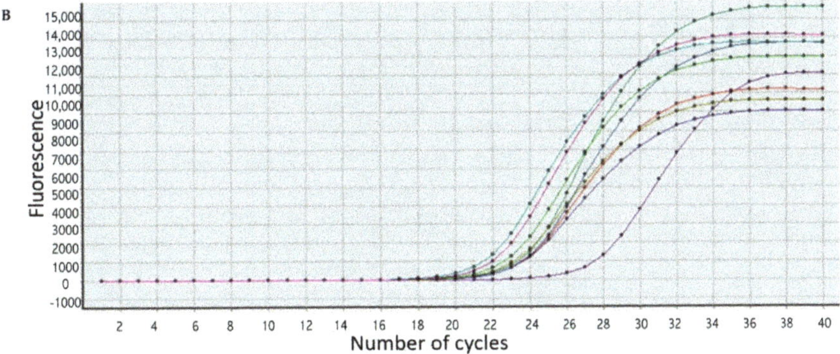

Figure 4. Expression of *acrA* gene in *E. coli* clinical isolates. (**A**) The level of expression without catechin treatment and with catechin treatment at concentration = 0.5 mg/mL. (**B**) The level of *acrA* expression in nine isolates represented in colored lines after catechin treatment. Data were expressed as mean ± SD with significant level **** $p < 0.001$.

Ct (thermal cycle) is the number of cycles required for the fluorescent signal to cross the thresholds and it is inversely proportional to the amount of the target gene (i.e., lower Ct indicates greater amount of target gene) [29]. Cts values of 29 indicate an abundance of target genes and strong positive reactions, according to the literature. Cts values ranging from 30–37 indicate a moderate number of target genes. Cts values of 38–40 are considered weak reactions with a small number of target genes [30,31]. In this study, *acrA* expression occurred in Ct < 29, reflecting a positive reaction. Moreover *E. coli* isolates that were treated with catechin showed a significant reduction in the target gene (*acrA*). By combining the results of the in silico and the in vitro studies, we suggested that catechin antibacterial action against multidrug-resistant uropathogenic *E. coli* is due to biofilm inhibition and efflux pump mechanism. According to the in silico result, catechin binds the substrate recognition site in the AcrAB-TolC of *E. coli*, which might indirectly affect the AcrA protein. Further in vitro tests on the catechin activity on the *E. coli* efflux pump system are warranted.

Table 4. The expression level of the 16S rRNA and *acrA* genes in *E. coli* clinical isolates.

Isolate Number	Mean of Ct of 16S rRNA (Untreated)	Mean of Ct of 16S rRNA (Treated)	Mean of Ct of *acrA* Gene (Untreated)	Mean of Ct of *acrA* Gene (Catechin Treated)	ΔCt for *acrA* Gene	Result
1	17	19.34	17.8	24.8	7	Down
2	16.9	18.45	17.2	24.7	7.5	Down
3	17.2	19.5	-	-	-	-
4	17.2	18.45	-	-	-	
5	17.3	21.5	16.8	24.7	7.9	Down
6	17.6	19.34	18.3	25	6.9	Down
7	18.2	19	-	-	-	-
8	17.7	21	-	-	-	-
9	18	20.41	17.8	24.8	7	Down
10	18.1	19.45	16.6	26.7	10.1	Down
11	17.6	18	-	-		-
12	17.7	18.2	18	27.2	9.2	Down
13	16.9	17.7	17.7	24.8	7.1	Down
14	16.7	17.9	16.9	24.7	7.8	Down
15	17.3	19.3	-	-	-	-

Ct, thermal cycle; -, acrA gene is not detected. ΔCt = Ct$_{acrA}$ (treated with catechin) − Ct$_{acrA}$ (untreated).

3. Discussion

Urinary tract infections (UTIs) are one of the most prevalent infectious diseases with a high recurrence rate worldwide [32]. Among several pathogens that cause UTIs, *E. coli* is considered as the major causative pathogen, accounting for 90% of community-acquired infections and 50% of nosocomial infections [33]. Recurrent UTIs with multidrug-resistant *E.coli* impose both an economic and health burden [34]. It has been reported that 78% of recurrent urinary tract infections are caused by multidrug-resistant, biofilm-forming *E. coli* [33]. Biofilm formation by many pathogens is considered as one of the indirect strategies for multidrug resistance [35]. Bacteria producing biofilm are difficult to eradicate and are able to transfer their resistance genes within their biofilm community, causing recurrent infections [36]. Many species have the ability to form biofilms, such as *Escherichia*, *Pseudomonas*, *Staphylococcus*, *Bacillus*, etc. [37].

A biofilm is a complex matrix composed of polysaccharides, nucleic acids, proteins, lipids, water, various ions, and other organic components in which cells bind together to survive harsh conditions such as host defense and accumulation of various noxious substances and antimicrobial agents [38]. According to the recent assessment by the National Institute of Health (NIH), more than 60% of in vivo infections are due to biofilm-forming microorganisms [39]. Currently, biofilm-producing bacteria have been designated as a major concern because of chronic infections [40]. It has been proven that biofilm formation makes bacteria 10–1000 times more antibiotic resistant. Several pathogens such as *Enterococcus faecium*, *Staphylococcus aureus*, *Klebsiella pneumoniae*, *Acinetobacter baumannii*, *Pseudomonas aeruginosa*, and *Enterobacter* spp. that are known as "ESKAPE" infections are distinguished by their strong biofilm formation [38]. Thus, the search for new antimicrobials with biofilm inhibition properties is urgently needed.

Plant secondary metabolites have acquired extra attention in the area of drug development as they are safe, derived from natural sources, and have high bioavailability [41]. Many of them have been proven to have antibiofilm activity [42]. *Canarium patentinervium* Miq. is a rare plant from the Burseraceae family, genus *Canarium* L., found in Asia [43]. Our team previously documented several medicinal activities of this plant, such as antibacterial, antioxidant, anti-inflammatory, anticholinesterase, and antitumor [16,44,45]. In this study, the antibacterial activity of catechin isolated from the ethanolic extract of the leaves of *Canarium patentinervium* Miq. was assessed against reference and multidrug-resistant uropathogenic *E. coli*. Based on the findings, catechin had significant antibacterial activity (MIC ranges of 1–2 mg/mL, MBC/MIC $\leq$ 4) against all the tested strains. In addition, catechin showed high synergism with the tetracycline combination. Catechin's antibacterial effect against *E. coli* has been reported in several studies [46–48]. It was shown that catechin inhibited *E. coli* growth in a dose-dependent manner [48,49].

In this study, the biofilm formation ability of *E. coli* strains was assessed quantitatively, and the result indicated that all the tested *E. coli* strains formed biofilm and were multidrug resistant. Catechin inhibited *E. coli* biofilm significantly, with a percent inhibition range of 60–100%. The antibiofilm activity of catechin has been reported in several studies [50,51]. Catechin at a concentration of 0.026 g/L showed significant biofilm inhibition in *MRSA* strains through downregulation of *fnbA* and *icaBC* genes in *MRSA* [50]. Similarly, green tea epigallocatechin gallate EGCG at sub-inhibitory concentration remarkably reduced the adhesion of *MRSA* by interfering with bacterial glucosyltransferase involved in biofilm formation [52]. In *E. coli*, EGCG showed antibiofilm activity by attenuating the expression or activity of several virulence factors such as Shiga toxin [53].

In the current study, catechin attenuated the expression of the *acrA* gene related to *E. coli* biofilm formation and multidrug resistance. In a previous study, the expression of *acrA* was significantly reduced in multidrug-resistant *E. coli* strains under the pressure of four aminoglycosides (streptomycin, gentamicin, amikacin, and apramycin) at sub-inhibitory concentration [11]. Molecular docking was performed to achieve a better understanding of catechin action on *E. coli* biofilm-related genes. Based on the result, catechin exhibited high binding affinity to the AcrB protein, which is one of the AcrAB-TolC efflux pump proteins that are responsible for *E. coli* multidrug resistance [21]. The AcrAB-TolC efflux pump is an RND-type tripartite efflux pump that is a major contributor to multidrug resistance in Gram-negative bacteria. It has three major proteins, one of which is the AcrB protein, which represents the ligand interaction site. Upon ligand binding to AcrB, quaternary structural changes occurred in AcrB that communicated with AcrA to trigger structural changes leading to the opening of the TolC channel from the sealing resting state [54]. Although the AcrA protein lacks a substrate binding site, it serves as a link between AcrB and TolC and plays an important role in the stability of the AcrAB-TolC pump [54]. Based on the in silico result, we suggested that catechin isolated from *Canarium patentinervium* Miq. might indirectly reduce the expression of the biofilm related *acrA* gene by binding to the AcrB domain of the *E. coli* AcrAB-TolC efflux pump. Further in vitro tests to confirm the efflux pump inhibition effect of catechin are required.

4. Materials and Methods

4.1. Plant Material

The leaves and bark of *Canarium patentinervium* Miq. were previously collected from one individual tree in Bukit Putih, Selangor, Malaysia (3°5′24″ N 101°46′0″ E). The plant was collected with the approval and assistance of the local Indigenous people. The plant was identified by Mr. Kamaruddin (Forest Research Institute of Malaysia). A herbarium sample (PID 251210-12) has been deposited at the Forest Research Institute of Malaysia. The leaves and bark were air dried and ground into small particles using an industrial grinder. Our team previously isolated catechin from the ethanolic extracts of the leaves through bioassay-guided fractionation using Sephadex LH-20 (30 cm $\times$ 60 cm) and Silica gel (4 cm $\times$ 90 cm) [55].

4.2. Chemicals

The following chemicals were purchased from different manufacturers: MacConkey agar and Mannitol salt agar (Himedia, Thane, India), Nutrient agar, EMB agar medium, Mueller–Hinton agar medium and Mueller–Hinton broth medium (Oxoid, Hampshire,

England), Brain Heart Infusion (BHI) broth and CCA medium (Condalab, Madrid, Spain), Gram stain solutions (Fluka, Buch, Switzerland), Glycerol (B.D.H, London, UK), Normal saline (Mediplast, Dubai, United Arab Emirates), Absolute ethanol 99% (Merck, Darmstadt, Germany), DMSO $\geq$ 99% (Merck, Darmstadt, Germany), EasyScript ®First-Strand cDNA Synthesis SuperMix (Transgen, Beijing, China), Real mod TM green (Takara, Maebashi, Japan), Quick-RNA Fungal/Bacterial Microprep Kit (Zymo, CA, USA).

4.3. Bacterial Strains

The clinical isolates of *E. coli* were obtained from the Bacteriology Unit, Department of Clinical Laboratory in the Medical City, Baghdad, Iraq. Samples were collected from UTI patients during the period from November 2021 to December 2021. *E. coli* ATCC reference bacteria (ATCC 25922, ATCC 8739, and ATCC 43895) were obtained from the Central Health Lab, Baghdad, Iraq. For all experiments except for the gene expression, three *E. coli* (ATCC reference) cultures and 15 clinical isolates were used.

Bacteria were cultured in a Brain Heart Infusion (BHI) broth medium and incubated at 37 °C for 24 h (Incubator IN55 Plus, Memmert GmBH, Schwabach, Germany) to promote bacterial growth. To identify *E. coli*, bacteria were streaked in general nutrient agar and differential culture media such as Chromogenic Coliforms Agar (CCA) medium and Eosin Methylene Blue agar (EMB). Antibiotic susceptibility (Table 5) was performed in VITEK®2 system (bioMérieux, Craponne, France) according to manufacturing company, *E. coli* was inoculated on MacConkey agar and incubated at 37 °C for 24 h.

Table 5. Bacterial source and antibiotic resistance profile.

Bacteria	Source	Resistance Profile
Escherichia coli ATCC 25922	Central Health Lab/Iraq	(R) TET, ERY, VAN, RIF, AZM, PIP (S) GEN, MIN, MEM
Escherichia coli ATCC 8739	Central Health Lab/Iraq	(R) TET, ERY, CLI, VAN (S) GEN, AZM, RIF
Escherichia coli ATCC 43895	Central Health Lab/Iraq	(R) RIF, TET, ERY, AZM, PIP, TIC, TOB (S) CFM, FEP, MIN
Escherichia coli	Urine from UTI samples	(R) TIC, PIP, GEN, TOB, CIP, SXT (S) TIM, TZP, CAZ, FEP, ATM, IPM, MEM, AMK, MIN

(R) resistant, (S) sensitive, TET tetracycline, ERY erythromycin, VAN vancomycin, RIF rifampin, AZM azithromycin, PIP piperacillin, GEN gentamycin, MIN minocycline, MEM meropenem, CLI clindamycin, TIC ticarcillin, TOB tobramycin, CFM cefixime, FEP cefepime, CIP ciprofloxacin, SXT trimethoprim-sulfamethoxazole, TIM ticarcillin- Clavulanic acid, TZP piperacillin-tazobactam, CAZ ceftazidime, ATM aztreonam, IPM imipenem, AMK amikacin.

4.4. Evaluation of Antibacterial Activity

4.4.1. Disc Diffusion Assay

This test was performed using the Kirby–Bauer technique for disc diffusion following the National Committee for Clinical Laboratory Standards methods (NCCLS) [56]. Seven antimicrobial discs have been used (azithromycin 15 µg, vancomycin 30 µg, clindamycin 2 µg, erythromycin 10 µg, gentamicin 10 µg, rifampin 5 µg, and tetracycline 10 µg) (Bioanalyse, Ankara, Turkey) based on recommendations given by the Clinical Laboratory Standards Institute (CLSI,2020) [57]. The inoculum was prepared by transferring at least 3–5 isolated colonies that were grown previously in CCA agar to a bijou bottle containing 3 mL of normal saline and incubated at 35 °C for 2 h to achieve the turbidity of growth equal to the normal turbidity standard of 0.5 McFarland/625 nm (inoculation of 1×10^8 CFU/mL).

The bacterial broth was used 15 min after the inoculum turbidity was adjusted. Then the inoculum was streaked into a petri dish with Mueller–Hinton agar medium with a thickness of 4 mm. The plate was dried at room temperature, then the antibiotic discs were added and incubated at 35 °C for 24 h. Catechin was used at a concentration of

100 mg/mL dissolved in DMSO $\geq$ 99%. The agar well diffusion method was performed for catechin based on the CLSI recommendation. A well with a diameter of 6 mm was punched aseptically into the petri dish with a sterile cork borer, and 20 μL of catechin solution was transferred into the well and incubated at 37 °C for 24 h [58]. Positive and negative controls were used, where the negative control included the solvent (DMSO $\geq$ 99%), and the positive control represented the antibiotic discs. The diameter of the zone of inhibition was measured using a digital vernier caliper to determine the microbial growth. The experiments were performed in triplicate and the mean values were presented.

4.4.2. Minimum Inhibitory Concentration (MIC)

The broth microdilution method was performed based on the guidelines given by CLSI as described by Eloff [59]. A stock solution of catechin dissolved in DMSO $\geq$ 99% was prepared in 1.5 mL microcentrifuge tubes (Eppendorff) to a final concentration of 64 mg/mL and filtered with a 0.22 Millipore filter. Likewise, each individual antibiotic powder (Panpharma S.A., Luitré, France) was dissolved in a suitable solvent, DMSO $\geq$ 99% for (azithromycin, tetracycline, erythromycin, and rifampin) and distilled water for (gentamicin, vancomycin, and clindamycin) with an initial concentration of 64 mg/mL for all antibiotics. Two-fold serial dilutions were made from the stock solution to obtain a concentration range from 32 mg/mL to 0.125 mg/mL using Mueller–Hinton broth in 96-well plates.

A standardized bacterial suspension (100 μL) at a concentration of 1×10^6 CFU/mL was added to each well containing the previously prepared 100 μL of diluted antimicrobial agents, resulting in a final volume of 200 μL in each well and final antibiotic concentrations ranging from 16 mg/mL to 0.125 mg/mL for a recommended final bacterial cell count of about 5×10^5 CFU/mL. To find out how sensitive the bacteria being tested were, a positive control was made up of broth, the antimicrobial agent's solvent (either distilled water or DMSO), and the bacteria. On the other hand, a broth without inoculum and an antimicrobial agent solvent served as the negative control. Then the microtiter plates were incubated at 37 °C for 24 h. After that, the MIC value was determined visually by recording the lowest concentration with no visible growth (the first clear well). MIC values were determined in triplicate and repeated to confirm activity.

4.4.3. Minimum Bactericidal Concentration (MBC)

The minimum bactericidal concentration (MBC) was recorded as the lowest concentration that resulted in 99.9% killing of bacteria growth [60]. The MBC assay was performed using Ozturk and Ercisli method [61] by which only the susceptible bacteria from the MIC assay were considered. Ten microliters were taken from the well obtained from the MIC experiment (MIC value) and two wells above the MIC value well and spread on MHA plates. The number of colonies was counted after 18–24 h of incubation at 37 °C. The concentration of a sample that produces < 10 colonies is considered as MBC value. Each experiment was performed in triplicate and the MBC/MIC ratio was determined. If the ratio MBC/MIC $\leq$ 4, the effect is considered bactericidal, but if the ratio MBC/MIC > 4, the effect is defined as bacteriostatic [19,62].

4.4.4. Synergistic Assay

In the presence of different antimicrobial agents (gentamicin, clindamycin, vancomycin, tetracycline, erythromycin, azithromycin, and Rifampin), the antimicrobial activity of catechin was tested. The bacterial strain was spread with a turbidity of 0.5 McFarland on Mueller–Hinton agar (MHA) plates. For the assessments of the synergistic effects, selected antibiotic discs were discretely impregnated with 5 μL of catechin (at the MBC value) and employed on the inoculated agar plates. Plates were incubated at 37 °C for 18 h. The zones of inhibition produced by catechin in combination with standard antibiotics after overnight incubation were estimated. If zones of combination treatment > 'more than' (zone of catechin + zone of the corresponding antibiotic), it was interpreted as synergism; if zone of combination treatment = (zone of catechin + zone of correspondence antibiotic), it was interpreted as additive; if zone of combination treatment < 'less than' (zone of catechin + zone of the corresponding antibiotic), it was interpreted as antagonism [63,64].

4.5. Evaluation of Antibiofilm Activity

4.5.1. Biofilm Formation by *E. coli*

Microtiter plate assay or tissue culture plate (TCP) was used for quantitative determination of *E. coli* biofilm formation in accordance with O'Toole with some modification [65]. In brief, a sterile polystyrene tissue culture plate (composed of 96 flat bottom wells) was filled with 200 µL of the diluted prepared bacterial suspension and incubated at 37 °C for 24 h. Then, the content of each well was gently removed by tapping the plates, and the wells were washed twice with 200 µL phosphate buffer saline (PBS) (pH 7.2) to remove free-floating 'planktonic' bacteria. Biofilms formed by adherent 'sessile' bacteria on plate were fixed by placing them in the incubator at 37 °C for 30 min. Then, the wells were stained with 200 µL of 1% crystal violet solution for 45 min. After that, the microtiter plates were rinsed three times with sterile distilled water to remove excess dye and left to dry for 45 min at room temperature, then de-stained by adding 200 µL ethanol. A sterile bacterial broth was used as a negative control to identify non-specific binding. A micro-ELISA reader (at 570/655 nm wavelength) was used to measure the optical densities (OD) of stained bacterial biofilms. The experiment was conducted in triplicate and the average OD values were considered. The results were graded into strong, moderate, and non or weak biofilm.

4.5.2. Biofilm Inhibition Assay

Catechin at MIC value was added to each well of a 96-well microplate except for the positive control well, which contained gentamicin and the negative control well, which contained bacterial broth only [66,67]. Bacterial culture (1×10^8 CFU/mL) in an amount of 100 µL was pipetted into each well and the well was then incubated at 37 °C for 18 h. After that, the content of each well was removed, and the wells were rinsed three times with distilled water and allow to dry at 60 °C for 45 min. Then the wells were stained with 200 µL of 1% crystal violet and incubated at room temperature for 30 min. Finally, the plates were rinsed with distilled water, de-stained with ethanol, and incubated at room temperature for 15 min. A microplate ELISA reader (model 680, Bio-Rad, Hercules, CA, USA) at 570/655 nm was used to measure the optical densities. The experiment was conducted in triplicate and the mean absorbance was considered. The percentage of biofilm inhibition was then compared with the positive control and calculated according to the following formula:

$$\% \ Of \ inhibition = \frac{OD \ control - OD \ treatment * 100}{OD \ control} \tag{1}$$

4.6. In Silico Study

Molecular Docking

A docking study was conducted to predict the target protein for catechin antibiofilm action. In a previous study, the *acrA* gene that is related to antibiotic resistance of *E. coli* biofilm was identified [11]. The crystal structure of the target protein was retrieved from the Protein Data Bank database (https://www.rcsb.org/, accessed on 25 April 2022). AcrA protein (PDB-ID: 5NG5) is a component of the *E. coli* AcrAB-TolC efflux pump and consists of an AcrA subunit with six chains and 373 sequence length, an AcrB subunit with three chains and 1049 sequence length, an AcrZ subunit with three chains and 55 sequence length, and a TolC protein with three chains and 493 sequence length. The target protein was prepared for docking by removing water, adding hydrogens and kollman charges and saved as pdb format using Autodock tools 1.5.1 [68]. Then the natural ligand was extracted from the protein using Discovery Studio Visualizer software (https://discover.3ds.com/discovery-studio-visualizer-download, accessed on 25 April 2022). The ligand and the protein were edited in Autodock tools 1.5.1. The grid box was centered on the ligand and control docking was run using Autodock Vina 1.1.2 software [69]. The same protocol was repeated with catechin as the ligand and the 3D structure of catechin (ID: 9064) was obtained from the PubChem data base (https://pubchem.ncbi.nlm.nih.gov/, accessed on 26 April 2022). The docking simulation was repeated three times, and the average binding affinity was considered. Several parameters were evaluated, such as the binding affinity of catechin toward the target protein, the ligand interaction site, the amino acids involved in the binding pocket, and the bonds formed at the interaction site. Results were analyzed using PyMOL [70] and visualized using Discovery Studio Visualizer (Dassault Systèmes, San Diego, CA, USA).

4.7. Gene Expression Using Quantitative Real-Time RT-PCR

To evaluate the effect of catechin on biofilm-related gene expression, quantitative real-time PCR (Mx3000P qPCR System, Agilent, Santa Clara, USA) was performed on *E. coli* clinical isolates with and without catechin (at the sub-inhibitory concentration = 0.5 mg/mL) [71]. *E. coli* clinical isolates (2×10^5 CFU) were inoculated in 250 mL of tryptic soy broth (TSB) [50]. Quick-RNA™ Fungal/Bacterial Microprep Kit was used to isolate the total RNA according to the manufacturer's instructions (Zymo, USA). The RNA was treated with DNase/RNase-Free water to remove genomic DNA. The concentration and optical absorbance of each extracted RNA were confirmed with Nanodrop (Epoch Microplate Spectrophotometer, Agilent, Santa Clara, USA) at 260 nm/280 nm and samples were kept at $-80\ ^\circ$C to elicit DNA contamination from total isolated RNA. EasyScript® First-Strand cDNA Synthesis SuperMix kit was used to synthesize the cDNA according to the manufacturer's instructions (Transgen/China). The cDNA reaction components including Random Primer (N9) in a volume of 1 μL, $2 \times$ ES Reaction Mix (10 μL), EasvScript®RT/RI Enzyme Mix (1 μL), RNase-free Water (Up to 20 μL), Eluted RNA (5 μL). qRT-PCR was performed by the SYBR Green gene expression assay [27]. In each sterile PCR tube, sample components of Real MODTM Green W2 2x qPCRmix in a volume of 10 μL, Forward Primer (10 μM) in a volume of 2.0 μL, Reverse Primer (10 μM) in a volume of 2.0 μL, Template DNA (4 μL), and DNase/RNase free water (up to 20 μL) were added. The thermal program was performed as follows: initial activation step, which was optimized at 95 $^\circ$C for 10 min, 1 cycle; denaturation step, which was optimized at 95 $^\circ$C, for 30 s; annealing was optimized at 60 $^\circ$C, for 30 s; extension was optimized at 72 $^\circ$C for 30 s. Denaturation, annealing, and extension steps were performed over 40 cycles. Then there was the final extension, which was optimized at 72 $^\circ$C, for 5 min, 1 cycle. The 16S rRNA gene was used as a housekeeping gene, while the *acrA* gene was used as a target gene for *E. coli*. The primers used in this study are listed in Table 6.

Table 6. The primer sequences for the real time qPCR analysis.

Genes	Type	Sequences (5′–3′)	Temperature (C)
16 sRNA (reference gene)	Forward	AGAGTTTGATCMTGGCTCAG	50
	Reverse	CTGCTGCSYCCCGTAG	52
*acrA*gene (target gene)	Forward	TTGAAATTCAGGAT	53
	Reverse	CTTAGCCCTAACAGGATGTG	57.2

4.8. Statistical Analysis

All experiments were performed in triplicate and data were expressed as mean ± standard deviation using GraphPad Prism 9 software for Mac, www.graphpad.com (GraphPad, San Diego, CA, USA). The data were analyzed using one-way ANOVA followed by Tukey test. For the gene expression, unpaired T-test was performed. A significant difference was considered at the level of $p < 0.01$. Data for the percentage of biofilm inhibiting activity of catechin were presented using Microsoft Excel.

5. Conclusions

Biofilm-forming *E. coli* places a significant burden on the health of the population. With a high rate of recurrence, *E. coli* becomes increasingly resistant to antimicrobial treatments and more difficult to eradicate. Catechin isolated from *Canarium patentinervium* Miq. exhibited strong biofilm inhibiting activity by reducing the expression of the biofilm-related *acrA* gene. This study highlights how important natural products are for treating infectious diseases that are resistant to the currently available antimicrobials.

Author Contributions: Conceptualization, M.R.; methodology, N.J. and Y.K.M.; software, A.F and N.J.; validation, M.R. and N.H.A.; formal analysis, A.F. and N.J. All authors have read and agreed to the published version of the manuscript.

Funding: This research received no external funding.

Data Availability Statement: The data used and/or analyzed in this study are available from the corresponding author on reasonable request.

References

1. Braz, V.S.; Melchior, K.; Moreira, C.G. *Escherichia coli* as a Multifaceted Pathogenic and Versatile Bacterium. *Front. Cell. Infect. Microbiol.* **2020**, *10*, 548492. [CrossRef]
2. Desvaux, M.; Dalmasso, G.; Beyrouthy, R.; Barnich, N.; Delmas, J.; Bonnet, R. Pathogenicity Factors of Genomic Islands in Intestinal and Extraintestinal *Escherichia coli*. *Front. Microbiol.* **2020**, *11*, 2065. [CrossRef]
3. Kaper, J.B.; Nataro, J.P.; Mobley, H.L.T. Pathogenic *Escherichia coli*. *Nat. Rev. Microbiol.* **2004**, *2*, 123–140. [CrossRef]
4. Öztürk, R.; Murt, A. Epidemiology of urological infections: A global burden. *World J. Urol.* **2020**, *38*, 2669–2679. [CrossRef]
5. Stamm, W.E.; Norrby, S.R. Urinary Tract Infections: Disease Panorama and Challenges. *J. Infect. Dis.* **2001**, *183*, S1–S4. [CrossRef]
6. Medina, M.; Castillo-Pino, E. An introduction to the epidemiology and burden of urinary tract infections. *Ther. Adv. Urol.* **2019**, *11*, 1756287219832172. [CrossRef] [PubMed]
7. Ramírez-Castillo, F.Y.; Moreno-Flores, A.C.; Avelar-González, F.J.; Márquez-Díaz, F.; Harel, J.; Guerrero-Barrera, A.L. An evaluation of multidrug-resistant *Escherichia coli* isolates in urinary tract infections from Aguascalientes, Mexico: Cross-sectional study. *Ann. Clin. Microbiol. Antimicrob.* **2018**, *17*, 34. [CrossRef] [PubMed]
8. Galindo-Méndez, M. Antimicrobial Resistance in *Escherichia coli*. *Microbiol. Spectr.* **2018**, *6*, 6–14. [CrossRef]
9. Bohnert, J.A.; Karamian, B.; Nikaido, H. Optimized Nile Red efflux assay of AcrAB-TolC multidrug efflux system shows competition between substrates. *Antimicrob. Agents Chemother.* **2010**, *54*, 3770–3775. [CrossRef] [PubMed]
10. Majumder, S.; Jung, D.; Ronholm, J.; George, S. Prevalence and mechanisms of antibiotic resistance in *Escherichia coli* isolated from mastitic dairy cattle in Canada. *BMC Microbiol.* **2021**, *21*, 222. [CrossRef]
11. Chen, C.; Liao, X.; Jiang, H.; Zhu, H.; Yue, L.; Li, S.; Fang, B.; Liu, Y. Characteristics of *Escherichia coli* biofilm production, genetic typing, drug resistance pattern and gene expression under aminoglycoside pressures. *Environ. Toxicol. Pharmacol.* **2010**, *30*, 5–10. [CrossRef] [PubMed]
12. Aziz, M.A.; Khan, A.H.; Adnan, M.; Ullah, H. Traditional uses of medicinal plants used by Indigenous communities for veterinary practices at Bajaur Agency, Pakistan. *J. Ethnobiol. Ethnomed.* **2018**, *14*, 11. [CrossRef]
13. Ahmad Khan, M.S.; Ahmad, I. Chapter 1—Herbal Medicine: Current Trends and Future Prospects. In *New Look to Phytomedicine*; Ahmad Khan, M.S., Ahmad, I., Chattopadhyay, D., Eds.; Academic Press: Cambridge, MA, USA, 2019; pp. 3–13. Available online: https://www.sciencedirect.com/science/article/pii/B978012814619400001X (accessed on 13 March 2022).
14. Leenhouts, P.W.; Kalkman, C.; Lam, H.J. Burseraceae. Flora Malesiana—Series 1. *Spermatophyta* **1955**, *5*, 209–296.
15. Mogana, R.; Teng-Jin, K.; Wiart, C. In Vitro Antimicrobial, Antioxidant Activities and Phytochemical Analysis of *Canarium patentinervium* Miq. from Malaysia. *Biotechnol. Res. Int.* **2011**, *2011*, 768673. [CrossRef] [PubMed]
16. Mogana, R.; Adhikari, A.; Tzar, M.N.; Ramliza, R.; Wiart, C. Antibacterial activities of the extracts, fractions and isolated compounds from *Canarium patentinervium* Miq. against bacterial clinical isolates. *BMC Complement. Med. Ther.* **2020**, *20*, 55. [CrossRef] [PubMed]
17. Andrews, J.M. Determination of minimum inhibitory concentrations. *J. Antimicrob. Chemother.* **2001**, *48*, 5–16. [CrossRef] [PubMed]
18. Sartoratto, A.; Machado, A.L.M.; Delarmelina, C.; Figueira, G.M.; Duarte, M.C.T.; Rehder, V.L.G. Composition and antimicrobial activity of essential oils from aromatic plants used in Brazil. *Braz. J. Microbiol.* **2004**, *35*, 275–280. [CrossRef]
19. Thomas, B.T.; Adeleke, A.J.; Raheem-Ademola, R.R.; Kolawole, R.; Musa, O.S. Efficiency of some disinfectants on bacterial wound pathogens. *Life Sci. J.* **2012**, *9*, 752–755.
20. Mathur, T.; Singhal, S.; Khan, S.; Upadhyay, D.J.; Fatma, T.; Rattan, A. Detection of biofilm formation among the clinical isolates of Staphylococci: An evaluation of three different screening methods. *Indian J. Med. Microbiol.* **2006**, *24*, 25–29. [CrossRef]

21. Rajapaksha, P.; Ojo, I.; Yang, L.; Pandeya, A.; Abeywansha, T.; Wei, Y. Insight into the AcrAB-TolC Complex Assembly Process Learned from Competition Studies. *Antibiotics* **2021**, *10*, 830. [CrossRef]
22. Eicher, T.; Seeger, M.A.; Anselmi, C.; Zhou, W.; Brandstätter, L.; Verrey, F.; Diederichs, K.; Faraldo-Gómez, J.D.; Pos, K.M. Coupling of remote alternating-access transport mechanisms for protons and substrates in the multidrug efflux pump AcrB. *eLife* **2014**, *3*, e03145. [CrossRef]
23. Daury, L.; Orange, F.; Taveau, J.-C.; Verchère, A.; Monlezun, L.; Gounou, C.; Marreddy, R.; Picard, M.; Broutin, I.; Pos, K.M.; et al. Tripartite assembly of RND multidrug efflux pumps. *Nat. Commun.* **2016**, *7*, 10731. [CrossRef] [PubMed]
24. Kobylka, J.; Kuth, M.S.; Müller, R.T.; Geertsma, E.R.; Pos, K.M. AcrB: A mean, keen, drug efflux machine. *Ann. N. Y. Acad. Sci.* **2020**, *1459*, 38–68. [CrossRef]
25. Zgurskaya, H.I.; Nikaido, H. AcrA is a highly asymmetric protein capable of spanning the periplasm. *J. Mol. Biol.* **1999**, *285*, 409–420. Available online: https://www.sciencedirect.com/science/article/pii/S0022283698923130 (accessed on 10 December 2021). [CrossRef]
26. Hobbs, E.C.; Yin, X.; Paul, B.J.; Astarita, J.L.; Storz, G. Conserved small protein associates with the multidrug efflux pump AcrB and differentially affects antibiotic resistance. *Proc. Natl. Acad. Sci. USA* **2012**, *109*, 16696–16701. [CrossRef]
27. Ponchel, F.; Toomes, C.; Bransfield, K.; Leong, F.T.; Douglas, S.H.; Field, S.L.; Bell, S.M.; Combaret, V.; Puisieux, A.; Mighell, A.J.; et al. Real-time PCR based on SYBR-Green I fluorescence: An alternative to the TaqMan assay for a relative quantification of gene rearrangements, gene amplifications and micro gene deletions. *BMC Biotechnol.* **2003**, *3*, 18. [CrossRef]
28. Woese, C.R. Bacterial evolution. *Microbiol. Rev.* **1987**, *51*, 221–271. [CrossRef]
29. Kim, Y.H.; Yang, I.; Bae, Y.-S.; Park, S.-R. Performance evaluation of thermal cyclers for PCR in a rapid cycling condition. *BioTechniques* **2008**, *44*, 495–505. [CrossRef]
30. de Oliveira, H.C.; Pinto Garcia Junior, A.A.; Gromboni, J.G.G.; Farias Filho, R.V.; do Nascimento, C.S.; Arias Wenceslau, A. Influence of heat stress, sex and genetic groups on reference genes stability in muscle tissue of chicken. *PLoS ONE* **2017**, *12*, e0176402.
31. Waldeisen, J.R.; Wang, T.; Mitra, D.; Lee, L.P. A Real-Time PCR Antibiogram for Drug-Resistant Sepsis. *PLoS ONE* **2011**, *6*, e28528. [CrossRef]
32. Murray, B.O.; Flores, C.; Williams, C.; Flusberg, D.A.; Marr, E.E.; Kwiatkowska, K.M.; Charest, J.L.; Isenberg, B.C.; Rohn, J.L. Recurrent Urinary Tract Infection: A Mystery in Search of Better Model Systems. *Front. Cell. Infect. Microbiol.* **2021**, *11*. Available online: https://www.frontiersin.org/article/10.3389/fcimb.2021.691210 (accessed on 2 January 2022). [CrossRef]
33. Mittal, S.; Sharma, M.; Chaudhary, U. Biofilm and multidrug resistance in uropathogenic *Escherichia coli*. *Pathog. Glob. Health* **2015**, *109*, 26–29. [CrossRef]
34. Foxman, B. The epidemiology of urinary tract infection. *Nat. Rev. Urol.* **2010**, *7*, 653–660. [CrossRef]
35. de la Fuente-Núñez, C.; Korolik, V.; Bains, M.; Nguyen, U.; Breidenstein, E.B.M.; Horsman, S.; Lewenza, S.; Burrows, L.; Hancock, R.E.W. Inhibition of bacterial biofilm formation and swarming motility by a small synthetic cationic peptide. *Antimicrob. Agents Chemother.* **2012**, *56*, 2696–2704. [CrossRef]
36. Famuyide, I.M.; Aro, A.O.; Fasina, F.O.; Eloff, J.N.; McGaw, L.J. Antibacterial and antibiofilm activity of acetone leaf extracts of nine under-investigated south African Eugenia and Syzygium (*Myrtaceae*) species and their selectivity indices. *BMC Complement. Altern. Med.* **2019**, *19*, 141. [CrossRef]
37. de la Fuente-Núñez, C.; Cardoso, M.H.; de Souza Cândido, E.; Franco, O.L.; Hancock, R.E.W. Synthetic antibiofilm peptides. *Biochim. Biophys. Acta (BBA)-Biomembr.* **2016**, *1858*, 1061–1069. Available online: https://www.sciencedirect.com/science/article/pii/S0005273615004137 (accessed on 13 December 2021). [CrossRef]
38. Gajdács, M.; Kárpáti, K.; Nagy, Á.L.; Gugolya, M.; Stájer, A.; Burián, K. Association between biofilm-production and antibiotic resistance in *Escherichia coli* isolates: A laboratory-based case study and a literature review. *Acta Microbiol. Immunol. Hung.* **2021**, *68*, 217–226. [CrossRef]
39. Bryers, J.D. Medical biofilm. *Biotechnol. Bioeng.* **2008**, *100*, 1–18. [CrossRef]
40. Sharma, D.; Misba, L.; Khan, A.U. Antibiotics versus biofilm: An emerging battleground in microbial communities. *Antimicrob. Resist. Infect. Control* **2019**, *8*, 76. [CrossRef]
41. Atanasov, A.G.; Zotchev, S.B.; Dirsch, V.M.; Supuran, C.T. Natural products in drug discovery: Advances and opportunities. *Nat. Rev. Drug Discov.* **2021**, *20*, 200–216. [CrossRef]
42. Adnan, M.; Patel, M.; Deshpande, S.; Alreshidi, M.; Siddiqui, A.J.; Reddy, M.N.; Emira, N.; De Feo, V. Effect of Adiantum philippense Extract on Biofilm Formation, Adhesion with Its Antibacterial Activities Against Foodborne Pathogens, and Characterization of Bioactive Metabolites: An in vitro-in silico Approach. *Front. Microbiol.* **2020**, *11*, 823. [CrossRef]
43. Burseraceae, P.W.; Leyden, H.J.L. Leenhouts Dioecious, Rarely Monoecious Trees the Outer Side by Distinct, Closed Twigs, Petioles with Those in Twigs Mostly Amphivasal Mainly Sclerenchymatic XYLEM, Those in the Petioles and Petiolules Collateral and Consisting of Abundant Imparipinnat. Available online: https://repository.naturalis.nl/pub/532548 (accessed on 10 May 2022).
44. Mogana, R.; Bradshaw, T.D.; Jin, K.T.; Wiart, C. In Vitro antitumor Potential of *Canarium patentinervium* Miq. *Acad J. Cancer Res.* **2011**, *4*, 1–4.
45. Mogana, R.; Teng-Jin, K.; Wiart, C. Anti-Inflammatory, Anticholinesterase, and Antioxidant Potential of Scopoletin Isolated from *Canarium patentinervium* Miq. (Burseraceae Kunth). *Evid.-Based Complementary Altern. Med.* **2013**, *2013*, 734824. [CrossRef]

46. Ma, Y.; Ding, S.; Fei, Y.; Liu, G.; Jang, H.; Fang, J. Antimicrobial activity of anthocyanins and catechins against foodborne pathogens *Escherichia coli* and Salmonella. *Food Control* **2019**, *106*, 106712. [CrossRef]

47. Jeon, J.; Kim, J.H.; Lee, C.K.; Oh, C.H.; Song, H.J. The Antimicrobial Activity of (−)-Epigallocatehin-3-Gallate and Green Tea Extracts against Pseudomonas aeruginosa and *Escherichia coli* Isolated from Skin Wounds. *Ann. Dermatol.* **2014**, *26*, 564–569. Available online: https://pubmed.ncbi.nlm.nih.gov/25324647 (accessed on 26 September 2014).

48. Chunmei, D.; Jiabo, W.; Weijun, K.; Cheng, P.; Xiaohe, X. Investigation of anti-microbial activity of catechin on *Escherichia coli* growth by microcalorimetry. *Environ. Toxicol. Pharmacol.* **2010**, *30*, 284–288. [CrossRef]

49. Díaz-Gómez, R.; Toledo-Araya, H.; López-Solís, R.; Obreque-Slier, E. Combined effect of gallic acid and catechin against *Escherichia coli*. *LWT* **2014**, *59*, 896–900. [CrossRef]

50. Zhao, Y.; Qu, Y.; Tang, J.; Chen, J.; Liu, J. Tea Catechin Inhibits Biofilm Formation of Methicillin-Resistant S. aureus. *J. Food Qual.* **2021**, *2021*, 8873091. [CrossRef]

51. Hengge, R. Targeting Bacterial Biofilms by the Green Tea Polyphenol EGCG. *Molecules* **2019**, *24*, 2403. [CrossRef]

52. Hamilton-Miller, J.T. Anti-cariogenic properties of tea (*Camellia sinensis*). *J. Med. Microbiol.* **2001**, *50*, 299–302. [CrossRef]

53. Lee, K.-M.; Kim, W.-S.; Lim, J.; Nam, S.; Youn, M.; Nam, S.-W.; Kim, Y.; Kim, S.-H.; Park, W.; Park, S. Antipathogenic Properties of Green Tea Polyphenol Epigallocatechin Gallate at Concentrations below the MIC against Enterohemorrhagic *Escherichia coli* O157:H7. *J. Food Prot.* **2009**, *72*, 325–331. [CrossRef]

54. Wang, Z.; Fan, G.; Hryc, C.F.; Blaza, J.N.; Serysheva, I.I.; Schmid, M.F.; Chiu, W.; Luisi, B.F.; Du, D. An allosteric transport mechanism for the AcrAB-TolC multidrug efflux pump. *eLife* **2017**, *6*, e24905. [CrossRef]

55. Mogana, R.; Adhikari, A.; Debnath, S.; Hazra, S.; Hazra, B.; Teng-Jin, K.; Wiart, C. The Antiacetylcholinesterase and Antileishmanial Activities of *Canarium patentinervium* Miq. *BioMed Res. Int.* **2014**, *2014*, 903529. [CrossRef]

56. Kiehlbauch, J.A.; Hannett, G.E.; Salfinger, M.; Archinal, W.; Monserrat, C.; Carlyn, C. Use of the National Committee for Clinical Laboratory Standards Guidelines for Disk Diffusion Susceptibility Testing in New York State Laboratories. *J. Clin. Microbiol.* **2000**, *38*, 3341–3348. [CrossRef]

57. CLSI. *2021 Performance Standards for Antimicrobial Susceptibility Testing*, 30th ed.; CLSI M100-ED29; Clinical and Laboratory Standards Institute: Wayne, PA, USA, 2020; Volume 40, pp. 50–51.

58. Balouiri, M.; Sadiki, M.; Ibnsouda, S.K. Methods for in vitro evaluating antimicrobial activity: A review. *J. Pharm. Anal.* **2015**, *6*, 71–79. [CrossRef]

59. Eloff, J.N. A Sensitive and Quick Microplate Method to Determine the Minimal Inhibitory Concentration of Plant Extracts for Bacteria. *Planta Med.* **1998**, *64*, 711–713. [CrossRef]

60. Parvekar, P.; Palaskar, J.; Metgud, S.; Maria, R.; Dutta, S. The minimum inhibitory concentration (MIC) and minimum bactericidal concentration (MBC) of silver nanoparticles against *Staphylococcus aureus*. *Biomater. Investig. Dent.* **2020**, *7*, 105–109. [CrossRef]

61. Ozturk, S.; Ercisli, S. Chemical composition and in vitro antibacterial activity of *Seseli libanotis*. *World J. Microbiol. Biotechnol.* **2006**, *22*, 261–265. [CrossRef]

62. Levison, M.E. Pharmacodynamics of antimicrobial drugs. *Infect. Dis. Clin.* **2004**, *18*, 451–465. Available online: https://www.sciencedirect.com/science/article/pii/S0891552004000753 (accessed on 15 May 2022).

63. Cantore, P.L.; Iacobellis, N.S.; De Marco, A.; Capasso, F.; Senatore, F. Antibacterial Activity of *Coriandrum sativum* L. and *Foeniculum vulgare* Miller Var. *vulgare* (Miller) Essential Oils. *J. Agric. Food Chem.* **2004**, *52*, 7862–7866. [CrossRef]

64. Saquib, S.A.; AlQahtani, N.A.; Ahmad, I.; Kader, M.A.; Al Shahrani, S.S.; Asiri, E.A. Evaluation and Comparison of Antibacterial Efficacy of Herbal Extracts in Combination with Antibiotics on Periodontal pathobionts: An in vitro Microbiological Study. *Antibiotics* **2019**, *8*, 89. [CrossRef]

65. O'Toole, G.A. Microtiter Dish Biofilm Formation Assay. *J. Vis. Exp.* **2011**, *47*, e2437. [CrossRef] [PubMed]

66. Sandasi, M.; Leonard, C.; Viljoen, A. The effect of five common essential oil components on Listeria monocytogenes biofilms. *Food Control* **2008**, *19*, 1070–1075. [CrossRef]

67. Djordjevic, D.; Wiedmann, M.; McLandsborough, L.A. Microtiter Plate Assay for Assessment of *Listeria monocytogenes* Biofilm Formation. *Appl. Environ. Microbiol.* **2002**, *68*, 2950–2958. [CrossRef] [PubMed]

68. Morris, G.M.; Huey, R.; Lindstrom, W.; Sanner, M.F.; Belew, R.K.; Goodsell, D.S.; Olson, A.J. AutoDock4 and AutoDockTools4: Automated docking with selective receptor flexibility. *J. Comput. Chem.* **2009**, *30*, 2785–2791. [CrossRef]

69. Danish, S.; Shakil, S.; Haneef, D. A simple click by click protocol to perform docking: AutoDock 4.2 made easy for non-bioinformaticians. *EXCLI J.* **2013**, *12*, 831–857.

70. Yuan, S.; Chan, H.S.; Hu, Z. Using PyMOL as a platform for computational drug design. *WIREs Comput. Mol. Sci.* **2017**, *7*, e1298. [CrossRef]

71. Kosari, F.; Taheri, M.; Moradi, A.; Alni, R.H.; Alikhani, M.Y. Evaluation of cinnamon extract effects on clbB gene expression and biofilm formation in *Escherichia coli* strains isolated from colon cancer patients. *BMC Cancer* **2020**, *20*, 267. [CrossRef]

antibiotics

Article

Antibiofilm Effect of Cinnamaldehyde-Chitosan Nanoparticles against the Biofilm of *Staphylococcus aureus*

Jiaman Xu [1,2], Quan Lin [1,2], Maokun Sheng [1,2], Ting Ding [1,2], Bing Li [3], Yan Gao [4] and Yulong Tan [1,2,*]

[1] Special Food Research Institute, Qingdao Agricultural University, Qingdao 266109, China
[2] Qingdao Special Food Research Institute, Qingdao 266109, China
[3] Marine Science and Engineering College, Qingdao Agriculture University, Qingdao 266109, China
[4] Marine Science Research Institute of Shandong Province (National Oceanographic Center of Qingdao), Qingdao 266071, China
* Correspondence: tanyulong@qau.edu.cn; Tel.: +86-532-5895-7918

Citation: Xu, J.; Lin, Q.; Sheng, M.; Ding, T.; Li, B.; Gao, Y.; Tan, Y. Antibiofilm Effect of Cinnamaldehyde-Chitosan Nanoparticles against the Biofilm of *Staphylococcus aureus*. *Antibiotics* 2022, 11, 1403. https://doi.org/10.3390/antibiotics11101403

Academic Editor: Wolf-Rainer Abraham

Received: 6 September 2022
Accepted: 2 October 2022
Published: 13 October 2022

Publisher's Note: MDPI stays neutral with regard to jurisdictional claims in published maps and institutional affiliations.

Abstract: Food contamination caused by food-spoilage bacteria and pathogenic bacteria seriously affects public health. *Staphylococcus aureus* is a typical foodborne pathogen which easily forms biofilm. Once biofilm is formed, it is difficult to remove. The use of nanotechnology for antibiofilm purposes is becoming more widespread because of its ability to increase the bioavailability and biosorption of many drugs. In this work, chitosan nanoparticles (CSNPs) were prepared by the ion–gel method with polyanionic sodium triphosphate (TPP). Cinnamaldehyde (CA) was loaded onto the CSNPs. The particle size, potential, morphology, encapsulation efficiency and in vitro release behavior of cinnamaldehyde–chitosan nanoparticles (CSNP-CAs) were studied, and the activity of CA against *S. aureus* biofilms was evaluated. The biofilm structure on the silicone surface was investigated by scanning electron microscopy (SEM). Confocal laser scanning microscopy (CLSM) was used to detect live/dead organisms within biofilms. The results showed that CSNP-CAs were dispersed in a circle with an average diameter of 298.1 nm and a zeta potential of +38.73 mV. The encapsulation efficiency of cinnamaldehyde (CA) reached 39.7%. In vitro release studies have shown that CA can be continuously released from the CSNPs. Compared with free drugs, CSNP-CAs have a higher efficacy in removing *S. aureus* biofilm, and the eradication rate of biofilm can reach 61%. The antibiofilm effects of CSNP-CAs are determined by their antibacterial properties. The minimum inhibitory concentration (MIC) of CA is 1.25 mg/mL; at this concentration the bacterial cell wall ruptures and the permeability of the cell membrane increases, which leads to leakage of the contents. At the same time, we verified that the MIC of CSNP-CAs is 2.5 mg/mL (drug concentration). The synergy between CA and CSNPs demonstrates the combinatorial application of a composite as an efficient novel therapeutic agent against antibiofilm. We can apply it in food preservation and other contexts, providing new ideas for food preservation.

Keywords: chitosan; nanoparticle; antibiofilm; cinnamaldehyde; *Staphylococcus aureus*

1. Introduction

Foodborne diseases are prominent health problems. Due to the long food chain and complex pathogen sources, food is easily contaminated. Meat, poultry, eggs, aquatic products, dairy products and other foods are very vulnerable to *Staphylococcus aureus* [1]. *S. aureus* often contaminates food in the following ways: via food-processing personnel, when food-processing staff or cooks' hands or clothes are contaminated with germs and are not cleaned; when the food itself is contaminated before processing or is contaminated during processing, resulting in enterotoxin and food poisoning; when the packaging of cooked food products is not sealed and the contents are contaminated during transportation; and when, e.g., dairy cows suffer from suppurative mastitis or there is local purulence in livestock, contamination can occur in other parts of the body, etc. Biofilm can also form on the surfaces of food and food-processing equipment [2]. Bacterial biofilm refers to a large number of aggregated, membrane-like substances formed by bacteria adhering to a contact surface and secreting polysaccharide matrix, fibrin, lipid protein, etc. [3,4]. The extracellular polymers on the surface of the biofilm are intricately aggregated, forming a complex and orderly overall structure, which effectively protects the stability of the biofilm on the carrier surface. The microorganisms in the biofilm occur together in the form of clusters and are firmly adsorbed on the surface of the carrier. Using common physicochemical methods is often difficult to completely remove biofilm. The physical methods include ultrasonic [5,6], low-current and other methods, such as applications of electromagnetic and shock waves. Although physical methods can meet the requirements for removing biofilms, it is difficult to achieve the desired effect for some large-scale processing equipment. The chemical methods include disinfectants and antibiotics, etc. [7]. These common chemical disinfectants, such as sodium hypochlorite and chlorine dioxide, and antibiotics, such as penicillin and tetracycline, cannot achieve ideal removal effects, and their use will also lead to the development of bacterial drug resistance [8]. Therefore, new drugs to overcome biofilm resistance and eliminate biofilm-protected bacteria need to be developed.

Cinnamaldehyde (CA) is an active component extracted from cinnamon bark, which has a variety of effects, such as anticancer, antifungal and antibacterial activities [9–11]. It has been reported that CA can repress bacteria, yeasts and filamentous molds by inhibiting ATPases, cell-wall biosynthesis and by changing membrane structure and integrity [11]. Albano et al. confirmed that CA had the ability to reduce the growth of *Staphylococcus epidermidis* in planktonic state, inhibit biofilm formation and eradicate formed biofilm [12]. Yu et al. found that *Campylobacter* strains treated with 15.63 µg/mL CA exhibited significantly decreased bacterial auto-aggregation, motility, exopolysaccharide production and soluble protein levels, which proved that it had the ability to remove biofilm [13]. CA also had an inhibitory effect on *Candida albicans* by inhibiting the release of its virulence factors [14]. Kot et al. found that trans-cinnamaldehyde is a promising antibiofilm agent for use in MRSA-biofilm-related infections [15].

Since the antibacterial properties of essential oils (EOs) are limited by their high volatility and water-insolubility, the stability of EOs has become a key issue in recent years [16]. Nanoparticles (NPs) have good biocompatibility and can be more stable for the release of EOs and have been widely used in encapsulating EOs [17]. NPs loaded with an EO can protect the EO from the external environment to prolong its inhibitory effect on microorganisms. Chitosan (CS) is an alkaline polysaccharide containing more free amino groups. It is used as a biological carrier, having good safety and biocompatibility, and has been identified as having an effective antibacterial membrane effect [18–22]. When the pH is lower than 10, the amino groups will be protonated and positively charged. The lipid layer on the surface of the biofilm is negatively charged and easy to electrostatically adsorb with positively charged chitosan particles [23]. As a pharmaceutical carrier, chitosan nanoparticles (CSNPs), with the advantages of slow or controlled drug release, can improve drug solubility and stability, thereby enhancing drug effects and reducing the side effects of drugs [24]. This paper mainly studied the antibiotic film effect of chitosan nanoparticles with cinnamaldehyde.

2. Results

2.1. Characterization of Nanoparticles

The surface morphologies and morphologies of CSNPs were shown by TEM. CSNP-CAs had smooth surfaces and nearly spherical shapes (Figure 1). The average diameters of the CSNPs and CSNP-CAs were estimated to be 167.3 nm and 298.1 nm, respectively. The surface zeta potentials of the CSNPs and CSNP-CAs were (+34.60) mV and (+38.73) mV, respectively. The average amount of CA on the carriers was 39.7%.

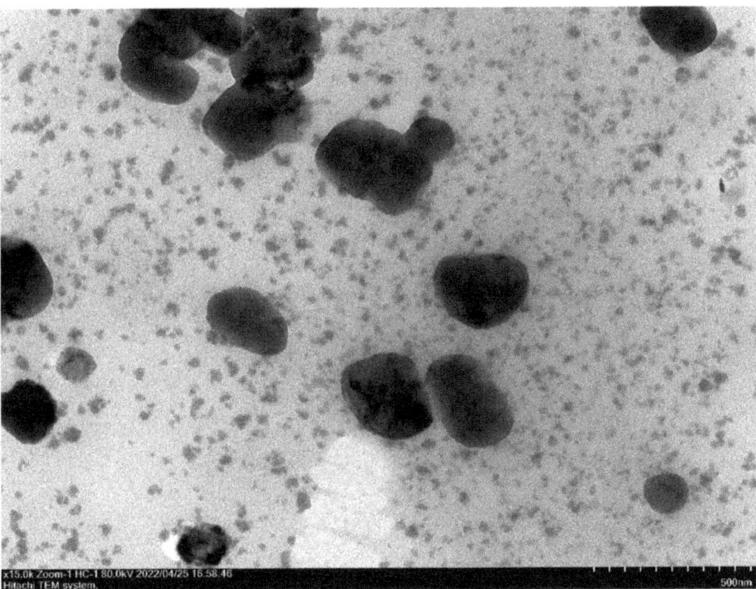

Figure 1. TEM image of CSNP-CAs.

2.2. In Vitro Release Studies

It can be seen from the figure (Figure 2) that the release effect of CSNP-CAs can be significantly divided into two stages. The drug release was rapid in the early stage, including the release of unencapsulated CA adsorbed on the surface of the nanoparticles. The drug release was slow in the later stage, which was due to the gradual release of the drug inside the NPs. The cumulative release amount was about 77.01% in 7 h, and the release tended to be gentle after 10 h. The cumulative release amount reached 86.00% in 48 h.

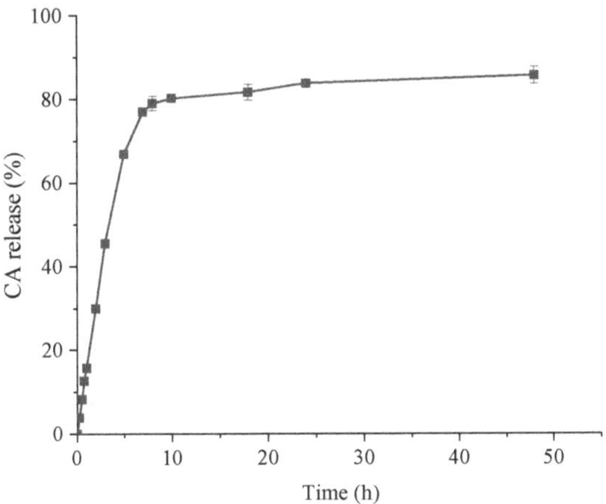

Figure 2. In vitro release profiles for CA from CA-loaded chitosan nanoparticles, using an initial weight ratio of chitosan to CA of 4:1. Data are presented as means ± SDs, *n* = 3.

2.3. Antibacterial Activity of CA and CSNP-CAs

The antibacterial activity of CA against *S. aureus* strains (NCTC 8325, RN6390, 15981, Col) were evaluated in vitro by measuring MIC and MBC values. The MIC and MBC values for this strain were 1.25 mg/mL and 2.5 mg/mL when the MIC of CSNP-CAs was 2.5 mg/mL.

2.4. Inhibition Activity on Biofilm Formation of CA and CSNP-CAs

As can be seen in Figure 3, biofilm reduction was achieved with all the concentrations of CA tested in this work. Both free CA and CSNP-CAs showed good biofilm inhibition. The activities of free CA and loaded CA both increased with the increase in CA concentration, indicating that the inhibition activities of free CA and loaded CA with respect to biofilm were concentration-dependent. For free CA, there was almost no biofilm formation when the concentration was greater than 1.25 mg/mL. In the same concentration range, the inhibitory activity of loaded CA (CSNP-CAs) was slightly lower than that of free CA.

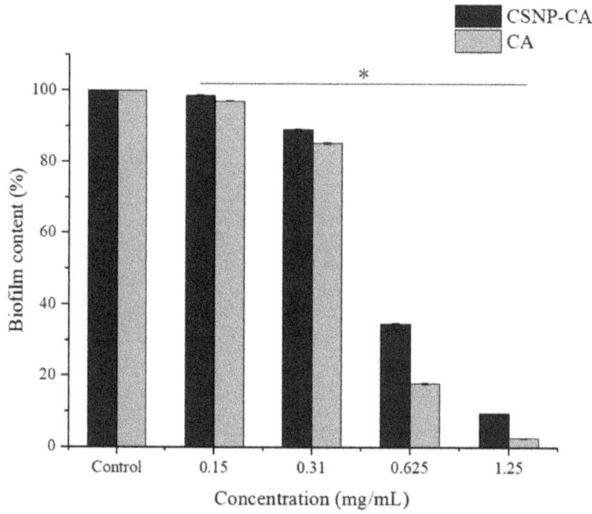

Figure 3. Inhibition effects of NPs on biofilm in 96-well microplates. The results represent the means and standard deviations (error bars) of three independent experiments. * $p < 0.05$ for comparison between the untreated and treated groups.

2.5. Antibiofilm Activity on Mature Biofilm of CA and CSNP-CAs

In order to evaluate the effect of CSNP-CAs on the established biofilm, we assessed the antibiofilm effect of CA and CSNP-CAs by repeated treatment for two days. Figure 4 shows the antibiofilm activity of CSNP-CAs and CA on mature biofilm at various CA concentrations after 48 h of treatment. As can be seen from the figure, the biofilm was removed at each concentration, but the effect was different. The removal effect of the CSNP-CAs was better than that of free CA. When the drug concentration reached 4 × MIC, the clearance rate of CSNP-CAs for the biofilm was 48.10% and that of free CA was 38.66%.

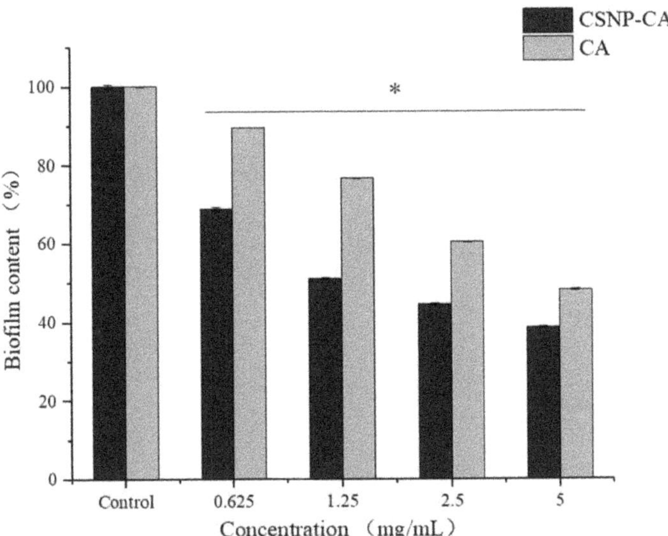

Figure 4. Disruption effects of NPs on mature biofilm in 96-well microplates. The results represent the means and standard deviations (error bars) of three independent experiments. * $p < 0.05$ for comparison between the untreated and treated groups.

2.6. Antibiofilm Effect of CSNP-CAs on Food-Grade Silicone

The biofilm grown on food-grade silicone platelets was observed by SEM (Figure 5). *S. aureus* biofilm grown without CA treatment exhibited the typical 3D morphology with dense structures and water channels (Figure 5A). Under the same conditions but treated with free CA and CSNP-CAs, the biofilm showed structural changes (Figure 5B–C). CA can be removed by bactericidal action to form biofilms (Figure 5B). The loading of CA and NPs enhanced the removal of biofilms (Figure 5C). The silicone surface showed more blank areas, with only single or no cells sticking to the surface.

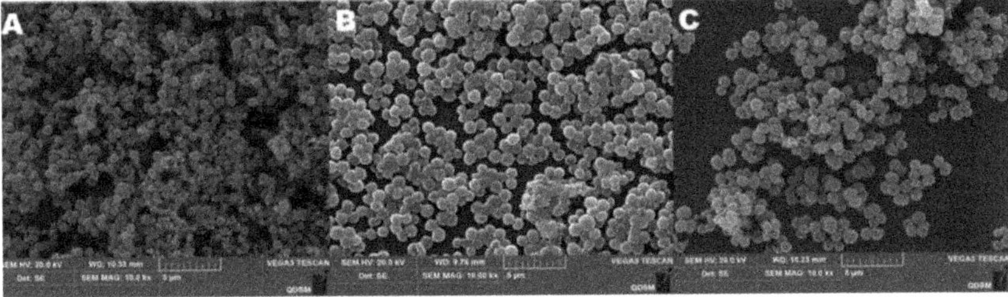

Figure 5. SEM images of *S. aureus* biofilm formations on medical-grade silicone surfaces with media without treatment (**A**) and with CA (**B**) and CSNP-CA treatments (**C**).

The images obtained with CLSM (Live/Dead staining) confirmed the results. In the control group (Figure 6A), a thick green structure (live cells) could be seen, indicating a mature biofilm structure with active cells. Free CA treatment (Figure 6B) resulted in a reduction in biofilm thickness. Biofilms had less biomass and reduced biofilm thickness, and more biofilms were stained red (dead cells). After the treatment with CSNP-CAs (Figure 6C), biofilms had less biomass and reduced biofilm thickness, the biofilm was almost completely red (dead cells) and the structure of the biofilm was destroyed. In conclusion, the CSNP-CAs could better destroy the structure of biofilm and kill the cells.

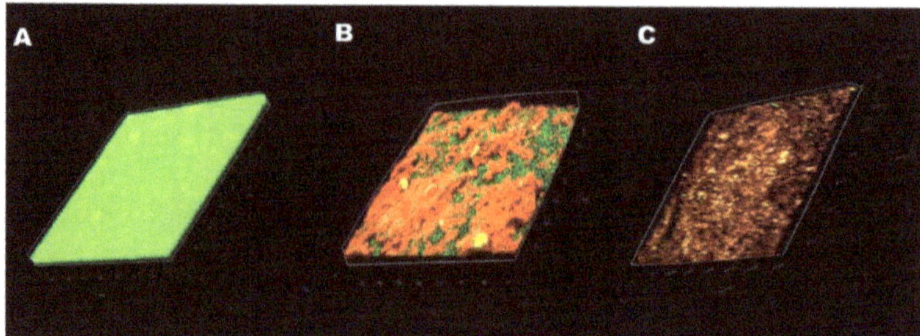

Figure 6. CLSM images of *S. aureus* biofilm formations on medical-grade silicone surfaces without treatment (**A**) and with CA (**B**) and CSNP-CA treatments (**C**). Biofilms were stained with the Live/Dead® BacLight™ Bacterial Viability and Counting Kit. CLSM reconstructions show the three-dimensional staining patterns for live cells (SYTO-9, green) and dead cells (propidium iodide, red). Magnification, ×10.

2.7. Effect of CA on Cell-Wall Integrity of S. aureus

As shown in Figure 7A, the content of extracellular alkaline phosphatase in the bacterial solution treated with CA was increased by more than threefold. The content of alkaline phosphatase reached the maximum at about 8 h, and it tended to be stable after 8 h. It could be inferred that CA could destroy the cell wall of *S. aureus* and increase the permeability of the cell wall.

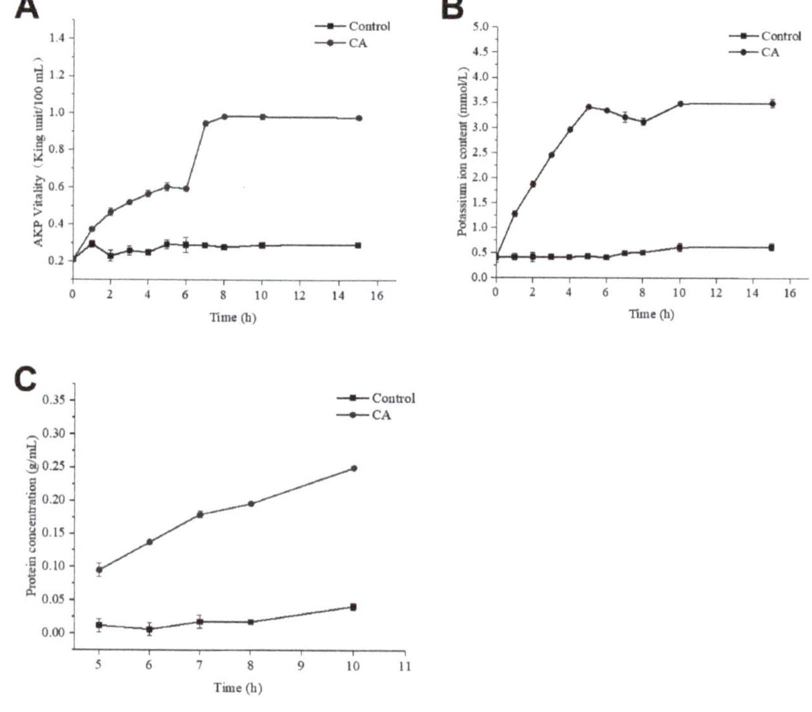

Figure 7. The changes in AKP concentration (**A**), K^+ concentration (**B**) and protein concentration (**C**) for *S. aureus* treated with CA.

2.8. Effect of CA on Membrane Permeability of S. aureus

Figure 8 clearly shows the effect of CA on the morphology of *S. aureus*. In the control group without CA, the bacterial cells were spherical and relatively intact, and the bacterial cell membrane was smooth and not broken (Figure 8A). When the concentration of CA was 1 × MIC, *S. aureus* lost its original spherical state, many cells collapsed and cell membranes even ruptured (Figure 8B). After adding CA, the permeability of cell membranes increased, leading to the outflow of potassium ions and proteins in cells. In Figure 7B, when CA was added to the bacterial solution, the concentration of extracellular potassium ions increased gradually and reached a peak at 5 h.

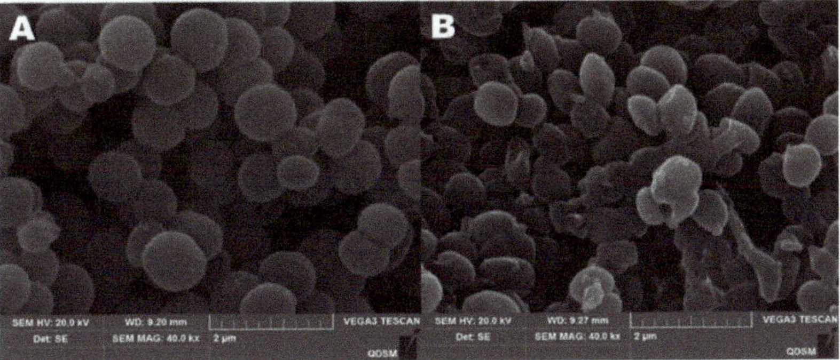

Figure 8. SEM images of *S. aureus* untreated (**A**) and treated (**B**) with cinnamaldehyde for 4 h.

Meanwhile, the concentration of protein in the bacterial suspension also increased gradually (Figure 7C). Compared with the control, the protein content increased six times. This result showed that the addition of CA could rupture the cells of *S. aureus* and cause cytoplasmic leakage.

The damaged membranes of *S. aureus* cells were observed by flow cytometry, as shown in Figure 9. The figure shows three different types of cells, namely, FITC-/PI- living cells (Q3), FITC+/PI+ dead cells (Q2) and FITC-/PI+ necrotic cells and debris (Q1). It can be seen from the figure (Figure 9A) that almost all control cells were clustered in the Q3 type, which proves that *S. aureus* has a complete cell-membrane structure in this state. However, after CA treatment (Figure 9B), the majority of cells changed from Q3 to Q1, which revealed that the integrity of *S. aureus* membranes was disrupted after CA treatment. This phenomenon is consistent with the results observed in SEM experiments. All experiments proved that cell membranes were damaged after CA treatment.

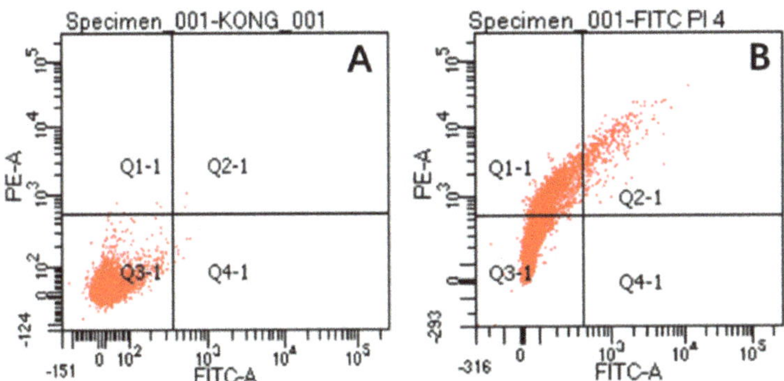

Figure 9. The permeability of *S. aureus* cell membranes before (**A**) and after CA treatment (**B**).

3. Discussion

Nowadays, there is an urgent need to develop a method that can efficiently solve the biofilm problem of foodborne pathogens without causing drug resistance in pathogens. NPs are defined as particles or materials at the nanometer scale [25]. The antimicrobial agent loaded on NPs can be protected from sequestering drugs by the biofilm matrix [26]. The antibacterial agent CA selected in this experiment is a safety additive recognized by the Food and Drug Administration (FDA) [27]. The maximum experimental amount used in this paper (5 mg/mL) was much lower than the maximum addition amount (286 mg/mL) specified by the FDA. CA can inhibit the synthesis of bacterial cell walls and affect membrane permeability and enzyme systems. In the experiment, the drug concentration of $1 \times$ MIC led to a significant increase in the content of alkaline phosphatase, which proved that CA can destroy the integrity of bacterial cell walls. The destruction of the bacterial cell walls led to the death of the bacteria, which was consistent with the conclusion of Shreaz et al. [11]. Therefore, this article supports the method of loading CA on CSNPs to clear biofilm and kill bacteria.

Regarding drug delivery, in vitro release profiles for CA from NPs were evaluated. CSNP-CAs showed a typical release profile, with rapid-burst release at the beginning, followed by sustained release. The initial burst may have been due to the drug loading on the surface of the material, after which a persistent pattern was established due to the release of the drug from the NPs. This phenomenon is more effective for biofilm clearance, as higher initial drug doses reduce the antibiotic resistance of surviving bacteria in biofilms [28].

CSNP-CAs were not as effective in inhibiting biofilm formation as free CA, precisely because the CSNPs had a sustained-release effect, the drug concentration released in the time period considered here was small and the antibiofilm effect was weak. As shown in this work, CSNP-CAs were more effective in removing biofilms than free CA. The unique antibiofilm properties of chitosan were mainly attributed to its polycationic nature, conferred by the amino functionality (NH_2) of the N-acetylglucosamine unit [29–31]. The positive charge of chitosan reacts electrostatically with negatively charged biofilm components, such as EPS, proteins and DNA, resulting in an inhibitory effect on bacterial biofilms [32,33]. At the same time, nanoparticles can also carry drugs into biofilm and release drugs to act on the bacteria themselves and kill bacteria. Therefore, CSNP-CAs are more effective in scavenging biofilms than free CA.

This experiment has confirmed that CSNP-CAs have the effect of removing biofilm and that the antibacterial effect of cinnamaldehyde plays an important role in it. CA destroyed the cell wall of *S. aureus* and changed the permeability of the cell membrane, resulting in the outflow of potassium ions, alkaline phosphatase, protein and other substances. Alkaline phosphatase is an enzyme between the cell wall and cell membrane. Under normal circumstances, its activity cannot be detected outside the cell. When the cell wall or cell membrane is damaged, alkaline phosphatase will penetrate the outside of the cell and threaten the integrity of the cell wall. Inoue et al. found that potassium-ion leakage provided direct evidence of membrane damage [34]. The content of extracellular potassium ions in the bacterial solution treated with $1 \times$ MIC of CA increased continuously, which clearly showed the damage to the bacteria caused by CA. Since most of the proteins in the cell are present in the cytoplasm, leakage of the cytoplasm leads to increased levels of extracellular proteins. Propidium iodide (PI) is a nucleic-acid dye that cannot penetrate an intact cell membrane. Since the membranes of injured cells and necrotic cells are destroyed, PI can enter the cell membranes of injured cells and necrotic cells [35], so that the cells can be stained with red fluorescence [36]. Phospholipid serine (PS) exists only on the inner side of the cell membrane in normal cells. When a specific injury occurs, the PS acid on the inner side of the plasma membrane is redistributed, flipped from the inside of the cell membrane to the outside of the cell membrane, and exposed on the outer surface of the cell. PS eversion can be detected by Annexin V. Double staining with FITC-labeled Annexin V and PI enables sensitive detection of live, injured and dead cells via flow cytometry.

Therefore, the intracellular PI content can verify the permeability of the cell membrane. The changes in the permeability of the cell membrane and plasma membrane of *S. aureus* lead to cell death.

In summary, we have verified that CA can destroy the cell membrane: on the one hand, it can change the cell-membrane potential and cause the leakage and death of potassium ions, proteins and other contents; on the other hand, CA acts on the cell membrane to redistribute PS inside the plasma membrane, which damages the cell membrane. SEM images can also clearly show changes in cell morphology and even cell rupture. In addition, the combination of CSNPs and CA makes it easier for CA to enter the mature biofilm, to achieve the effect of clearing biofilm.

4. Materials and Methods

4.1. Bacterial Strains and Growth Media

At present, most *S. aureus* strains have drug resistance. We used four different strains (NCTC 8325, RN6390, 15981, Col) for the determination of MIC. *S. aureus* RN6390 can easily form biofilm and has been wildly used for bovine mastitis infection assays in experiments [37,38], so we used RN6390 to conduct the following detailed experiments. The strain was cultured overnight in Tryptic Soy Broth (TSB) medium from Solarbio (Beijing, China) at 37 °C, 220 rpm.

CS (low-molecular-weight; degree of deacetylation: 75–85%) and CA ($\geq$95%) were purchased from Macklin (Shanghai, China). Other reagents were purchased from Solarbio (Beijing, China).

4.2. Preparation of Nanoparticles

CSNPs were prepared by the ion crosslinking method with polyanionic sodium triphosphate (TPP). In brief, CS was dissolved in acetic acid (1%) and stirred overnight to bring the concentration of chitosan to 1 mg/mL. Then, TPP was dissolved in deionized water for 2 h. TPP solution was mixed dropwise with the CS solution under stirring conditions (volume ratio CS: TPP = 3:1). CSNPs were separated by ultracentrifugation (12,000 rpm, 30 min) and then resuspended in PBS.

4.3. Preparation of CSNP-CAs

For the preparation of the CSNP-CAs, CA (1%, *v/v*) was dissolved in ethanol (25%, *v/v*) and slowly dropped into the prepared chitosan solution under mechanical agitation to ensure that the volume ratio of CA to CS solution was 1:4. Then, the TPP solution was added to the mixed solution to ensure that the volume ratio of CS solution to TPP solution was 3:1. The mixed solution continued to be stirred at 600 rpm for 4 h. The prepared liquid was centrifuged at 12,000 rpm, and the precipitate was washed with distilled water. The CSNP-CA suspensions were freeze-dried and stored at −20 °C before use.

4.4. Particle Size and Zeta Potential of NPs

The size and surface charge of the NPs were measured with a ZetaSizer Nano ZS90 (Malvern Instruments, Malvern, UK). CSNP suspensions were analyzed by dynamic light scattering (DLS) for size and potential determinations. The morphological characteristics were confirmed with transmission electron microscopy (TEM, JEM-1200EX, 80kv, Tokyo, Japan).

4.5. Drug Encapsulation

The standard curve of CA was determined by UV spectroscopy (Persee, TU-1810, Beijing, China) at 291 nm. The amounts of CA encapsulated in the NPs were determined by centrifugation. The CSNP-CA solution was centrifuged at 12,000 rpm for 30 min. Then, the absorbance of the supernatant was detected by UV spectroscopy. The CA loading capacity (LC) was calculated according to:

$$LC = (A - B)/C \tag{1}$$

where A = total amount of CA, B = total amount of non-loaded CA and C = weight of the NPs.

4.6. In Vitro Release Test

The in vitro drug-release effect of the CSNP-CAs was passed by dynamic dialysis. The CSNP-CAs were transferred into a dialysis bag (with a retained relative molecular weight of 3500) and 50 mL of PBS buffer was used as the release medium. The dialysis bag was kept at a constant temperature of 37 °C and agitated at 70 rpm. At predetermined time intervals, 5 mL of dialysate was taken and supplemented with the same amount of release medium. The samples were measured at 291 nm with a UV spectrophotometer, 3 times for each time point, and an in vitro drug-release curve for the CSNP-CAs was drawn.

4.7. Determinations of Minimum Inhibitory Concentration (MIC) and Minimum Bactericidal Concentration (MBC)

The determination of the MIC was carried out according to the method described by Tan et al [3]. Stock solutions of CA were kept in 25% (v/v) ethanol to enhance their solubility in suspension. Antimicrobial effects of CA were evaluated using a serial twofold dilution method. CA was initially diluted in 25% (v/v) ethanol (1:100 v/v) and then in TSB medium. Serial twofold dilutions for CA were prepared to obtain concentrations ranging from 10 to 0.15 mg/mL. A quantity of 100 μL of bacteria at a final concentration of 1×10^5 CFU/mL in TSB was added into the wells of 96-well microplates. Free CA (10, 5, 2.5, 1.25, 0.62, 0.31, 0.15 or 0 mg/mL) was added to each well. The microplate was incubated at 37 °C for 24 h at 150 rpm. MIC refers to the lowest drug concentration that inhibits the growth of bacteria in culture medium after 18 to 24 hours of culture in vitro. MBC refers to the minimum drug concentration capable of killing cultured bacteria after 18 to 24 hours in vitro.

4.8. Growth of Biofilm

Biofilm formation in 96-well microplates was supported and assayed as described previously [39]. The microplates were incubated at 37 °C for 24 h without shaking. After treatment with different concentrations of CA and CSNP-CAs, the wells were washed 3 times with PBS to remove impurities and plankton. Then, each well was stained with 100 μL of 0.1% (w/v) crystal violet (CV) solution for 15 min. A quantity of 100 μL of 30% (v/v) acetic acid was used for extraction. After shaking, the absorbance was measured at 570 nm.

4.9. Removal of Biofilm

Biofilm was grown in a 96-well microplate without CSNP-CAs for 48 h, as described above. Then, CSNP-CAs and free CA solution with different concentrations were added to each well to remove the biofilm. After 24 h, the microplates were washed with PBS, and the removal ability was detected by CV staining.

4.10. Antibiofilm Effect on Food-Grade Silicone Surfaces

Scanning electron microscopy (SEM) and biofilm-formation analyses were conducted according to the method of Tan et al. [4]. Biofilms were formed and treated with CA and CSNP-Cas, as described above, on silicone (7 mm in diameter; Hongxiang, Jiaxing, China) as the surface substrate. Afterwards, the biofilm was fixed with 3% glutaraldehyde at 4 °C overnight and then dehydrated in a series of ethanol solutions for 20 min each (70%, 80%, 96% and 100%). After chemical dehydration with tert-butanol (Macklin, Shanghai, China), the samples were coated with gold and analyzed via SEM (VEGA3, TESCAN, Brno, Czech Republic), using a 10 kV accelerating voltage.

4.11. Live/Dead Assay

Biofilm cells were stained with a Live/Dead® BacLight™ Bacterial Staining Kit (40274, Yeasen, Shanghai, China), according to the manufacturer's protocol, for 30 min at 37 °C in the dark. Images of biofilm were analyzed by confocal laser scanning microscopy (CLSM; ZEISS, LSM900, Oberkohen, Germany).

4.12. Effect of CA on Membrane Permeability of S. aureus

To the bacterial suspension, 1 × MIC CA was added at 37 °C, and PBS buffer was used as the control. Samples were taken hourly to measure potassium-ion concentration. The determination steps for potassium-ion contents were performed according to the instructions provided with the Nanjing Jiancheng potassium ion kit, and the absorbance was measured at 440 nm (Persee, TU-1810, Beijing, China). All the experiments were carried out in triplicate.

4.13. Effect of CA on Cell Wall Integrity of S. aureus

To the bacterial suspension, 1 × MIC CA was added at 37 °C, and PBS buffer was used as the control. Samples were periodically sampled by hourly sampling and centrifuged at 1000 rpm for 10 min, and the supernatant was saved for assay. The instructions provided with the Nanjing Jiancheng alkaline phosphatase determination kit (AKP) were referred to for the determination steps, and absorbance was measured at 520 nm.

4.14. The Morphologies of the Bacteria Observed by SEM

The bacterial suspension was cultured to the logarithmic phase, centrifuged at 4000 rpm for 10 min and the supernatant was discarded. Then, the precipitation was fixed with 2.5% (v/v) glutaraldehyde at 4 °C overnight, followed by dehydration in a series of ethanol solutions (10%, 30%, 50%, 70%, 80%, 90%, 95%, 100%) for 20 min. Then, the samples were placed in 50% and 100% tert-butanol solutions (Macklin, China) for chemically dehydration for 15 min and vacuum dried for 48 h. Then, the samples were sprayed with gold and analyzed via SEM (VEGA3, TESCAN, Brno, Czech Republic).

4.15. Flow Cytometry Analysis

S. aureus was collected by centrifugation (10,000 rpm, 10 min, 4 °C). The bacterial suspension was washed 3 times with sterile PBS buffer and diluted to bacterial suspension OD_{600} = 0.01. Then, 2 × MIC CA was added to the bacterial suspension prior to incubation for 4 h. Flow cytometry analysis was conducted according to the method of Deng et al., with minor modifications [40]. A quantity of 1 mL *S. aureus* bacterial suspension was added to 5 μL of FITC and 5 μL of PI staining solution (Meilunbio, Dalian, China), placed in the dark at 37 °C for 30 min and then detected by flow cytometry (Facs AriaIII, BD, New York, NY, USA).

4.16. Statistical Analysis

All the experiments were carried out in triplicate. Statistical analyses were performed with SPSS 19.0 (SPSS Inc., Chicago, IL, USA). Means ± standard deviations (SDs) were calculated for each experiment. Statistical significance was determined by t-test analysis, with significant differences determined at $p < 0.05$.

5. Conclusions

The results showed that the antibiofilm activity of CA against *S. aureus* was enhanced when loaded on CSNPs. The antibiofilm activity of CSNP-CAs was determined by biofilm inhibition, biofilm disruption and biofilm live/dead bacterial changes on food-grade silicone surfaces. Synthetic CSNP-CAs can be used as potential therapeutics to control *S. aureus* biofilm formation in the future. Furthermore, CSNPs can be used as platforms to design more chemically modified agents or be loaded with other antibiofilm agents for more functional drug-delivery systems.

Author Contributions: Conceptualization, Y.T. and J.X.; methodology, J.X.; software, J.X.; validation, J.X., Q.L. and M.S.; formal analysis, B.L.; investigation, Y.G.; resources, Y.T.; data curation, J.X.; writing—original draft preparation, J.X.; writing—review and editing, Y.T. and T.D.; visualization, M.S.; supervision, Y.T.; project administration, Y.T.; funding acquisition, Y.T. All authors have read and agreed to the published version of the manuscript.

Funding: The authors gratefully acknowledge the Young Taishan Scholars Program of Shandong Province (tsqn202103094) and the Talent Research Foundation of Qingdao Agricultural University (665/1121020).

Conflicts of Interest: The authors declare no conflict of interest.

References

1. Bortolaia, V.; Espinosa-Gongora, C.; Guardabassi, L. Human health risks associated with antimicrobial-resistant enterococci and *Staphylococcus aureus* on poultry meat. *Clin. Microbiol. Infect* **2016**, *22*, 130–140. [CrossRef] [PubMed]

2. Hall-Stoodley, L.; Costerton, J.W.; Stoodley, P. Bacterial biofilms: From the Natural environment to infectious diseases. *Nat. Rev. Microbiol.* **2004**, *2*, 95–108. [CrossRef] [PubMed]

3. Furukawa, S.; Akiyoshi, Y.; O'Toole, G.A.; Ogihara, H.; Morinaga, Y. Sugar fatty acid esters inhibit biofilm formation by food-borne pathogenic bacteria. *Int. J. Food Microbiol.* **2010**, *138*, 176–180. [CrossRef] [PubMed]

4. Tan, Y.; Ma, S.; Leonhard, M.; Moser, D.; Haselmann, G.M.; Wang, J.; Eder, D.; Schneider-Stickler, B. Enhancing antibiofilm activity with functional chitosan nanoparticles targeting biofilm cells and biofilm matrix. *Carbohydr Polym.* **2018**, *200*, 35–42. [CrossRef] [PubMed]

5. Vyas, N.; Wang, Q.X.; Walmsley, A.D. Improved biofilm removal using cavitation from a dental ultrasonic scaler vibrating in carbonated water. *Ultrason Sonochem.* **2021**, *70*, 105338. [CrossRef]

6. Shi, S.F.; Zhang, X.L.; Zhu, C.; Chen, D.S.; Guo, Y.Y. Ultrasonically enhanced rifampin activity against internalized *Staphylococcus aureus*. *Exp. The. Med.* **2013**, *5*, 257–262. [CrossRef]

7. Sadekuzzaman, M.; Yang, S.; Mizan, M.F.R.; Ha, S.D. Current and recent advanced strategies for combating biofilms. *Com. Rev. Food Sci. F* **2015**, *14*, 491–509. [CrossRef]

8. Srinivasan, S.; Harrington, G.W.; Xagoraraki, I.; Goel, R. Factors affecting bulk to total bacteria ratio in drinking water distribution systems. *Water Res.* **2008**, *42*, 3393–3404. [CrossRef]

9. Doyle, A.A.; Stephens, J.C. A review of cinnamaldehyde and its derivatives as antibacterial agents. *Fitoterapia* **2019**, *139*, 104405. [CrossRef]

10. Zhu, R.; Liu, H.; Liu, C.; Wang, L.; Ma, R.; Chen, B.; Li, L.; Niu, J.; Fu, M.; Zhang, D.; et al. Cinnamaldehyde in diabetes: A review of pharmacology, pharmacokinetics and safety. *Pharmacol. Res.* **2017**, *122*, 78–89. [CrossRef]

11. Shreaz, S.; Wani, W.A.; Behbehani, J.M.; Raja, V.; Irshad, M.; Karched, M.; Ali, I.; Siddiqi, W.A.; Hun, L.T. Cinnamaldehyde and its derivatives, a novel class of antifungal agents. *Fitoterapia* **2016**, *112*, 116–131. [CrossRef] [PubMed]

12. Albano, M.; Crulhas, B.P.; Alves, F.C.B.; Pereira, A.F.M.; Andrade, B.F.M.B.; Barbosa, L.N.; Furlanetto, A.; Lyra, L.P.D.S.; Rall, V.L.M.; Júnior, A.F. Antibacterial and anti-biofilm activities of cinnamaldehyde against *S. epidermidis*. *Microb. Pathog.* **2019**, *126*, 231–238. [CrossRef]

13. Yu, H.H.; Song, Y.J.; Yu, H.S.; Lee, N.K.; Paik, H.D. Investigating the antimicrobial and antibiofilm effects of cinnamaldehyde against *Campylobacter* spp. using cell surface characteristics. *J. Food Sci.* **2020**, *85*, 157–164. [CrossRef]

14. Kim, J.; Bao, T.H.Q.; Shin, Y.K.; Kim, K.Y. Antifungal activity of magnoflorine against *Candida* strains. *World J. Microb. Biot.* **2018**, *34*, 167. [CrossRef]

15. Kot, B.; Sytykiewicz, H.; Sprawka, I.; Witeska, M. Effect of *trans*-cinnamaldehyde on methicillin-resistant *Staphylococcus aureus* biofilm formation: Metabolic activity assessment and analysis of the biofilm-associated genes expression. *Int. J. Mol. Sci.* **2020**, *21*, 102. [CrossRef] [PubMed]

16. Benjemaa, M.; Neves, M.A.; Falleh, H.; Lsoda, H.; Ksouri, R.; Nakajima, M. Nanoencapsulation of Thymus capitatus essential oil: Formulation process, physical stability characterization and antibacterial efficiency monitoring. *Ind. Crop. Prod.* **2018**, *113*, 414–421. [CrossRef]

17. Taghipour, Y.D.; Hajialyani, M.; Naseri, R.; Hesari, M.; Mohammadi, P.; Stefanucci, A.; Mollica, A.; Farzaei, M.H.; Abdollahi, M. Nanoformulations of natural products for management of metabolic syndrome. *Int. J. Nanomed.* **2019**, *14*, 5303–5321. [CrossRef]

18. Lodhi, G.; Kim, Y.S.; Hwang, J.W.; Kim, S.K.; Jeon, Y.J.; Je, J.Y.; Ahn, C.B.; Moon, S.H.; Jeon, B.T.; Park, P.J. Chitooligosaccharide and its derivatives: Preparation and biological applications. *BioMed Res. Int.* **2014**, *2014*, 654913. [CrossRef]

19. Kashyap, P.L.; Xiang, X.; Heiden, P. Chitosan nanoparticle based delivery systems for sustainable agriculture. *Int. J. Biol. Macromol.* **2015**, *77*, 36–51. [CrossRef]

20. Youssef, A.M.; Abou-Yousef, H.; El-Sayed, S.M.; Kamel, S. Mechanical and antibacterial properties of novel high performance chitosan/nanocomposite films. *Int. J. Biol. Macromol.* **2015**, *76*, 25–32. [CrossRef]

21. Haldorai, Y.; Shim, J.J. Chitosan-Zinc oxide hybrid composite for enhanced dye degradation and antibacterial activity. *Compos. Interface* **2013**, *20*, 365–377. [CrossRef]

22. Liang, J.; Yan, H.; Puligundla, P.; Gao, X.; Zhou, Y.; Wan, X. Applications of chitosan nanoparticles to enhance absorption and bioavailability of tea polyphenols: A review. *Food Hydrocolloid.* **2017**, *69*, 286–292. [CrossRef]

23. Karthik, C.S.; Chethana, M.H.; Manukumar, H.M.; Ananda, A.P.; Sandeep, S.; Nagashree, S.; Mallesha, L.; Mallu, P.; Jayanth, H.S.; Dayananda, B.P. Synthesis and characterization of chitosan silver nanoparticle decorated with benzodioxane coupled piperazine as an effective anti-biofilm agent against MRSA: A validation of molecular docking and dynamics. *Int. J. Biol. Macromol.* **2021**, *181*, 540–551. [CrossRef]

24. Elgadir, M.A.; Uddin, M.S.; Ferdosh, S.; Adam, A.; Chowdhury, A.J.K.; Sarker, M.Z.I. Impact of chitosan composites and chitosan nanoparticle composites on various drug delivery systems: A review. *J. Food Drug Anal.* **2015**, *23*, 619–629. [CrossRef]

25. Zamborini, F.P.; Bao, L.; Dasari, R. Nanoparticles in measurement science. *Anal. Chem.* **2012**, *84*, 541–576. [CrossRef]

26. Tan, Y.; Ma, S.; Leonhard, M.; Moser, D.; Schneider-Stickler, B. β-1,3-glucanase disrupts biofilm formation and increases antifungal susceptibility of *Candida albicans* DAY185. *Int. J. Biol. Macromol.* **2018**, *108*, 942–946. [CrossRef]

27. Ye, H.; Shen, S.; Xu, J.; Lin, S.; Yuan, Y.; Jones, G.S. Synergistic interactions of cinnamaldehyde in combination with carvacrol against food-borne bacteria. *Food Control* **2013**, *34*, 619–623. [CrossRef]

28. Forier, K.; Raemdonck, K.; De Smedt, S.C.; Demeester, J.; Coenye, T.; Braeckmans, K. Lipid and polymer nanoparticles for drug delivery to bacterial biofilms. *J. Control. Release* **2014**, *190*, 607–623. [CrossRef] [PubMed]

29. Bellich, B.; D'Agostino, I.; Semeraro, S.; Gamini, A.; Cesaro, A. "The good, the bad and the ugly" of chitosans. *Mar. Drugs* **2016**, *14*, 99. [CrossRef]

30. Cheung, R.C.F.; Ng, T.B.; Wong, J.H.; Chan, W.Y. Chitosan: An update on potential biomedical and pharmaceutical applications. *Mar. Drugs* **2015**, *13*, 5156–5186. [CrossRef]

31. Mu, H.; Guo, F.; Niu, H.; Liu, Q.; Wang, S.; Duan, J. Chitosan improves anti-biofilm efficacy of gentamicin through facilitating antibiotic penetration. *Int. J. Mol. Sci.* **2014**, *15*, 22296–22308. [CrossRef] [PubMed]

32. Jiang, F.; Deng, Y.; Yeh, C.K.; Sun, Y. Quaternized chitosans bind onto preexisting biofilms and eradicate pre-attached microorganisms. *J. Mater Chem. B* **2014**, *2*, 8518–8527. [CrossRef] [PubMed]

33. Shrestha, A.; Hamblin, M.R.; Kishen, A. Characterization of a conjugate between Rose Bengal and chitosan for targeted antibiofilm and tissue stabilization effects as a potential treatment of infected dentin. *Antimicrob. Agents Ch.* **2012**, *56*, 4876–4884. [CrossRef]

34. Inoue, Y.; Shiraishi, A.; Hada, T.; Hirose, K.; Hamashima, H.; Shimada, J. The antibacterial effects of terpene alcohols on *Staphylococcus aureus* and their mode of action. *FEMS Microbiol. Lett.* **2006**, *237*, 325–331. [CrossRef]

35. Frohling, A.; Baier, M.; Ehlbeck, J.; Knorr, D.; Schlüter, O. Atmospheric pressure plasma treatment of *Listeria innocua* and *Escherichia coli* at polysaccharide surfaces: Inactivation kinetics and flow cytometric characterization. *Innov. Food Sci. Emerg.* **2012**, *13*, 142–150. [CrossRef]

36. Mukherjee, P.K.; Chandra, J.; Kuhn, D.M.; Ghannoum, M.A. Mechanism of fluconazole resistance in *Candida albicans* biofilms: Phase-specific role of efflux pumps and membrane sterols. *Infect. Immun.* **2003**, *71*, 4333–4340. [CrossRef] [PubMed]

37. Borezée-Durant, E.; Hiron, A.; Piard, J.C.; Juillard, V. Dual role of the oligopeptide permease opp3 during growth of staphylococcus aureus in milk. *Appl. Environ. Microb.* **2009**, *75*, 3355–3357. [CrossRef]

38. Wesson, C.A.; Liou, L.E.; Todd, K.M.; Bohach, G.A.; Trumble, W.R.; Bayles, K.W. Staphylococcus aureus agr and sar global regulators influence internalization and induction of apoptosis. *Infect. Immun.* **1998**, *66*, 5238–5243. [CrossRef]

39. Klinger-Strobel, M.; Ernst, J.; Lautenschlaeger, C.; Pletz, M.W.; Fischer, D.; Makarewicz, O. A blue fluorescent labeling technique utilizing micro-and nanoparticles for tracking in Live/Dead® stained pathogenic biofilms of *Staphylococcus aureus* and *Burkholderia cepacia*. *Int. J. Nanomed.* **2016**, *11*, 575–583. [CrossRef]

40. Ji, D.B.; Zhang, L.Y.; Li, C.L.; Ye, J.; Zhu, H.B. Effect of Hydroxysafflor yellow A on human umbilical vein endothelial cells under hypoxia. *Vasc. Pharmacol.* **2009**, *50*, 137–145. [CrossRef]

antibiotics

MDPI

Article

Effect of Extracts, Fractions, and Isolated Molecules of *Casearia sylvestris* to Control *Streptococcus mutans* Cariogenic Biofilm

Sabrina M. Ribeiro [1,†]**, Paula C. P. Bueno** [2,3,4]**, Alberto José Cavalheiro** [2,†] **and Marlise I. Klein** [1,5,*]

1   Department of Dental Materials and Prosthodontics, School of Dentistry, São Paulo State University (UNESP), Araraquara 14801-903, SP, Brazil
2   Department of Organic Chemistry, Institute of Chemistry, São Paulo State University (UNESP), Araraquara 14800-900, SP, Brazil
3   Department of Physics and Chemistry, Faculty of Pharmaceutical Sciences of Ribeirão Preto, University of São Paulo (USP), Ribeirão Preto 14040-903, SP, Brazil
4   Department of Organic Chemistry, Institute of Chemistry, Federal University of Alfenas, Alfenas 37130-001, MG, Brazil
5   Department of Oral Diagnosis, Piracicaba Dental School, State University of Campinas (UNICAMP), Piracicaba 13414-903, SP, Brazil
*   Correspondence: marlise@unicamp.br
†   These authors contributed equally to this work.

Abstract: The effects of extracts, fractions, and molecules of *Casearia sylvestris* to control the cariogenic biofilm of *Streptococcus mutans* were evaluated. First, the antimicrobial and antibiofilm (initial and pre-formed biofilms) in prolonged exposure (24 h) models were investigated. Second, formulations (with and without fluoride) were assessed for topical effects (brief exposure) on biofilms. Third, selected treatments were evaluated via bacterium growth inhibition curves associated with gene expression and scanning electron microscopy. In initial biofilms, the ethyl acetate (AcOEt) and ethanolic (EtOH) fractions from Brasília (BRA/DF; 250 µg/mL) and Presidente Venceslau/SP (Water/EtOH 60:40 and Water/EtOH 40:60; 500 µg/mL) reduced ≥6-logs vs. vehicle. Only the molecule Caseargrewiin F (CsF; 125 µg/mL) reduced the viable cell count of pre-formed biofilms (5 logs vs. vehicle). For topical effects, no formulation affected biofilm components. For the growth inhibition assay, CsF yielded a constant recovery of surviving cells (≅3.5 logs) until 24 h (i.e., bacteriostatic), and AcOEt_BRA/DF caused progressive cell death, without cells at 24 h (i.e., bactericidal). CsF and AcOEt_BRA/DF damaged *S. mutans* cells and influenced the expression of virulence genes. Thus, an effect against biofilms occurred after prolonged exposure due to the bacteriostatic and/or bactericidal capacity of a fraction and a molecule from *C. sylvestris*.

Keywords: *Streptococcus mutans*; biofilm; extracellular matrix; dental caries; *Casearia sylvestris*

Citation: Ribeiro, S.M.; Bueno, P.C.P.; Cavalheiro, A.J.; Klein, M.I. Effect of Extracts, Fractions, and Isolated Molecules of *Casearia sylvestris* to Control *Streptococcus mutans* Cariogenic Biofilm. *Antibiotics* **2023**, *12*, 329. https://doi.org/10.3390/antibiotics12020329

Academic Editors: Ding-Qiang Chen, Yulong Tan, Ren-You Gan, Guanggang Qu, Zhenbo Xu and Junyan Liu

Received: 13 December 2022
Revised: 31 January 2023
Accepted: 1 February 2023
Published: 4 February 2023

1. Introduction

Dental caries is a chronic and multifactorial condition that results from the formation of a polymicrobial biofilm and dynamic interactions between microorganisms present in this biofilm, salivary constituents, and dietary carbohydrates (e.g., sucrose) [1,2]. Despite attempts to raise awareness, this oral condition is a worrying public health problem and impairs the quality of life of millions of people [1]. *Streptococcus mutans* is the bacterium associated with dental caries etiology (although other microorganisms may be associated) [2,3]. *S. mutans* is acidogenic, aciduric, and the leading producer of extracellular matrix in biofilms, known as dental plaque [4]. This trait occurs because *S. mutans* encodes multiple exoenzymes (e.g., glycosyltransferases or Gtfs), and in the presence of sucrose (and starch), they produce copious amounts of exopolysaccharides (i.e., mainly glucans but also fructans) [4,5].

The extracellular matrix provides a cohesive and acidic environment of limited diffusion [6], restricting access to buffering saliva and antimicrobial agents [6,7]. In cariogenic biofilms, exopolysaccharides are primordial components in the organization of the extracellular matrix [4,5] and are determinants of virulence [6]. In addition to exopolysaccharides, extracellular DNA (eDNA) and lipoteichoic acids (LTA) are found in large amounts in cariogenic biofilms [8] and contribute to the matrix structural organization and properties [9,10]. These virulence factors modulate the pathogenesis of dental caries and, thus, are selective therapeutic targets for preventing this disease.

Fluoride, in its various modalities of administration, is the basis for caries prevention; however, its current delivery forms are insufficient to overcome the cariogenic challenges in many individuals; therefore, additional approaches are needed to increase its effectiveness [11]. Chemical agents (such as chlorhexidine?CHX) are widely used to control cariogenic biofilms [12]. Although CHX can suppress mutans group streptococci levels, its efficacy is reduced against mature biofilms, mainly because the exopolysaccharides of the matrix have a negative charge and affect the penetration of CHX (a cationic substance) into the biofilm, compromising its antimicrobial activity in these biofilms [6,12]. In addition, CHX eliminates oral bacteria that convert nitrate to nitrite, which can raise systolic blood pressure [13]. Therefore, CHX cannot be used for a continuous and prolonged period [13]. Natural products are a vast source of structurally diverse molecules with various biological properties. Therefore, natural antibacterial substances are useful for developing alternative or adjunctive anticaries therapies. For example, plant extracts have recently been incorporated into these products to improve their antimicrobial properties [14].

Casearia sylvestris Sw. (Salicaceae) is one of the most promising species from the genus *Casearia* due to its biological properties and uses in folk medicine. *C. sylvestris* has a high adaptive capacity and is widely disseminated in Central and South America, and in Brazil, it occurs in practically all biomes [15]. Chemically, extracts from leaves of *C. sylvestris* var. *sylvestris* (Atlantic Forest) are rich in diterpenes (taxonomic markers for the genus) [16], while phenolic compounds (flavonoids) predominate in var. *lingua* (Cerrado) [17]. The range of variations around a basic skeleton found in the diterpenes and flavonoids of *C. sylvestris* provides very interesting models for studies on the relationship between chemical structure and biological activity of these compounds, which have already demonstrated potential activity in the control of cariogenic biofilms [18].

Studies about different *C. sylvestris* varieties and their chemical constituents have pointed out a plethora of biological activities such as cytotoxic, anti-inflammatory, and anti-tumor effects, among others [18–21]. However, few studies have been carried out at a deep level regarding its antimicrobial potential [22,23], and little is clarified about its biological activity against pathogenic microorganisms found in the oral cavity [24]. Thus, prospective studies of plant extracts and/or isolated molecules with antimicrobial and antibiofilm properties are relevant for dentistry and other areas. Therefore, this in vitro study evaluated the effect of extracts, fractions, and isolated molecules of *C. sylvestris* (Atlantic and Cerrado Biomes; *sylvestris*, *lingua*, and intermediate varieties) to control the cariogenic biofilm of *S. mutans*.

2. Results

2.1. Antimicrobial Activity of Crude Extracts

The measure was considered effective when the $\log_{10}$ CFU/mL count was reduced by 3 logs (vs. vehicle) [25,26]. Extracts caused mean reductions of 0.5 logs vs. vehicle control (Figure 1); therefore, none inhibited > 3 logs of viable cell counts. Thus, the observed effect is not biologically significant [27]. These data differ from those obtained previously when adequate reduction occurred [18]. These findings demonstrate the requirement to test all batches of extracts to ensure that minimal activity standards are achieved.

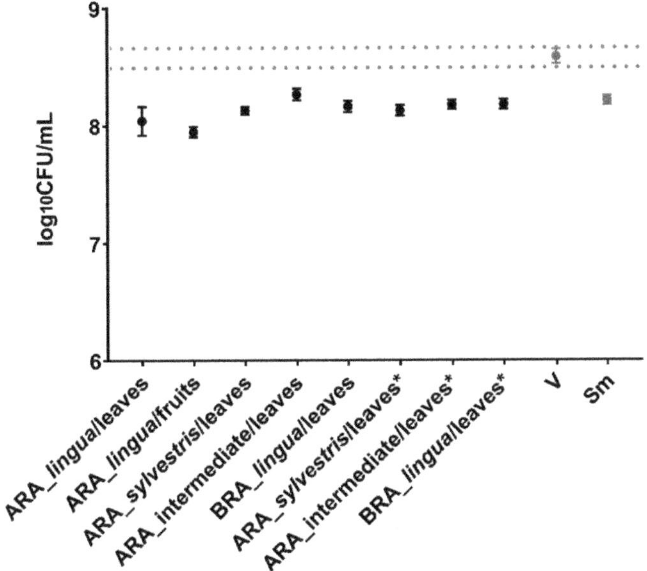

Figure 1. Antimicrobial activity of *C. sylvestris* crude extracts against *S. mutans*. Log$_{10}$ CFU data from planktonic cells treated by crude extracts (500 µg/mL). The central data are the mean, and the error bars are the confidence interval (95% CI). The dotted lines use vehicle data to compare the effectiveness of tested treatments. The growth control (no treatment) is represented as Sm for *S. mutans* and V for the vehicle control (with the concentration in each well being 1.75% EtOH and 0.31% DMSO). The asterisk indicates the raw extracts with the pH adjusted to a value close to the vehicle. The experiments were performed in triplicate on two separate occasions (n = 2).

2.2. Antibiofilm Activity Using the Polystyrene Microplate Model (Prolonged Exposure)

The CFU/mL data obtained in the antibiofilm assays were converted to log$_{10}$ to verify the log reduction. As for the antimicrobial assay, the effect was considered adequate when the log$_{10}$ CFU/mL count was reduced by 3 logs (vs. vehicle) for 24 h old and 48 h old biofilms [25,26]. In early (24 h) biofilms, no extract inhibited > 3 logs of viable cell counts (Figure 2A). Therefore, these extracts did not present good biological activity. All fractions obtained through methodology 1 reduced the viable population count of *S. mutans*. However, considerable reductions occurred for the AcOEt and EtOH fractions of BRA/DF, which decreased 8 and 6 logs of viable cell counts, respectively (vs. vehicle; Figure 2B). For the fractions of methodology 2, AcOEt_BRA/DF reduced 8 logs of viable cell count (vs. vehicle; Figure 2C). None of the Hex fractions caused a considerable reduction (Tables S1 and S2 in the Supplementary Materials). Another attempt to characterize the active fraction was made through the fractionation of extracts from leaves (F) and branches (G) of *C. sylvestris* collected in Presidente Venceslau (PRE) by SPE-C18. The fractions Water/EtOH 60:40 (F4.2b) and Water/EtOH 40:60 (G3.3a), both at 500 µg/mL, reduced, respectively, 6 and 8 logs of the viable cell count of the biofilms (vs. vehicle; Figure 2D). Their chromatographic profiles indicate that they are fractions rich in clerodane diterpenes and possibly tannins (data shown in Figure A1).

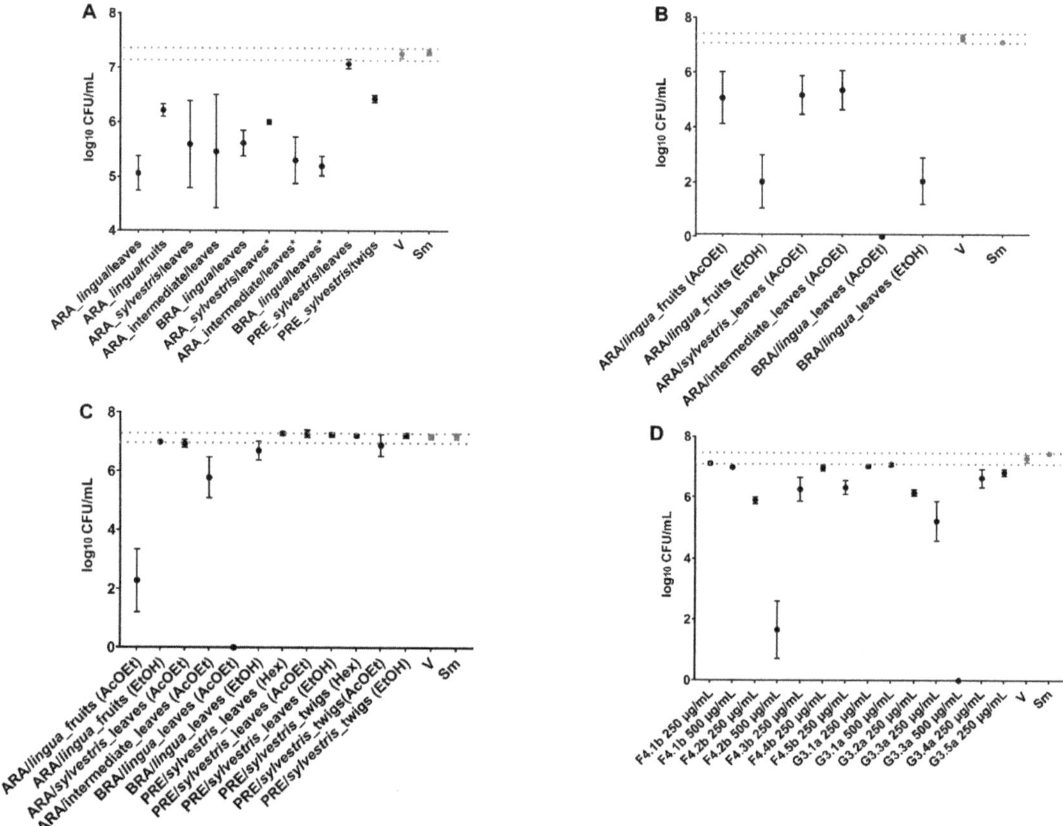

Figure 2. The activity of crude extracts, fractions, and molecules of *C. sylvestris* against initial biofilms. The graphs show in (**A**) $\log_{10}$ CFU of the early (24 h) biofilms treated by the crude extracts (500 µg/mL); (**B**) $\log_{10}$ CFU of the early (24 h) biofilms treated by fractions obtained via methodology 1 (dry fractions in sample concentrator, 250 µg/mL); (**C**) $\log_{10}$ CFU of biofilms treated by fractions obtained via methodology 2 (dry fractions in the fume hood, 250 µg/mL); (**D**) $\log_{10}$ CFU of the early (24 h) biofilms treated by SPE-C18 fractions of PRE/SP (250 µg/mL and 500 µg/mL). The central data are the mean, and the error bars are the confidence interval (95%CI). The dotted lines use vehicle data to compare the effectiveness of tested treatments. The growth control (no treatment) is represented as Sm for *S. mutans* and V for the vehicle control (with the concentration in each well being 1.75% EtOH and 0.31% DMSO for the crude extracts, while for the AcOEt, EtOH, and Hex fractions, it was 5.26% EtOH and 0.94% DMSO, finally for the SPE-C18 fractions, 2.63% EtOH). In (**A**), the asterisk indicates the raw extracts with the pH adjusted to the value close to that of the vehicle. The experiments were performed in triplicate on two separate occasions (*n* = 2).

Among the four isolated molecules tested, only CsF (125 µg/mL) reduced the viable cell count of pre-formed biofilms vs. vehicle (reduced $\cong$ 5 logs; Figure 3). The BRA/DF fractions (500 µg/mL) did not interfere adequately with the viability of *S. mutans* in the treated biofilms. For the samples of *C. sylvestris* obtained by SPE-Si/C, the chemical profile showed bands that could not be related to the main secondary metabolites already described for this species, that is, flavonoids and clerodane diterpenes, except for the AcOEt fraction obtained from the extract of leaves of *C. sylvestris* collected in Brasilia-DF (BRA/DF). This fraction is enriched in a constituent whose UV spectrum is characteristic of casearins, with λmax at 241 nm. The relationship of clerodane diterpenes with antibiofilm activity was confirmed by the data presented in Figure 3 for CsF, where the activities of the AcOEt_BRA/DF fraction and patterns of glycosylated flavonoids and casearins are confronted. For the AcOEt _BRA/DF fraction, chromatography identified the main peak, indicating a casearin. However, this peak differs at this time from the peaks of the other casearins used in this study (F, X, and J; chromatographic data shown in Figure A2). Therefore, we would hypothesize that in AcOEt_BRA/DF, there is the presence of an unidentified casearin.

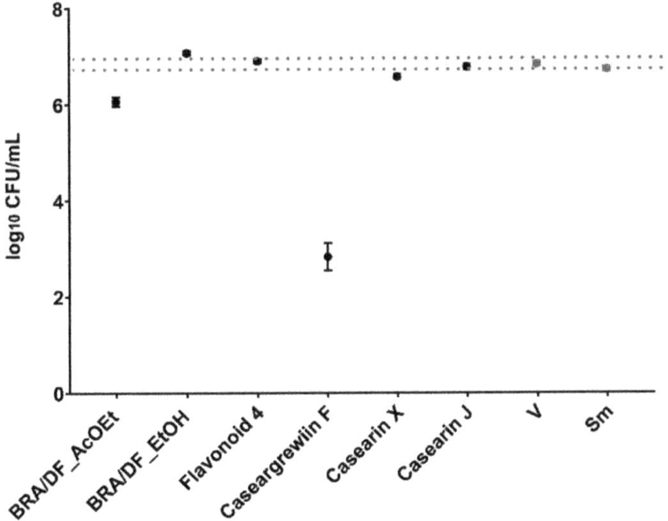

Figure 3. Log$_{10}$ CFU of pre-formed biofilms (48 h old biofilms) treated by fractions (500 µg/mL), flavonoid (125 µg/mL), and casearins (125 µg/mL). The central data are the mean, and the error bars are the confidence interval (95% CI). The dotted lines use vehicle data to compare the effectiveness of tested treatments. The growth control (no treatment) is represented as Sm for *S. mutans* and V for the vehicle control (with the concentration in each well being 1.75% EtOH and 0.31% DMSO). The experiments were performed in triplicate on two separate occasions ($n = 2$).

2.3. Effect of Topical Treatments on Biofilms Grown on Saliva-Coated Hydroxyapatite (sHA) Discs

2.3.1. pH of the Spent Culture Medium

The pH data are represented in Figure 4. These values reflect acidogenicity, measuring the concentration of free hydrogen ions in the medium. However, they do not show whether acid concentration occurred at any specific location within the biofilm or at the interface between the biofilm and the substrate (HA disc). The mean values for 43, 51, and 67 h old biofilms are lower or more acidic compared to 19 h old biofilms. All values are below 5.5, a pH considered critical for demineralizing dental enamel. However, in general, there is a similarity in the pH values between the evaluated groups over time, although the fluoride formulations present higher values.

SPE - C18

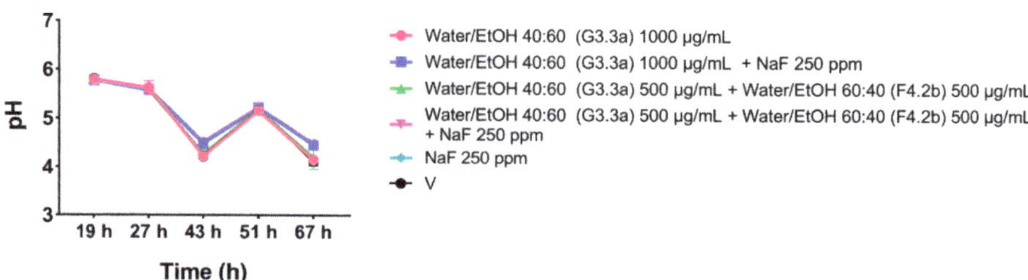

AcOEt_BRA/DF

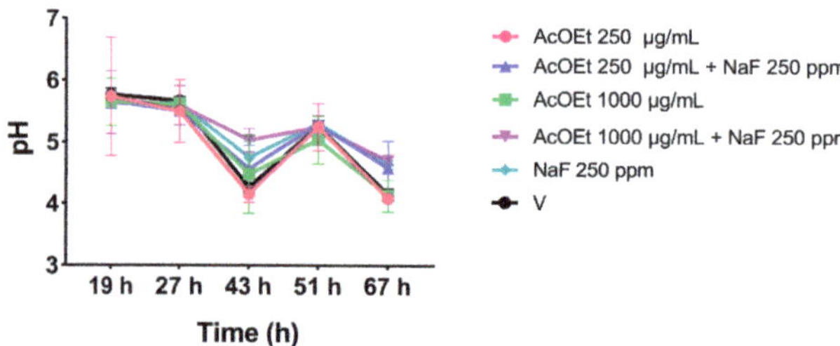

Caseargrewiin F

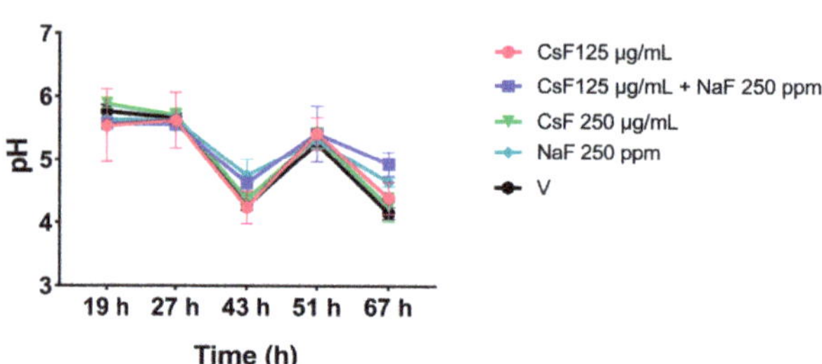

Figure 4. pH of spent culture medium from topically treated biofilms at different developmental stages. The spent biofilm culture media were analyzed at 19, 27, 43, 51, and 67 h. The central data are the mean, and the error bars are the confidence interval (95% CI). The vehicle control is represented as V (with the concentration being 10.52 % EtOH for the SPE-C18 fractions, 21.04% EtOH and 3.75% DMSO for the AcOEt_BRA/DF fractions, and 10.52% EtOH and 1.87% DMSO for CsF). The experiments for SPE-C18, AcOEt_BRA/DF, and CsF 125 µg/mL were performed in duplicate on one experimental occasion ($n = 1$) and in duplicate on two separate occasions ($n = 2$) for CsF 250 µg/mL.

2.3.2. Bacterial Population and Biofilm Dry Weight (Insoluble Dry Weight)

All groups were compared to the vehicle control. Topical exposure (1.5 min for SPE-C18 fractions and 10 min for AcOEt_BRA/DF and CsF) affected only the bacterial population, with an increase for Water/EtOH 40:60 (G3.3a) 1000 µg/mL + NaF250 ppm, Water/EtOH 40:60 (G3.3a) 500 µg/mL + Water/EtOH 60:40 (F4.2b) 500 µg/mL + NaF250 ppm (Figure 5A) and at the lowest concentrations tested for the AcOEt_BRA/DF 250 fraction µg/mL and 125 µg/mL CsF combined with NaF (Figure 5B); this also occurred for the control NaF vs. vehicle (Figure 5A,B). There was no pronounced effect on insoluble dry weight (Figure 5C,D).

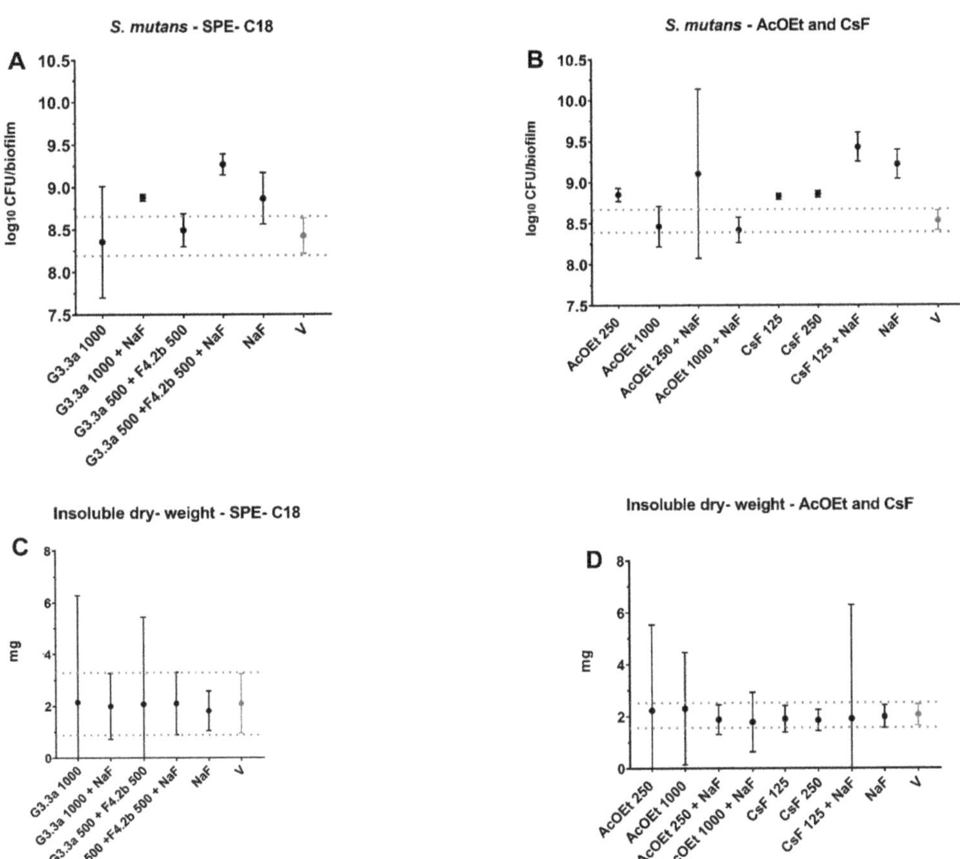

Figure 5. Characterization of biofilms after topical treatments. (**A**) Bacterial population ($\log_{10}$ CFU/biofilm) of the SPE-C18 fractions; (**B**) bacterial population ($\log_{10}$ CFU/biofilm) of BRA/DF_AcOEt and Caseargrewiin F (CsF) fractions; (**C**) dry weight (mg) for biofilms treated with SPE-C18 fractions; (**D**) dry weight (mg) for biofilms treated with BRA/DF_AcOEt and Caseargrewiin F. The central data are the mean, and the error bars are the confidence interval (95% CI). The dotted lines use vehicle data to compare the effectiveness of tested treatments. The vehicle control is represented as V (with the concentration being 10.52% EtOH for the SPE-C18 fractions, 21.04% EtOH and 3.75% DMSO for the AcOEt_BRA/DF fractions, and 10.52% EtOH and 1.87% DMSO for CsF). The experiments for SPE-C18, AcOEt_BRA/DF, and CsF 125 µg/mL were performed in duplicate on one experimental occasion ($n = 1$) and in duplicate on two separate occasions ($n = 2$) for CsF 250 µg/mL.

2.3.3. Components of the Extracellular Matrix of Biofilms

The amount of insoluble (ASP, recovered in the insoluble portion of the biofilm matrix) and soluble (WSP) exopolysaccharides and eDNA (recovered in the soluble portion of the biofilm matrix) are represented in Figure 6. There was no pronounced effect for these extracellular matrix components. Therefore, the data indicate that the antibiofilm effect found in the plate-bottom model may have been mostly due to prolonged exposure (24 h of exposure during initial biofilm formation or 24 h after initial biofilm formation), and this effect may be due to an antibacterial action (cell death; see data in Section 2.4).

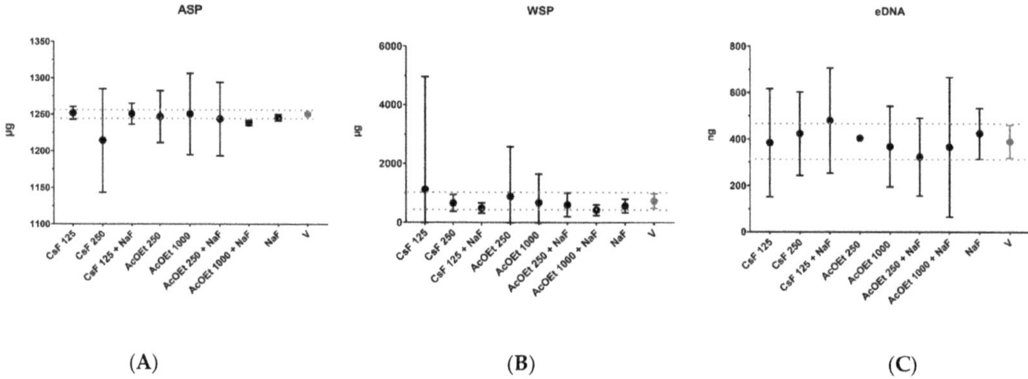

Figure 6. Components of the extracellular matrix of *S. mutans* biofilms after topical treatments. The graphs show the quantification of (**A**) ASP (µg), (**B**) WSP (µg), and (**C**) eDNA (ηg) of biofilms (formed on sHA discs) after treatment with selected agents and controls. The central data are the mean, and the error bars are the confidence interval (95% CI). The dotted lines use vehicle data to compare the effectiveness of tested treatments. The vehicle control is represented as V (with the concentration being 21.04% EtOH and 3.75% DMSO for the AcOEt_BRA/DF fractions and 10.52% EtOH and 1.87% DMSO for CsF). The experiments for SPE-C18, AcOEt_BRA/DF, and CsF 125 µg/mL formulations were performed in duplicate on one experimental occasion ($n = 1$) and duplicate on two separate occasions ($n = 2$) for CsF 250 µg/mL.

2.3.4. Three-Dimensional Structure of Biofilms Using Confocal Microscopy

Figure 7 shows representations of the 3D structure of biofilms on the surface of sHA discs. In the images of biofilms treated with the vehicle control, there are large and defined clusters of microcolonies (in green) protected by exopolysaccharides in the extracellular matrix (in red). The experimental treatments (alone or in combination with NaF) showed similar structural conformation. However, in addition to large microcolonies, there are small microcolonies (among the largest) scattered on the surface of the disc. Finally, for the fluorine-treated (NaF) biofilms, clusters of microcolonies similar to the vehicle-treated biofilms were observed. Therefore, the treatments did not interfere with the three-dimensional organization associated with biofilm virulence.

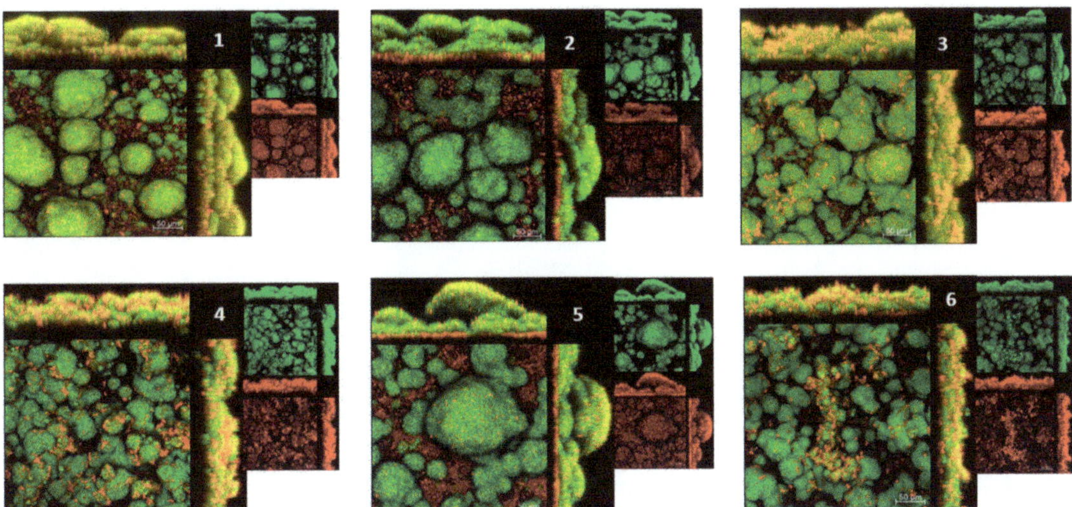

Figure 7. 3D architecture of *S. mutans* biofilm grown on sHA discs topically treated with SPE-C18 fractions. Representative confocal images of 67 h old biofilms are displayed in this figure. The red color represents exopolysaccharides produced by *S. mutans* (Alexa Fluor 647), and the green color represents bacterial cells (SYTO 9). The larger image in each set represents the overlay images of the smaller red and green channels (50 µm scale bars), which are shown separately at a smaller size. 1. Water/EtOH 40:60 (G3.3a)—1000 µg/mL; 2. Water/EtOH 40:60 (G3.3a)—1000 µg/mL + NaF 250 ppm; 3. Water/EtOH 60:40 (F4.2b)—500 µg/mL + Water/EtOH 40:60 (G3.3a)—500 µg/mL; 4. Water/EtOH 60:40 (F4.2b)—500 µg/mL + Water/EtOH 40:60 (G3.3a)—500 µg/mL + NaF 250 ppm; 5. NaF 250 ppm; 6. Vehicle (10.52% EtOH).

2.4. Growth Inhibition Curve of CsF and AcOEt_BRA/DF and Three-Dimensional Structure of Planktonic Cultures Using SEM

The data obtained for the growth inhibition curve in planktonic culture are represented in Figure 8A. The effect of AcOEt 250 µg/mL is more pronounced (effective in killing *S. mutans*), with substantial bacterial death after 1 h of exposure (2 logs vs. vehicle), with complete elimination of bacteria within 4 to 24 h of exposure. This effect of AcOEt 250 µg/mL may be due to its composition (presence of an unidentified casearin; Figure A2) or even the concentration used. Importantly, AcOEt 250 µg/mL resulted in 1 log reduction in bacterial population vs. the vehicle in a 48 h biofilm (pre-formed biofilm) and complete elimination of the bacteria in an initial 24 h biofilm. In contrast, there was gradual death from 1 h for CsF 125 µg/mL, and in 4 h, the mean reduction compared to the vehicle control was 1.71 logs. The average recovery of surviving colonies (cells) for CsF was constant from 4 to 24 h, the average difference in 24 h being 3.5 logs compared to the control (vehicle). These data show that the effect of CsF was bacteriostatic, while that of AcOEt was bacteriocidal. Thus, CsF and AcOEt need to be in contact with *S. mutans* cells for much longer than the duration of topical applications (experimentally in this study and recommended for formulations for chemical control of oral biofilms). SEM analysis showed that both AcOEt and CsF promoted the disruption of planktonic *S. mutans* cells (Figure 8B). The images show the presence of amorphous material originating from the ruptured cells, with remnants of the cell wall and released intracellular content (arrows). The surface topography of vehicle-control-treated cultures shows that the vehicle used did not affect *S. mutans* cell morphology.

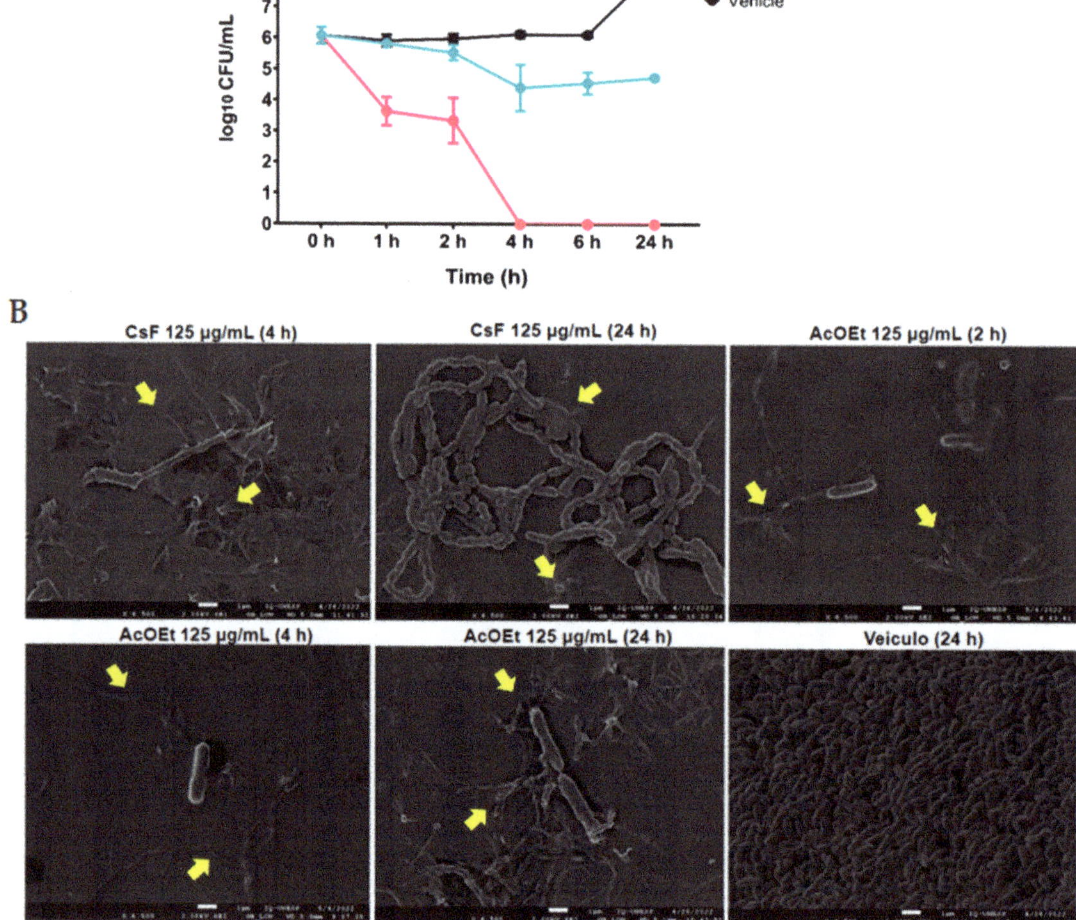

Figure 8. Effects of AcOEt and CsF on planktonic cultures of *S. mutans*. (**A**) The central data for each time, for each evaluated treatment, are the means and the confidence interval (CI 95%) for error bars; (**B**) representative SEM images of *S. mutans* planktonic cells after different exposure times to CsF and AcOEt. The arrows indicate amorphous material from ruptured cells. Scale bars are 1 µm. The concentration of the control vehicle in each well was 5.26% EtOH and 0.94% DMSO.

2.5. Gene Expression

The effect of seven selected treatments on *S. mutans* gene expression is depicted in Figure 9. Besides choosing the AcOEt fraction and molecule CsF, other compounds with information on possible targets in *S. mutans* cells were tested. Data were obtained on one experimental occasion and quantified by qPCR in triplicate. Genes were not normalized by *16S rRNA* because there was a difference in expression between flavonoid myricetin (J10595) and NaF vs. vehicle. Thus, the normalization was performed by using the same amount of RNA used for cDNA synthesis. *16S rRNA* presented a higher expression after treatment with myricetin (J10595) and NaF. For the *nox1*, the expression was repressed by hydroxychalcone C135 and induced by AcOEt and myricetin (J10595) (vs. vehicle). For *eno*, a substantial reduction in the expression occurred after AcOEt, *tt*-farnesol (Far; a terpenoid), and NaF, while myricetin (J10595) increased the expression of this gene. Exposure to C135, myricetin (J10595), and NaF increased *atpD* expression, while exposure to *tt*-farnesol repressed *atpD* (vs. vehicle). The *gtfB* gene, which encodes an exoenzyme that synthesizes insoluble exopolysaccharides (glucans), showed higher expression for C135, *tt*-farnesol, CsF, myricetin (J10595), and NaF (vs. vehicle). In contrast, it appears that AcOEt reduces the expression of this gene. The gene *lrgA* was induced after treatment with C135, compound 1771 (an inhibitor of LTA metabolism), and myricetin (J10595). This finding demonstrates that the product of this gene, LrgA, which is responsible for coordinating the remodeling of the cytoplasmic membrane (a process that releases DNA to the extracellular milieu), can be affected directly or indirectly because the agents may have interfered with the fluidity of this membrane.

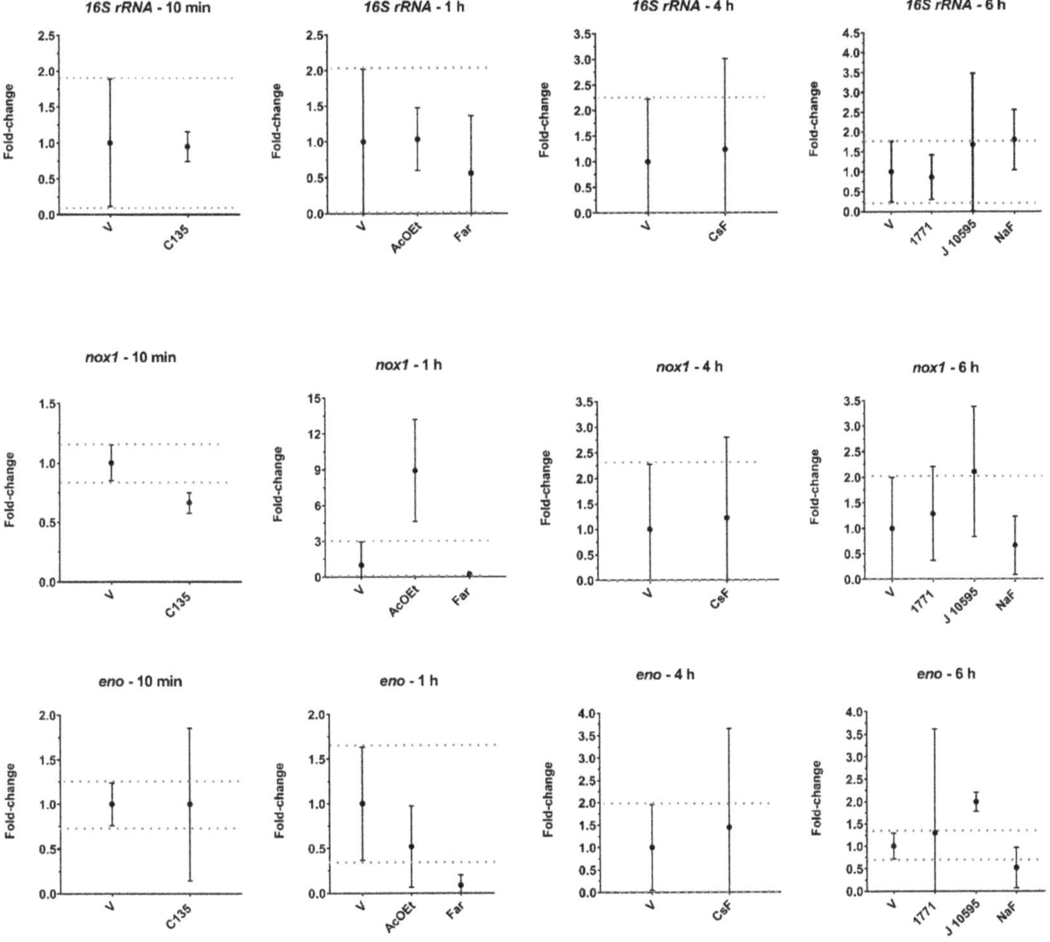

Figure 9. *Cont.*

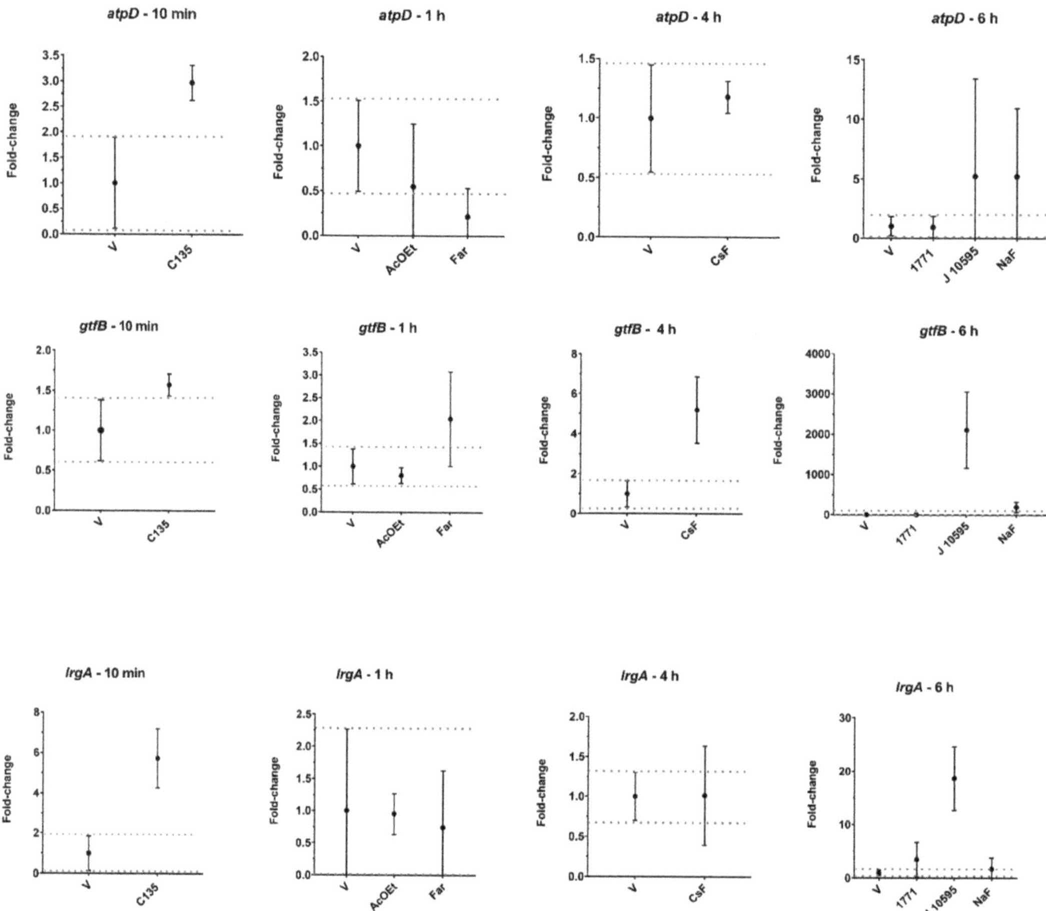

Figure 9. *S. mutans* gene expression of planktonic cultures treated with distinct agent. The central data are the mean, and the error bars are the confidence interval (95% CI). The dotted lines use vehicle data to compare the effectiveness of tested treatments. For detailed information, see the graph of *gtfB* at 6 h for 1771 vs. V in Figure A3.

3. Discussion

C. sylvestris has a phytochemical composition marked by clerodane-types diterpenes and glycosylated flavonoids [17,18,20,28,29]. Prospection studies of this plant have shown activity against cariogenic biofilms. Therefore, it is of interest to the community that its biological properties be explored. Our results indicate that Caseargrewiin F (CsF) and the AcOEt fraction obtained from the leaves of sample BRA/DF (corresponding to variety *lingua* from the Cerrado biome) are effective in inhibiting *S. mutans* through irreversible damage to its structure and changes in the expression of specific virulence genes. However, the topical application does not exert any activity on *S. mutans* biofilm components (viable population, biomass, extracellular matrix components, and structural organization). These findings indicate that the antibiofilm effect in the polystyrene plate-bottom models may have been mainly due to prolonged exposure (24 h) during (the initial biofilm) or after biofilm formation (on pre-formed biofilms). Moreover, the activity of these treatments was primarily due to antibacterial action, verified in the growth inhibition assay. In contrast to previous findings [18], here, no crude extract of *C. sylvestris* showed activity against the long-exposure (24 h) models (Figure 2), which demonstrates that the chemical composition of secondary metabolites is closely related to geographic location, seasonal effects, and biological diversity and drastically influences the biological activity of samples [18,30].

In this long-exposure model, the AcOEt_BRA/DF and SPE-C18 fractions of PRE/SP (Water/EtOH 60:40-F4.2b and Water/EtOH 40:60-G3.3a) reduced the population of the initial biofilms. Additionally, of the isolated molecules evaluated, only CsF showed activity against pre-formed biofilms. CsF is a clerodane-type diterpene with the molecular formula $C_{28}H_{40}O_8$, which was previously isolated from the ethanolic extract of *C. sylvestris* leaf [20]. The chemical structure of CsF seems relevant for its biological activity against *S. mutans*. The presence of OAc at the R2 and R3 terminals of CsF could influence its activity since the exchange for OBu (Casearin X-$C_{32}H_{46}O_9$) and OMe (Casearin J-$C_{31}H_{44}O_9$) in R1 and OBu in R2 (Casearin X and J) seems to significantly inhibit casearin activity. No effect was observed for Casearin X and Casearin J, consistent with the hypothesis of radical terminal placement and molecule configuration for CsF exerting its impact on the bacterium cell. Nonetheless, when tested for topical effects (brief exposure) on *S. mutans* biofilms, these treatments (SPE-C18, AcOEt_BRA/DF, and CsF) did not inhibit biofilm components (Figures 5–7). Thus, for such agents to exert their effect, it would be necessary to prepare formulations capable of retaining their active principle in the oral cavity for the time required for their action. Therefore, in the future, these treatments could be used for loading in drug delivery systems (suitable for the oral cavity) to prolong the exposure time at an adequate concentration [31,32].

In addition to the need for a prolonged exposure time for potential activity (see Figures 2, 3 and 8A), some substances/molecules with antimicrobial activity may not be an antibiofilm agent, and a compound with antibiofilm activity (e.g., effect on inhibiting microbial adhesion and/or extracellular matrix build-up) may not be an antimicrobial per se [33]. Here, this behavior was observed for AcOEt_BRA/DF and CsF. Although brief exposure to these treatments had no activity, when evaluated through the *S. mutans* growth inhibition curve, CsF interfered with the viability of *S. mutans* after 1 h of contact, and the mean recovery of surviving cells was constant until the 24 h test, which demonstrates a bacteriostatic effect. However, the bacterial survival profile obtained for AcOEt_BRA/DF showed cell death after 1 h of exposure, with a bactericidal effect (Figure 8A). Therefore, the isolated molecule would be expected to be more effective in eradicating the microorganisms than the fraction. However, isolated molecules are often not as effective as active fractions because the biological activity of the whole natural product results from the synergistic or additive interactions of different compounds in the mixture and not of a single active molecule [30]. SEM analysis demonstrates that these treatments disrupt *S. mutans* cells (in planktonic culture; Figure 8B), which indicates that the antimicrobial mechanism is mainly related to irreversible damage to the microbial cytoplasmic membrane.

Thus, planktonic cells of *S. mutans* were also treated with different agents with recognized targets (compound 1771, C135, myricetin, *tt*-farnesol, and NaF) and, after, evaluated for the expression of virulence genes via the gene expression profile to understand the possible mechanisms of action involved in the biological activity observed for CsF and AcOEt_BRA/DF. Exposure to the flavanoid myricetin (J10595) induced *eno* and *atpD* gene expression. This induction may have occurred because *S. mutans* has developed adaptive acid tolerance responses through the induction of multiple cellular pathways to tolerate the acidification of the acidic environment it produces during glucose consumption [34]. One of these defense pathways against environmental challenges, such as acid shock, is the bacterial cytoplasmatic membrane itself [34]. The *atpD* gene encodes a functional subunit of the F_0F_1-ATPase system that is membrane-bound and important for the survival of *S. mutans* under acidic stress. Therefore, induction of *atpD* expression demonstrates that cells reacted to intracellular (cytoplasmic) acidification caused by increased glycolytic activity (increased eno expression), and this induced *atpD* expression [35].

The gene *gtfB* encodes the GtfB enzyme, which metabolizes sucrose into water-insoluble glucans [36]. Here, we observed that CsF, *tt*-farnesol, and myricetin (J10595) induced the expression of this gene. This result was unexpected because the medium contained glucose and the substrate of the GtfB enzyme is sucrose to synthesize glucans, demonstrating that the microorganism was expressing a crucial gene for its survival in a cariogenic biofilm. It may be that the agents promote stress to bacterial cells in their free form (planktonic), and in response to this condition, the microorganism responds with an increase in the gene expression of *gtfB* as an attempt to produce glucans on the cell surface or to adhere to a surface to protect the cell from the stressor. However, myricetin (J10595) is an effective inhibitor of *gtfB* expression in solution, whereas *tt*-farnesol targets the cytoplasmatic membrane, decreasing acid tolerance and the Gtfs enzymes of *S. mutans* [37,38]. Thus, it appears that cells in biofilm and cells in their free form are affected differently by agents. Therefore, the *gtfB* expression profile here (Figure 9) differs from previous ones, where the agents were tested on biofilms through brief exposure [37,38]. Furthermore, inhibition/induction of gene expression and enzyme activity are not always correlated because several post-transcriptional regulatory processes can occur after mRNA is produced (as reviewed before [39].

For *lrgA*, there was an increase in gene expression after treatment with hydroxychalcone C135, compound 1771 (LTA metabolism inhibitor that affects the composition of the Gram-positive cell wall), and myricetin (J10595). The induction of *lrgA* expression and in the expression of genes related to membrane alterations (e.g., induction of *atpD* expression, observed for C135) demonstrate that these agents compromise the ability of *S. mutans* since LrgA (product of *lrgA*) is a membrane-associated protein and controls autolysis and cell death by modulating bacterial cell wall permeability [40,41]. Furthermore, changes in fatty acid profiles affect F-ATPase function and overall membrane permeability, altering the ability of *S. mutans* to maintain intracellular ΔpH, greatly impairing acid tolerance; this induces *atpD* expression, as observed for C135 and myricetin (J10595).

Changes in *nox1* gene expression alter the fatty acid composition of the microbial membrane and interfere with the activity of its product (Nox) [42]. This finding demonstrates that the Nox product can be directly or indirectly affected by C135, AcOEt_BRA/DF, and myricetin (J10595) through changes in membrane physiology, normal function of enzymes involved in glycolysis (enolase), and exopolysaccharides (GtfB) production because these agents downregulated *eno* expression. Alterations in the expression of this gene have a potentially lethal effect on S. *mutans* [43], as observed by the structural damage caused by AcOEt_BRA/DF (SEM images). Repression of *eno* by *tt*-farnesol (a terpenoid) and NaF was already expected since *tt*-farnesol reduces the glycolytic activity of *S. mutans* [30,37], and it is well established that enolase is a target of fluoride [44]. The *atpD* gene was downregulated after treatment with *tt*-farnesol, and there seems to be a reduction for AcOEt, although the magnitude was not relevant at this exposure time. This finding would reinforce that the treatments cause cytoplasmic acidification, which in turn impairs the normal function of the enzymes involved in glycolysis (enolase), as already demonstrated for *tt*-farnesol [30,37]. This outcome would explain the bactericidal effect of AcOEt and the structural damage, mainly in the cell wall of *S. mutans*, observed by SEM analysis.

CsF is a clerodane-type diterpene, and AcOEt fractions are rich sources of this secondary metabolite (also called casearins) [18]. Therefore, we propose that the findings of this study are closely correlated with this class of secondary metabolites. Studies regarding the chemistry of different diterpenes and the relationship of their molecular geometry versus their effectiveness in inhibiting the growth of *S. mutans* demonstrate that the structural characteristics are fundamental for the antimicrobial activity observed for these metabolites [45]. The antimicrobial properties of diterpenes are associated with their potential to promote bacterial lysis and rupture of the cytoplasmic membrane. This activity occurs through the structural characteristics of diterpenes, which include a lipophilic structure capable of insertion into the cell membrane and a hydrophilic fragment having a hydrogen-bonding donor group, which interacts with phosphorylated groups in the membrane [23]. Therefore, we hypothesize that the biological effect obtained for CsF and AcOEt_BRA/DF is associated with bacteriostatic and/or bactericidal capacity due to alterations in the membrane of *S. mutans* by the terpenoids present in these formulations. Furthermore, agent–membrane interactions may occur during cell division or cell wall remodeling, as these events facilitate the entry and/or interaction of the molecule with the cytoplasmic membrane. Unlike CsF which has bacteriostatic activity, the reason that AcOEt works as a bactericidal agent seems to be related to the presence of a new casearin not yet identified in the literature. Therefore, we are working on additional analyses to identify this metabolite.

The ability to effectively disrupt biofilm-specific and lifestyle-essential pathways of bacterial pathogens, all without affecting the viability of normal flora, is an attractive approach to the prevention and/or reduction of biofilm-related diseases, especially those that occur in complex microenvironments, such as the human mouth [30]. Considering the findings of this study, CsF and AcOEt_BRA/DF bring new and significant perspectives for developing selective antibiofilm/antimicrobial agents for potential applications in the prevention of cavities. Although the details of the cytotoxicity of these agents have not been investigated here, our previous study demonstrated low/moderate toxicity at higher concentrations of crude extracts of *C. sylvestris* in long-term exposure models [18]. In the future, these treatments will be evaluated for their cytotoxic activity through three-dimensional culture models.

Interpretation of in vitro results requires caution to avoid overestimation of observed effects. The rapid screening model on polystyrene plates was used here to select agents with potential antimicrobial and/or antibiofilm activity; however, they do not reflect the complex polymicrobial, ecological, and environmental influences found in the oral cavity. Thus, to verify the properties shown in our conditions, future in vitro tests will be carried out for drug delivery systems to prolong the exposure time at an adequate concentration using a polymicrobial (microcosms) model. Additionally, recognizing the common ability to interact with the cell membrane of *S. mutans*, we propose as future investigations the simulation of the insertion of these agents in a phospholipid membrane model to validate our conclusions about the structure–activity relationships of this class of compounds [23].

4. Materials and Methods

4.1. Plant Material

Samples of *C. sylvetris* were collected from four trees from three Brazilian regions belonging to the biomes: Cerrado and Atlantic Forest between September/October 2019 and December 2020 (Table 1; Figure S1). All samples were sent to Instituto Agronômico de Campinas (State of São Paulo, Brazil) for identity confirmation and variety assignment by Profa. Dr. Roseli B. Torres. The plant is registered in the National System for the Management of Genetic Resources and Associated Traditional Knowledge (SisGen; Registration n° A00892A), and the collections were authorized by the Brazilian Institute for the Environment and Renewable Natural Resources (IBAMA) through the Authorization and Biodiversity Information (SISBIO; registration n° 33429-1).

Table 1. Samples of *C. sylvestris* from different Brazilian biomes.

Population [a]	Sample Code	Variety [b]	Biome	Plant Part	Geographic Location
Araraquara/SP	ARA/SP	L	Cerrado	Leaves	Associação Servidores Campus Araraquara–Ascar (21°49′08.7″ S 48°11′48.1″ W)
Araraquara/SP	ARA/SP	L	Cerrado	Fruits	Associação Servidores Campus Araraquara–Ascar (21°49′08.7″ S 48°11′48.1″ W)
Araraquara/SP	ARA/SP	S	Cerrado	Leaves	Associação Servidores Campus Araraquara–Ascar (21°49′08.7″ S 48°11′48.1″ W)
Araraquara/SP	ARA/SP	I	Cerrado	Leaves	Institute of Chemistry, UNESP (21°48′25.0″ S 48°11′33.2″ W)
Brasília/DF	BRA/DF	L	Cerrado	Leaves	University of Brasília–UNB (15°45′49.4″ S 47°54′23.4″ W)
Presidente Venceslau/SP	PRE/SP	S	Atlantic Forest	Leaves	Km 622 of the Raposo Tavares Highway (21°51′36.0″ S 51°56′29.8″ W)
Presidente Venceslau/SP	PRE/SP	S	Atlantic Forest	Twigs	Km 622 of the Raposo Tavares Highway (21°51′36.0″ S 51°56′29.8″ W)

[a] Four individuals sampled; [b] varieties based on botanical classification: L: var. *lingua*; S: var. *sylvestris*; I: intermediate morphology.

4.2. Sample Preparation

The plant materials (fruits, leaves, and/or twigs) were dehydrated (40 °C in a circulating air oven) and then stored protected from light and at room temperature until use. Then, the samples were individually crushed, and 20 g of each sample was used to prepare the crude extracts, as described before [18]. First, the extracts were combined and lyophilized, producing seven lyophilized crude extracts. Then, these extracts were solubilized with 84.15% ethanol (EtOH; Sigma-Aldrich Co. St. Louis, MO, USA) and 15% dimethyl sulfoxide (DMSO; Sigma-Aldrich Co. St. Louis, MO, USA) and stored at −80 °C until biological assays.

The same plant material was used for a different approach to prepare crude extracts (combined and subjected to Speed Vac model SPD-Thermo Scientific to remove the extracting solvent) to optimize fraction yields. Then, two methodologies were adopted for fractionation (SPE-Si/C fractions). In the first method, fractionation was carried out from the lyophilized crude extracts, and then the resulting fractions were dried with a Speed Vac (methodology 1). In the second method, fractionation was performed from the crude extracts dried under a Speed Vac, and then the fractions were dried in a fume hood (methodology 2). The fractionation of crude extracts was conducted as described previously [18]. Briefly, a mixture of 40–63 μm, 60 Å silica gel (Merck, Darmstadt, Germany), and activated carbon (Labsynth, Diadema, Brazil) (1:1) was added to solid phase extraction (SPE) cartridges. Columns were preconditioned with 95:5 ethyl acetate (both from J.T. Baker, HPLC grade), and then 150 mg of samples were applied. First, fractions were eluted with 10 mL of 95:5 (% *v/v*) hexane/ethyl acetate (Hex fraction), 100% ethyl acetate (AcOEt fraction), and 100% ethanol (EtOH fraction), respectively. Then, the solvents were evaporated with a Speed Vac or under the fume hood, resulting in their respective dry fractions. Next, the fractions were solubilized with 84.15% EtOH and 15% DMSO and stored at −80 °C until the biological assays.

For the PRE/SP crude extracts, solid phase fractionation was performed using reversed-phase silica (Si-C18) (SPE-C18 fractions). The stationary phase consisted of 30 g of C18 silica, dry packed in a polypropylene tube with an internal diameter of 3.7 cm (Polygoprep® Silica 60–50 C18, 50 μm; Macherey-Nagel). For sample application and elution, the dry crude extract was dispersed in C18 and deposited on the adsorbent. First, the elution was carried out with Water/EtOH in the proportions 95:05, 60:40, 40:60, and 20:80 and then with pure EtOH, with the aid of a vacuum, using about 150 mL of each eluent. Next, the solvents were evaporated with a Speed Vac, and then, fractions were solubilized with 84.15% EtOH and 15% DMSO and stored at −80 °C until biological assays. After the elution of each crude extract and fraction, the working solution of each one was evaluated for their pH value to ensure that all treatments and vehicle control showed no difference concerning this parameter (Table S3 in the Supplementary Materials).

Four compounds were isolated from leaves extracts screened in a previous study [18]: the flavonoid 4 (rutin), and the clerodane diterpenes Caseargrewiin F, Casearins X and J (Table 2). The isolation of molecules was performed as described before [17,28]. Casearin substitutes are represented in Table 3.

Table 2. Structure of flavonoids and clerodane diterpenes from *C. sylvestris*.

Molecules	Molecular Geometry	Reference
Flavonoid 4 (Quercetin-3-O-rutinoside–Rutin)		[28]
Caseargrewiin F		[17]
Casearin X		
Casearin J		

Table 3. *C. sylvetris* clerodane diterpenes and their substituents R1 to R5.

Casearins	R1	R2	R3	R4	R5
Caseargrewiin F	OBu	OAc	OAc	OH	H
Casearin X	OBu	OBu	OAc	OH	H
Casearin J	OMe	OBu	OAc	OH	OBu

Casearin substitutes represented in Table 2 and used in the study, where: OBu = n-$C_3H_7CO_2$ group; OMe = OCH_3 (methoxy) group; OAc group = CH_3CO_2 (acetate); OH = O-H (hydroxyl) group; H = hydrogen (adapted from [17]).

4.3. Bacterium Strain and Growth Conditions

Stocks of *S. mutans* strain UA159 (ATCC 700610) stored at $-80\,^{\circ}$C (tryptic soy broth containing 25% glycerol; Synth, Diadema, SP, Brazil) were thawed and seeded on blood agar plates (5% blood of sheep; Laborclin, Pinhais, PR, Brazil) and incubated at 37 $^{\circ}$C, 5% CO_2 for 48 h (Thermo Scientific, Waltham, MA, USA). After, starter cultures were prepared using 10 colonies that were inoculated in tryptone-yeast extract broth (TY; 2.5% tryptone, 1.5% yeast extract, pH 7.0; Becton Dickinson and Company, Sparks, MD, USA) containing 1% glucose (Synth, Diadema, SP, Brazil), followed by incubation for 16 h (37 $^{\circ}$C, 5% CO_2). Then, the starter cultures were diluted 1:20 in fresh TY + 1% glucose and incubated until the middle of the logarithmic growth phase (optical density or $OD5_{40nm}$ 0.847 ($\pm$0.273) and colony forming units per milliliter (CFU/mL) 1.37 $\times$ 10^9 ($\pm$6.10 $\times$ 10^7); Kasvi spectrophotometer, Beijing, China)). Inoculums for the assays described here were prepared with a defined population of 2 $\times$ 10^6 CFU/mL in TY + 1% glucose for antimicrobial assays and TY + 1% sucrose (Synth, Diadema, SP, Brazil) for biofilm assays.

4.4. Antimicrobial Activity through Prolonged Exposure Model (24 h)

Antimicrobial activity was evaluated for crude extracts (500 μg/mL). These concentrations were used due to the yield of extracts and fractions and the scientific literature, which considers adequate studies with 1 mg/mL for extracts or 0.1 mg/mL for isolated molecules [46]. First, 100 μL of *S. mutans* cultures (2 $\times$ 10^6 CFU/mL) were transferred to 96-well microplates (Kasvi, Beijing, China) containing test concentrations of treatments or vehicle (control with diluent of treatments) and culture medium (TY + 1% glucose), totalizing 200 μL (resulting in 1 $\times$ 10^6 CFU/mL). All experiments contained the following controls: wells containing culture medium only, wells containing only the experiment inoculum (microbial growth control), and wells containing the inoculum plus vehicle or 0 μg/mL). Then, the microplates were incubated (24 h, 37 $^{\circ}$C, 5% CO_2). After, a visual analysis of the wells was performed (turbidity: microbial growth; clear: no growth), followed by the reading of OD_{562nm} readings (ELISA plate reader, Biochrom Ez, Cambourne, UK). Furthermore, to determine the viability of the microbial cells, an aliquot from each well was used for a serial 10-fold dilution (10^{-1} to 10^{-5}) in microtubes containing saline solution (0.89% NaCl; Química Moderna, Barueri, SP, Brazil) and 10 μL aliquots of each dilution plus undiluted culture were used for plating in duplicate on BHI agar plates (Himedia, Dindhori, Nashik, India) and incubated (48, 37 $^{\circ}$C, 5% CO_2) followed by colony counting. CFU data were transformed into $\log_{10}$ and analyzed versus the vehicle control. Each treatment was performed in triplicate on two separate occasions ($n = 2$) [18,29].

4.5. Antibiofilm Activity through Prolonged Exposure Models (24 h)

The antibiofilm activity of crude extracts 500 μg/mL, fractions (250 μg/mL and 500 μg/mL), and molecules (125 μg/mL) was investigated [18,46]. This analysis was performed using two treatment exposure settings: (I) activity against initial biofilm formation (incubation of agents with cells from the beginning of biofilm formation until analysis after 24 h) for crude extracts, Hex; AcOEt and EtOH fractions, and SPE-C18 fractions; (II) activity against pre-formed biofilm (24 h old biofilms were exposed to treatments for 24 h, yielding

48 h old biofilms) for AcOEt_BRA/DF and EtOH_BRA/DF, Flavonoid 4, Caseargrewiin F (or CsF), and Casearins X and J.

4.5.1. Activity against Initial Biofilm Formation (24 h Old Biofilms)

For 24 h old biofilms, treatments were introduced at 0 h, and biofilms were evaluated at 24 h of development to assess inhibition of biofilm formation. Biofilms were formed in polystyrene microplate wells to verify the viable population (CFU/biofilm) of bacterium cells in biofilms treated by crude extracts or fractions, as described before [29]. A 96-well plate was prepared as described in item 4.4, including the set of controls. However, here, the culture medium used was TY + 1% sucrose. The plate was incubated (24 h, 37 °C, 5% CO_2). Then, a visual analysis of wells was performed, and the plate was subjected to orbital shaking (5 min, 75 rpm, 37 °C; Quimis, G816 M20, São Paulo, Brazil). The culture medium with loose cells was aspirated and discarded. The biofilms remaining in the wells were washed (three times) with a pipette and 200 µL of 0.89% NaCl solution to remove non-adhered cells. Next, these biofilms were scraped with pipet tips five times with 200 µL of 0.89% NaCl, totalizing 1 mL of biofilm suspension (from each well). This biofilm suspension was placed in a microtube and subjected to serial dilutions (10^{-1} to 10^{-5}), which were plated, as were the undiluted biofilm suspensions. The BHI plates were incubated (48 h, 37 °C, 5% CO_2), followed by colony counting. Next, data CFU were transformed into $\log_{10}$ and analyzed compared to the vehicle control. Two independent experiments were performed in triplicate ($n = 2$).

4.5.2. Activity against Pre-Formed Biofilms (48 h Old Biofilms)

In this setting, the biofilms were formed in polystyrene microplate wells without the addition of any treatment or vehicle control. After 24 h, the formed biofilms were exposed to treatments for 24 h to determine the prevention of biofilm accumulation (48 h biofilms). Then, the biofilms were evaluated at 48 h of development to verify the inhibition of biofilm formation via viable population analysis (CFU/biofilm) [33]. For this evaluation, 50 µL of final inoculum of *S. mutans* (2×10^6 CFU/mL) and 50 µL of TY + 1% sucrose (to obtain 1×10^6 CFU/mL), and 50 µL of TY + 1% sucrose (to reach 1×10^6 CFU/mL) were added to wells of 96-well plates. The microplate was incubated (24 h, 37 °C, 5% CO_2) without any treatment or vehicle control. After incubation and biofilm formation, visual analysis was performed, followed by culture medium removal and washing of the remaining biofilms (three times with 0.89% NaCl solution). Next, fresh culture medium TY + 1% sucrose and test concentrations of treatments or the vehicle were added. For each experiment, the same controls described in item 4.4 were included. The microplate was incubated again (24 h, 37 °C, 5% CO_2). After incubation (when biofilms were 48 h old), the same processing protocol applied for 24 h old biofilms was conducted until obtaining 1 mL of biofilm suspension. The biofilm suspensions were sonicated (30 s, 7 w, Sonicator QSonica, Q125, Newtown, CT, USA) and subjected to serial dilutions (10^{-1} to 10^{-5}), which were plated in BHI agar (in addition to undiluted biofilm suspensions), followed by incubation (48 h, 37 °C, 5% CO_2) and colony counting. Next, data CFU were transformed into $\log_{10}$ and analyzed compared to the vehicle control. Two independent experiments were performed in triplicate ($n = 2$).

4.6. Analysis of the Effect of Topical Application (Brief Exposure Time) of Formulations on Biofilms Formed on sHA Discs

In this step, selected treatments ("promising" data for antibiofilm activity on initial and pre-formed biofilms) were used to prepare formulations (combined with and without sodium fluoride-NaF). These formulations were then tested on sHA discs and biofilms grown on them [47]. The selected fractions were AcOEt_BRA/DF, SPE-C18 Water/EtOH 40:60 (G3.3a), and Water/EtOH 60:40 (F4.2b), and the molecule was CsF.

Biofilm Formation on sHA Discs and Topical Treatments

Biofilms of *S. mutans* UA159 were grown on sHA discs (surface area of 2.93 ± 0.2 cm^2, Clarkson Chromatography Products Inc., South Williamsport, PA, USA) in batch cultures for 67 h, as detailed elsewhere [9]. Saliva and pellicle preparation were performed as described before [48]. Saliva was donated by two volunteers who did not undergo antibiotic therapy in the last three months. The volunteers rinsed their mouths with 5 mL of Milli-Q water, then masticated a portion of parafilm, collecting 5 mL of saliva into a collection tube, which was then discarded. Next, the volunteers continued masticating the parafilm (Parafilm M; Sigma-Aldrich Co., St. Louis, MO, USA) and collected saliva into an ice-chilled Falcon tube. The saliva samples from all volunteers were combined and diluted 1:1 with adsorption buffer (AB buffer: 50 mM KCl, 1 mM KPO$_4$, 1 mM CaCl$_2$, 1 mM MgCl$_2$, 0.1 mM PMSF, in dd-H$_2$O, pH 6.5). The samples were centrifuged ($3220 \times g$, 20 min, 4 °C; Centrifuge 5810R, Eppendorf), and the clarified portion was filtered-sterilized (0.22 µm low protein binding polyethersulfone membrane filter, Rapid-flow Nalgene Thermo Scientific, Monterrey, Mexico). Fresh saliva was used for pellicle formation and culture medium preparation at the start of the experiment. Aliquots of the prepared saliva samples were stored at -80 °C until use for culture media preparation. Then, sHA discs were vertically positioned in wells of a 24-well microtiter plate (Kasvi, Beijing, China) with the help of custom-made disc holders [6]. Next, the sHA discs were dip-washed into wells containing AB buffer and topically treated with the tested treatments or vehicle control, as summarized in Table 4.

Table 4. Formulations used to topically treat biofilms formed on sHA discs.

Treatments	Topical Treatment Time (min)
G3.3a—1000 µg/mL G3.3a—1000 µg/mL + NaF 250 G3.3a—500 µg/mL + F4.2b—500 µg/mL G3.3a—500 µg/mL + F4.2b—500 µg/mL + NaF 250 NaF 250 Vehicle	1.5
AcOEt_BRA/DF—250 µg/mL AcOEt_BRA/DF—250 µg/mL + NaF 250 AcOEt_BRA/DF—1000 µ/mL AcOEt_BRA/DF—1000 µ/mL + NaF 250 NaF 250	10
CsF—125 µg/mL CsF—125 µg/mL + NaF 250 CsF—250 µg/mL CsF—250 µg/mL + NaF 250 NaF 250 Vehicle	10

G3.3a denotes the fraction SPE-C18 Water/EtOH 40:60; F4.2b denotes the fraction SPE-C18 Water/EtOH 60:40; NaF 250 is 250 ppm sodium fluoride and diluent carrier 42.075% EtOH, 7.5% DMSO and 50% pH 6.0 buffer.

The surfaces of each sHA disc were subjected to topical treatment for 1.5 min or 10 min by dripping 300 µL of each treatment or control (two discs per treatment and experiment). Then, the discs were washed by immersion in wells containing AB buffer (to remove excess treatment) and transferred to wells containing *S. mutans* culture for biofilm formation (0 h biofilm formation). The biofilm inoculum was prepared with the cultures of *S. mutans*. After reaching the late exponential growth phase, the cultures were diluted in TY + 0.1% sucrose and 20% saliva (to obtain 2×10^6 CFU/mL). The biofilms were incubated for 6 h (37 °C, 5% CO$_2$), when the discs with initial biofilm were rinsed with 0.89% NaCl, treated with each corresponding treatment (as above), rinsed with 0.89% NaCl (to remove excess treatment), and transferred back to the corresponding culture media until biofilms were 19 h old, when culture medium was changed.

The culture medium was changed twice a day until the end of each experimental occasion: at 8 a.m. (TY + 0.1% sucrose and 20% saliva) and 4 p.m. (0.5% sucrose + 1% starch (Sigma-Aldrich Co., St. Louis, MO, USA) and 20% saliva). After each media change, the pH of the spent media was measured. The biofilms were topically treated two hours after each culture change (Figure S2). Biofilms were grown until 67 h for evaluation of bacterial population, biomass, biochemical characteristics of the extracellular matrix, and structure (confocal microscopy).

4.7. Biofilm Analyses

4.7.1. Biofilm Processing and Standard Methods of Biochemical Analysis (Colorimetric) and Microbial Culture Method

When they reached 67 h, the biofilms were processed for analyses following previously described protocols [48]. Standard methods of biochemical analysis (colorimetric) were used to determine total protein content (in the soluble portion and the insoluble portion; Figure A4), exopolysaccharides content (water-soluble and -insoluble) and a bacterial culture method to determine the biofilm biomass (dry weight) and population. Furthermore, the amount of eDNA in the matrix was evaluated [47]. Briefly, after 67 h of formation, the biofilms were washed by immersion in wells containing sterile 0.89% NaCl solution. Each biofilm (disc) was transferred to a glass tube containing 1 mL of saline. Then, the walls of each tube were washed with 1 mL of saline solution. The glass tubes with biofilms/discs were placed in a Beaker, and the set was sonicated in a water bath for 10 min (model CD-4820, Kondentech Digital, São Carlos, Brazil). Afterward, the surfaces of each disc were scraped with the aid of a sterile metal spatula, taking care to remove any remaining biofilm from each disc. The volume of each biofilm suspension (2 mL) was transferred to a new 15 mL tube. Then, each glass tube was washed with 3 mL of saline solution, which was transferred to the tube containing the initial 2 mL, totaling 5 mL of biofilm suspension per biofilm/disc. Each biofilm suspension (5 mL) was sonicated through a probe at 7 w for 30 s. An aliquot of each suspension (100 µL) was used for serial dilutions (10^{-1} to 10^{-5}) to determine the number of CFU/biofilm by plating on BHI agar plates (48 h, 37 °C, 5% CO_2). Next, data CFU were transformed into $\log_{10}$ and analyzed compared to the vehicle control. One or two independent experiments were performed in duplicate.

The remaining volume (4.9 mL) was centrifuged (3220× *g*, 20 min, 4 °C). The supernatant (with soluble extracellular matrix components) was transferred to a new tube (15 mL Falcon tube–supernatant), and the pellet (precipitate with the microbial cells and insoluble matrix components) was washed twice with 2.6 mL sterile Milli-Q water (3220× *g*, 20 min, 4 °C). The supernatants generated during the two washes were combined with the first supernatant obtained, totaling 10 mL, which was used to isolate and quantify water-soluble exopolysaccharides (6 mL plus 18 mL of 99% EtOH to precipitate the polysaccharides, followed by phenol-sulfuric colorimetric assay) [49], eDNA (500 µL) and proteins (500 µL) [9]. The pellet was suspended in 1 mL of Milli-Q water, of which 50 µL was used to quantify proteins and 950 µL to quantify insoluble dry weight (biomass), followed by the isolation and quantitation of water-insoluble exopolysaccharides (or alkali-soluble polysaccharides [48].

4.7.2. Laser Scanning Confocal Fluorescence Microscopy

For confocal microscopy analyses, biofilms were formed and treated with SPE-C18 (as described above), except that here, 1 µM Alexa Fluor™ 647-labeled dextran conjugate (absorbance/fluorescence emission maxima of 647/668 nm; Molecular Probes, Carlsbad, CA, USA) was added to the culture medium at the beginning of and during the development of the biofilms [50]. This method allows for the incorporation of labeled dextrans into exopolysaccharides during its synthesis process and matrix build-up. At 67 h of development, biofilms/discs were dip-washed into wells containing 0.89% NaCl and transferred to wells containing 0.89% NaCl solution and SYTO™ 9 (485/498 nm; Molecular Probes, Carlsbad, CA, USA), which is a green, fluorescent nucleic acid marker for detecting bacteria [50]. Each biofilm was scanned at three randomly chosen positions, and optical sectioning at each of these positions generated a series of confocal images. The image of the three-dimensional structure of these biofilms was performed using a Zeiss LSM 780 microscope (Zeiss, Jena, Germany) equipped with a Multialkali-PMT detector, 488 nm (SYTO9) and 561 nm (Alexa Fluor 647) laser, EC Plan-Neofluar objective of 20x, with a scale of 0.312 × 0.312 μm per pixel and increments of 1.5 μm. Images were acquired and analyzed using ZEN Blue 2.3 software for 3D reconstruction.

4.8. Growth Inhibition Curve for Selected Treatments Associated with Gene Expression Analysis

The fraction AcOEt_BRA/DF (250 μg/mL) showed activity against initial biofilm formation (24 h), and the molecule CsF (125 μg/mL) showed activity against pre-formed biofilm (48 h). Thus, they were tested in planktonic culture to determine *S. mutans* growth inhibition curve, using the cultures at the mid-log growth phase [51].

4.8.1. Growth Inhibition Curve

In a 48-well polystyrene microplate, 150 μL of *S. mutans* inoculum (prepared as described above) was added to each well of the plate containing the treatments at the concentrations to be evaluated. Two wells were prepared to contain the inoculum and culture medium (TY + 1% glucose) and one containing only the culture medium (control for visual analysis of bacterial growth). An aliquot of the inoculum was seeded to determine the amount of CFU/mL at 0 h (before starting incubation and adding treatments-inoculum control). After incubation for 1 h, 2 h, 4 h, 6 h, and 24 h (37 °C, 5% CO_2), visual observation and seeding of cultures on BHI agar plates were performed. For that, a 10 μL aliquot of the pure culture was seeded on a BHI agar plate, and 40 μL was removed from each well and transferred to microtubes containing 360 μL of 0.89% NaCl (dilution 1:10 *v/v*), followed by serial dilution and seeding on BHI agar plates. The plates were incubated (48 h, 37 °C, 5% CO_2). After that, colony counts were performed for the treatment and vehicle control, and the calculation of the log number of CFU/mL of the treatment was compared to the vehicle control. Two experimental occasions were carried out, in duplicate, for the treatments and control tested (*n* = 2). At 2 h, 4 h and 24 h an aliquot of AcOEt_BRA/DF was diluted 1:1 (*v/v*) in 2.5% glutaraldehyde, and the same procedure was performed for CsF at 4 h and 24 h. The samples were stored in a refrigerator for scanning electron microscopy (SEM) analysis.

4.8.2. Preparation of Cultures for SEM Analysis

Samples were diluted 1:11 *v/v* in 2.5% glutaraldehyde and were centrifuged (5 min, 15,294× *g*, 4 °C). Then, the supernatant was discarded, taking care not to detach the pellet, and 500 μL of 70% EtOH was added to the pellets, and the samples were incubated for 1 h at room temperature. Afterward, the samples were centrifuged (5 min, 15,294× *g*, 4 °C; Centrifugue 5430R, Eppendorf, Hamburg, Germany), the supernatant discarded, 500 μL of 90% EtOH was added to the pellets, and the samples were incubated for another 1 h at room temperature. Afterward, the samples were centrifuged (5 min, 15,294× *g*, 4 °C) and the supernatant was discarded. Then, the pellets were resuspended in 10 μL of absolute EtOH, and this volume was transferred to clean coverslips and placed in a 24-well plate. After complete evaporation of absolute EtOH, the plate was kept in a glass desiccator with silica until the analysis. Each sample was fixed with double-sided tape on the sample holder, and after, the samples were covered with carbon. The analysis using a high-resolution field emission electron microscope (MEV-FEG; JEOL, model JSM-7500F) with PC operating software PC-SEM v 2,1,0,3, equipped with secondary electron detectors, backscattered and chemical analysis (energy dispersive spectroscopy-EDS; Thermo Scientific, Ultra Dry model, USA) with NSS 2.3 operating software.

4.9. Inhibition of Growth by Different Compounds for Gene Expression Analysis

Lastly, to explain the difference in survival profile between the compounds, *S. mutans* planktonic cells were treated with different agents with recognized targets to know which possible targets were involved in the observed biological activity. The following agents were used: compound 1771 [(5-phenyl-1,3,4-oxadiazol-2-yl)carbamoyl]methyl 2-{naphtho [2,1-b]furan-1-yl}acetate) (UkrOrgSynthesis, Ltd., Kiev, Ukraine, catalog n° PB25353228; purity not available), 4′ hydroxychalcone (C135) [(2E)-1-(4-hydroxyphenyl)-3-phenylprop-2-en-1-one) (AK Scientific, Inc., Union City, USA, catalog n° C135; 98% purity), myricetin (J10595) [3,5,7-trihydroxy-2-(3,4,5-trihydroxyphenyl)-4H-chromen-4-one] (AK Scientific, Inc., catalog n° J10595; 95% purity), *tt*-farnesol [(E,E)-3,7,11-trimethyl-2,6,10-dodecatrien-1-ol, trans,trans-3,7,11-trimethyl-2,6,10-dodecatrien-1-ol] (Sigma-Aldrich Co., St. Louis, MO, USA, catalog n° 46,193; 96% purity)], sodium fluoride (Sigma-Aldrich, catalog n° 71519), chlorhexidine digluconate solution (Sigma-Aldrich, catalog n° C9394).

The agents and their concentrations were selected based on antimicrobial activity data previously tested in the laboratory [33,47,51]. The concentration of CHX was the same in mouthwashes commercialized for the control of biofilms, and sodium fluoride is the most seen in mouthwashes [52]. A starter culture was prepared as described above. For the inoculum, two 50 mL Falcon tubes were used to dilute the starter culture (1:20) using 2 mL of starter culture plus 38 mL TY + 1% glucose. When the appropriate OD was reached, the cultures were centrifuged (3220× *g*, 20 min, 4 °C), and the supernatants were discarded. Next, the pellet was resuspended with half the initial volume of TY + 1% glucose. This volume was divided into 12 tubes, then treatments were added, followed by incubation, but the time was different based on the survival curve (Table S4 in the Supplementary Materials). After incubation for the described time, an aliquot was removed for plating (to confirm the reduction in cell viability; Table S5 in the Supplementary Materials). The tubes were placed on ice for 15 min, centrifuged (3220× *g*, 20 min, 4 °C), the supernatants were discarded, and the pellets were resuspended with 1 mL of RNAlater (Ambion, Molecular Probes, Austin, TX, USA). Samples were frozen at −80 °C until RNA was isolated.

Gene Expression of *S. mutans*

The RT-qPCR (Reverse Transcription–quantitative Polymerase chain reaction) methodology included RNA isolation, cDNA synthesis, and gene expression analysis via qPCR of selected genes. Five specific genes were selected for expression profile analysis: genes associated with exopolysaccharides (*gtfB*, synthesis of insoluble glucans) and eDNA (*lrgA*) metabolism, acid stress tolerance (*atpD*), and acid and oxidative stress tolerance (*nox1*), with glycolysis (*eno*, enolase enzyme, fluoride target; Table 5) [6,51,53,54]. The 16S rRNA gene was included as an expression control (as a normalizer for the expression of specific genes) [55].

Table 5. Primers used for RT-qPCR.

Gene	GenBank Locus Tag	Sequence (forward and reverse)	Primer Concentration (ηM)	Product Size (bp)	Reference
16 S rRNA		ACCAGAAAGGGACGGCTAAC TAGCCTTTTACTCCAGACTTTCCTG	200	122	[6]
gtfB	SMU_1004	AAACAACCGAAGCTGATAC CAATTTCTTTTACATTGGGAAG	250	90	
nox1	SMU.765	GGACAAGAATCTGGTGTTGA CAATATCAGTCTCTACCTTAGGC	250	91	[54]
atpD	SMU.1528c	GGCGACAAGTCTCAAAGAATTG AACCATCAGTTGACTCCATAGC	250	115	
eno	SMU_1247	GTTGAACTTCGCGATGGAGAT GTCAAGTGCGATCATTGCTTTAT	250	150	[51]
lrgA	SMU_575c	GTCTATCTATGCTGCTATT AAGGACATACATGAGAAC	300	109	[53]

An optimized methodology for *S. mutans* was used to isolate the RNA, following a phenol-chloroform separation method, and the purification through the treatment with DNAse in column (Rneasy Micro Kit, Qiagen) and DNAse in solution (TURBO DNAse; Ambion) [56]. Afterward, spectrophotometry was used to evaluate the amount (ηg/μL-OD260nm) and purity (OD 260/280 ratio) of the RNA samples (Nano-spectrophotometer DS-11+, Denovix). After purification, the RNA was evaluated for integrity through 1% agarose gel electrophoresis (Ultra Pure Invitrogen). There was no adequate RNA yield for the CHX-treated culture, and this sample was excluded from the analyses. Samples were kept at $-80\ ^\circ$C until cDNA synthesis for RT-qPCR.

cDNA was synthesized (in triplicate per sample) using 0.25 μg of total RNA, and the High-Capacity cDNA Reverse Transcription kit (Thermo Fisher; catalog n° 4368814). Reactions containing all kit ingredients except the reverse transcriptase enzyme served as negative controls, determining whether there was DNA contamination. Reactions were incubated using the CFX96 Touch™ Real-Time PCR Detection System (Bio-Rad Laboratories, Hercules, CA, USA), using the protocol determined by the manufacturer, following the cycle: 10 min/25 $^\circ$C, 120 min/37 $^\circ$C, 5 min/85 $^\circ$C, ∞/4 $^\circ$C. Then, the samples were stored in a freezer at $-20\ ^\circ$C until their dilution (1:5 for specific genes; 1:1000 for 16S rRNA gene and cDNA negative controls were not diluted) and used to quantify gene expression. The cDNA dilutions and negative controls were also stored in a $-20\ ^\circ$C freezer.

The cDNA and negative controls were amplified by the CFX96 using specific primers from the literature and 2× SYBR PowerUp Green Master Mix (Thermo Fisher; catalog n° A25776), following previously determined protocols [54]. For each gene, a standard curve based on the purified PCR product of the target gene was included [57]. The reactions were incubated in the CFX96 thermocycler, using the following amplification cycle: 2 min/50 $^\circ$C; 2 min/95 $^\circ$C; 39 times: 0:15 min/95 $^\circ$C; 0:30 min/58 $^\circ$C and 1 min/72 $^\circ$C; 0:15 min/95 $^\circ$C; Melt Curve 60.0 $^\circ$C to 95.0 $^\circ$C: Increment 0.5 $^\circ$C 0:05. It was observed that for the *gtfB* and *lrgA* genes, the amplification was inadequate using the SYBR Thermo (i.e., the efficiency was lower than 90%), and so, the reaction was carried out with the iQ™ SYBR® Green Supermix, Bio-rad (catalog n° 170-8882) and its cycle. This amplification was adequate, as per the MIQE246 guidelines [58]. The standard curves were used to transform the Quantification Cycle (Cq) values to relative numbers of cDNA molecules. Next, a fold difference was calculated per agent/treatment vs. the vehicle control for each gene.

4.10. Data Analyses

The data obtained were submitted to descriptive statistical analysis to compare groups with the vehicle. Data were organized in a database (Excel). The graphical representation of the results was performed using the statistical software Prism 8 (GraphPad Software). Data from antimicrobial activity, antibiofilm activity (long exposure time), and growth inhibition curve were analyzed by comparing the evaluated groups with vehicle control (independent variables) as CFU/mL count (dependent variable) and gene expression (fold-change). The biofilm characterization data included the response variables: biomass (mg), population (CFU/biofilm), and extracellular matrix components (ASP (μg), WSP (μg), and eDNA (ηg)).

Supplementary Materials: The following supporting information can be downloaded at: https://www.mdpi.com/article/10.3390/antibiotics12020329/s1, Table S1. Activity against initial biofilm formation (24 h) of fractions of *C. sylvestris* against *S. mutans*; Table S2. Activity against initial biofilm formation (24 h) of fractions of *C. sylvestris* against *S. mutans*; Figure S1. Samples of leaves and fruits of specimens collected from *C. sylvestris* showing anatomical details; Table S3. pH values of crude extracts of *C. sylvestris*; Table S4. The treatment agents used and their respective incubation were different based on the survival curve; Table S5. Cell viability reduction ($\log_{10}$ CFU/mL) of *S. mutans* cultures, Figure S2. Experimental design for topical treatment regimen in biofilms formed on sHA discs.

Author Contributions: Conceptualization, M.I.K. and A.J.C.; methodology, S.M.R., M.I.K. and A.J.C.; software, S.M.R., M.I.K. and A.J.C.; validation, S.M.R., M.I.K. and A.J.C.; formal analysis, S.M.R., M.I.K., A.J.C. and P.C.P.B.; investigation, S.M.R., M.I.K. and A.J.C.; resources, S.M.R., M.I.K. and

A.J.C.; data curation; M.I.K. and A.J.C.; writing—original draft preparation, S.M.R. and M.I.K.; writing—review and editing, S.M.R., M.I.K., A.J.C. and P.C.P.B.; supervision, M.I.K. and A.J.C.; project administration, M.I.K.; funding acquisition, M.I.K. and A.J.C. All authors have read and agreed to the published version of the manuscript.

Funding: This research was funded by a research grant and overhead funds from the São Paulo Research Foundation (FAPESP) grant number 2019/23175-7 to S.R.M. and 2011/21440-3 to P.C.P.B. The Coordination for the Improvement of Higher Education Personnel (CAPES)–Finance Code 001 provided additional support.

Institutional Review Board Statement: The study was conducted in accordance with the Institutional Ethical Committee at São Paulo State University (Unesp), School of Dentistry, Araraquara (CAAE: 26015019.3.0000.5416; approved on 13 December 2019).

Informed Consent Statement: Written informed consent has been obtained from the participants that donated saliva to publish this paper.

Data Availability Statement: The data presented in this study are available on request from the corresponding author.

Acknowledgments: The authors would like to thank the volunteers who donated saliva for this study, the LMA-IQ for the availability to use the scanning electron microscope at the São Paulo State University (UNESP), Institute of Chemistry, Araraquara, and Glaucius Oliva, coordinator of CIBFar-Center for Innovation in Biodiversity and Pharmaceuticals (CEPID/FAPESP), and Vanderlan da Silva Bolzani, Coordinator of INCT BioNat.

Conflicts of Interest: The authors declare no conflict of interest. The funders had no role in the design of the study; in the collection, analyses, or interpretation of data; in the writing of the manuscript; or in the decision to publish the results.

Appendix A

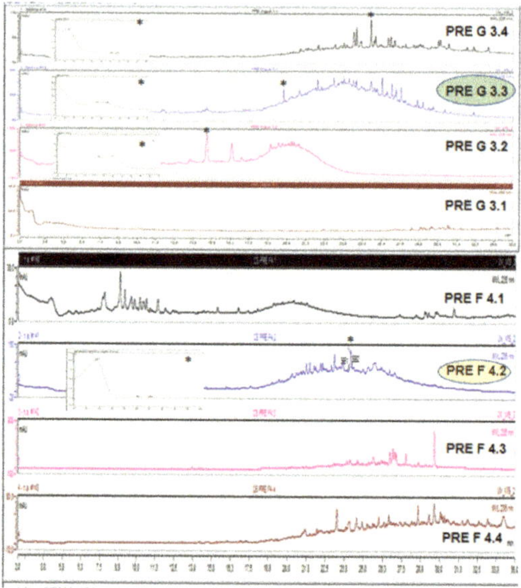

Figure A1. Chemical profile of the SPE-C18 fractions obtained from *C. sylvestris*. The chromatographic profiles of the SPE-C18 fractions indicate that they are rich in clerodane diterpenes and possibly tannins. Asterisks (*) indicate which chromatographic band corresponds to the UV spectrum.

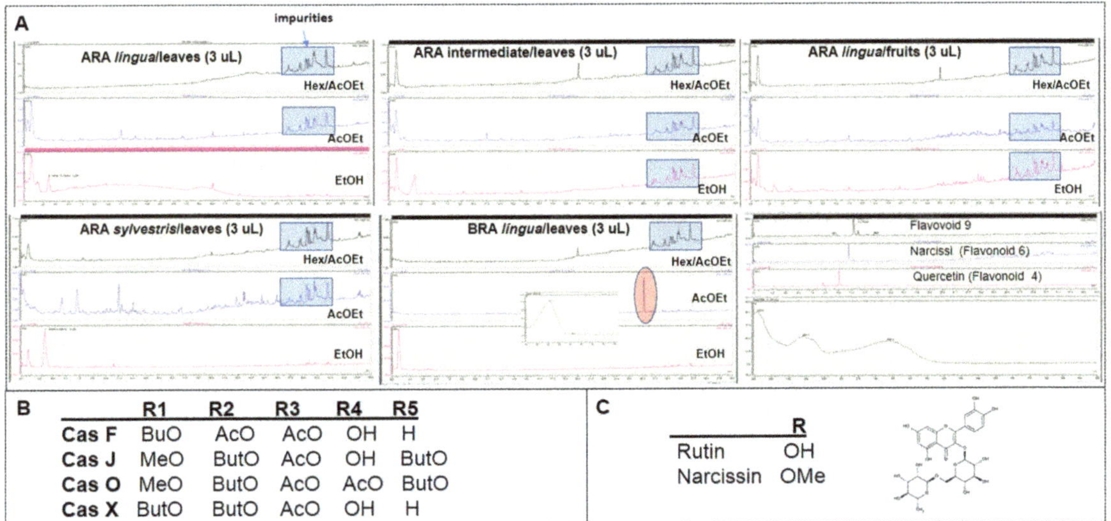

B

	R1	R2	R3	R4	R5
Cas F	BuO	AcO	AcO	OH	H
Cas J	MeO	ButO	AcO	OH	ButO
Cas O	MeO	ButO	AcO	AcO	ButO
Cas X	ButO	ButO	AcO	OH	H

C

	R
Rutin	OH
Narcissin	OMe

Figure A2. (**A**) The chemical profile of AcOEt fractions from *C. sylvestris* samples. (**B**) Substitutes of casearins, where: BuO = n-$C_3H_7CO_2$; MeO = OCH_3 (methoxy); AcO = CH_3CO_2 (acetate); OH = O-H (hydroxyl); H = hydrogen. (**C**) Structure of *C. sylvestris* flavonoids and their substituents. The chemical profile of AcOEt was obtained from *C. sylvestris*. The chemical profile of the AcOEt fractions for most samples of *C. sylvestris* showed bands that could not be related to the main secondary metabolites already described for this species, that is, flavonoids and clerodane diterpenes, with the exception of the AcOEt fraction obtained from the AcOEt extract leaves of *C. sylvestris* collected in Brasilia/DF. This fraction is enriched in a constituent whose UV spectrum is characteristic of casearins, with λmax at 241 nm.

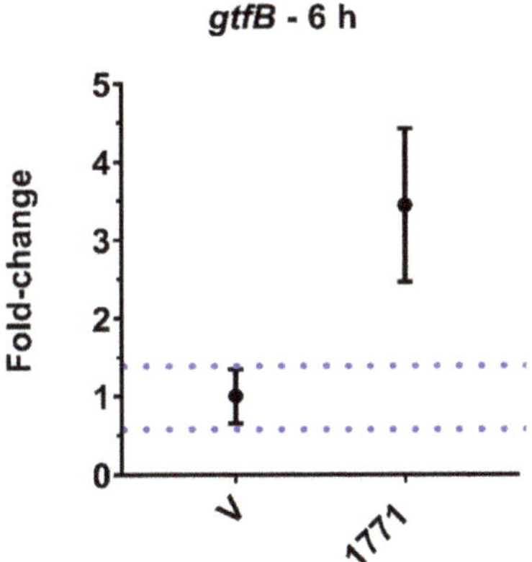

Figure A3. *S. mutans gtfB* expression of planktonic cultures. The central data are the mean, and the error bars are the confidence interval (95% CI). The dotted lines use vehicle data to compare the effectiveness of tested treatments.

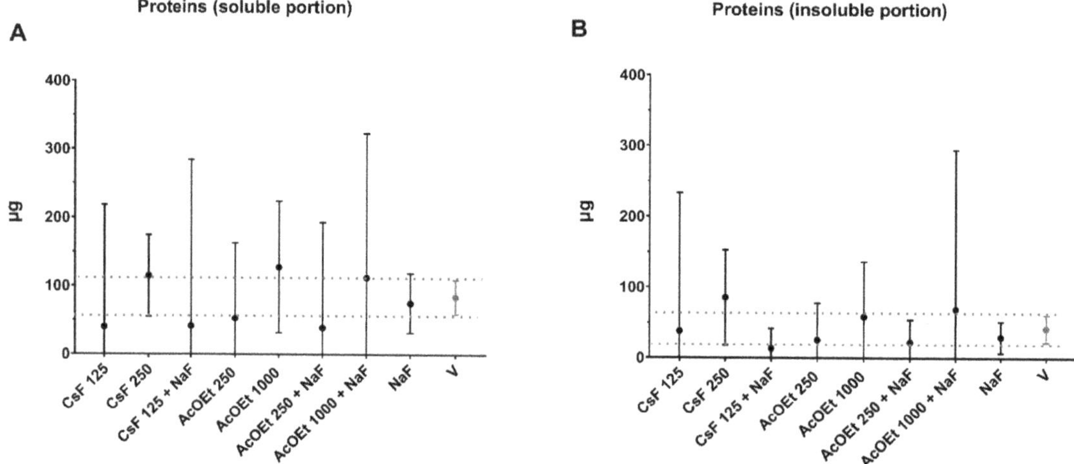

Figure A4. Total protein content (soluble and insoluble portions) of topically treated biofilms. The graphs show the total protein content in the soluble (**A**) and the insoluble portions (**B**) of topically treated biofilms. The central data are the mean, and the error bars are the confidence interval (95% CI). The blue dotted lines use vehicle data to compare the effectiveness of tested treatments. The vehicle control is represented as V (with the concentration being 21.04% EtOH and 3.75% DMSO for the AcOEt_BRA/DF fractions and 10.52% EtOH and 1.87% DMSO for CsF). The experiments for AcOEt_BRA/DF and CsF 125 µg/mL were performed in duplicate on one experimental occasion (n = 2) and in duplicate on two separate occasions (n = 4) for CsF 250 µg/mL.

References

1. Kassebaum, N.J.; Smith, A.G.C.; Bernabé, E.; Fleming, T.D.; Reynolds, A.E.; Vos, T.; Murray, C.J.L.; Marcenes, W.; GBD 2015 Oral Health Collaborators. Global, Regional, and National Prevalence, Incidence, and Disability-Adjusted Life Years for Oral Conditions for 195 Countries, 1990–2015: A Systematic Analysis for the Global Burden of Diseases, Injuries, and Risk Factors. *J. Dent. Res.* **2017**, *96*, 380–387. [CrossRef] [PubMed]
2. Bowen, W.H.; Burne, R.A.; Wu, H.; Koo, H. Oral Biofilms: Pathogens, Matrix, and Polymicrobial Interactions in Microenvironments. *Trends Microbiol.* **2018**, *26*, 229–242. [CrossRef] [PubMed]
3. Kolenbrander, P.E.; Palmer, R.J., Jr.; Periasamy, S.; Jakubovics, N.S. Oral multispecies biofilm development and the key role of cell–cell distance. *Nat. Rev. Microbiol.* **2010**, *8*, 471–480. [CrossRef]
4. Bowen, W.H.; Koo, H. Biology of *Streptococcus mutans*-Derived Glucosyltransferases: Role in Extracellular Matrix Formation of Cariogenic Biofilms. *Caries Res.* **2011**, *45*, 69–86. [CrossRef] [PubMed]
5. Leme, A.P.; Koo, H.; Bellato, C.M.; Bedi, G.; Cury, J.A. The Role of Sucrose in Cariogenic Dental Biofilm Formation—New Insight. *J. Dent. Res.* **2006**, *85*, 878–887. [CrossRef]
6. Xiao, J.; Klein, M.I.; Falsetta, M.L.; Lu, B.; Delahunty, C.M.; Yates, J.R.; Heydorn, A.; Koo, H. The Exopolysaccharide Matrix Modulates the Interaction between 3D Architecture and Virulence of a Mixed-Species Oral Biofilm. *PLoS Pathog.* **2012**, *8*, e1002623. [CrossRef]
7. Flemming, H.-C.; Wingender, J. The biofilm matrix. *Nat. Rev. Microbiol.* **2010**, *8*, 623–633. [CrossRef]
8. Klein, M.I.; Hwang, G.; Santos, P.H.S.; Campanella, O.H.; Koo, H. *Streptococcus mutans*-derived extracellular matrix in cariogenic oral biofilms. *Front. Cell. Infect. Microbiol.* **2015**, *5*, 10. [CrossRef]
9. Pedraza, M.C.C.; Novais, T.F.; Faustoferri, R.; Quivey, R.G., Jr.; Terekhov, A.; Hamaker, B.R.; Klein, M.I. Extracellular DNA and lipoteichoic acids interact with exopolysaccharides in the extracellular matrix of *Streptococcus mutans* biofilms. *Biofouling* **2017**, *33*, 722–740. [CrossRef]
10. Castillo Pedraza, M.C.; Rosalen, P.L.; de Castilho, A.R.F.; Freires, I.A.; de Sales Leite, L.; Faustoferri, R.C.; Quivey, R.G., Jr.; Klein, M.I. Inactivation of *Streptococcus mutans* genes lytST and dltAD impairs its pathogenicity in vivo. *J. Oral Microbiol.* **2019**, *11*, 1607505. [CrossRef]
11. Cury, J.A.; de Oliveira, B.H.; dos Santos, A.P.P.; Tenuta, L.M.A. Are fluoride releasing dental materials clinically effective on caries control? *Dent. Mater.* **2016**, *32*, 323–333. [CrossRef]
12. Mattos-Graner, R.O.; Klein, M.I.; Smith, D.J. Lessons Learned from Clinical Studies: Roles of Mutans Streptococci in the Pathogenesis of Dental Caries. *Curr. Oral Health Rep.* **2013**, *1*, 70–78. [CrossRef]

13. Tribble, G.D.; Angelov, N.; Weltman, R.; Wang, B.-Y.; Eswaran, S.V.; Gay, I.C.; Parthasarathy, K.; Dao, D.-H.V.; Richardson, K.N.; Ismail, N.M.; et al. Frequency of Tongue Cleaning Impacts the Human Tongue Microbiome Composition and Enterosalivary Circulation of Nitrate. *Front. Cell. Infect. Microbiol.* **2019**, *9*, 39. [CrossRef]
14. Freires, I.A.; Rosalen, P.L. How Natural Product Research has Contributed to Oral Care Product Development? A Critical View. *Pharm. Res.* **2016**, *33*, 1311–1317. [CrossRef]
15. Ferreira, P.M.P.; Costa-Lotufo, L.V.; Moraes, M.O.; Barros, F.W.; Martins, A.M.; Cavalheiro, A.J.; Bolzani, V.S.; Santos, A.G.; Pessoa, C. Folk uses and pharmacological properties of *Casearia sylvestris*: A medicinal review. *An. Acad. Bras. Cienc.* **2011**, *83*, 1373–1384. [CrossRef]
16. Xia, L.; Guo, Q.; Tu, P.; Chai, X. The genus Casearia: A phytochemical and pharmacological overview. *Phytochem. Rev.* **2014**, *14*, 99–135. [CrossRef]
17. Bueno, P.C.P.; Pereira, F.M.V.; Torres, R.B.; Cavalheiro, A.J. Development of a comprehensive method for analyzing clerodane-type diterpenes and phenolic compounds from *Casearia sylvestris* Swartz (Salicaceae) based on ultra high performance liquid chromatography combined with chemometric tools. *J. Sep. Sci.* **2015**, *38*, 1649–1656. [CrossRef]
18. Ribeiro, S.M.; Fratucelli, D.O.; Bueno, P.C.P.; De Castro, M.K.V.; Francisco, A.A.; Cavalheiro, A.J.; Klein, M.I. Antimicrobial and antibiofilm activities of *Casearia sylvestris* extracts from distinct Brazilian biomes against *Streptococcus mutans* and Candida albicans. *BMC Complement. Altern. Med.* **2019**, *19*, 308. [CrossRef]
19. Spósito, L.; Oda, F.B.; Vieira, J.H.; Carvalho, F.A.; Ramos, M.A.D.S.; de Castro, R.C.; Crevelin, E.J.; Crotti, A.E.M.; Santos, A.G.; da Silva, P.B.; et al. In vitro and in vivo anti-*Helicobacter pylori* activity of *Casearia sylvestris* leaf derivatives. *J. Ethnopharmacol.* **2018**, *233*, 1–12. [CrossRef]
20. Bueno, P.C.P.; Abarca, L.F.S.; Anhesine, N.B.; Giffoni, M.S.; Pereira, F.M.V.; Torres, R.B.; de Sousa, R.W.R.; Ferreira, P.M.P.; Pessoa, C.; Cavalheiro, A.J. Infraspecific Chemical Variability and Biological Activity of *Casearia sylvestris* from Different Brazilian Biomes. *Planta Med.* **2020**, *87*, 148–159. [CrossRef]
21. dos Santos, A.L.; Amaral, M.; Hasegawa, F.R.; Lago, J.H.G.; Tempone, A.G.; Sartorelli, P. (−)-T-Cadinol—A Sesquiterpene Isolated From *Casearia sylvestris* (Salicaceae)—Displayed In Vitro Activity and Causes Hyperpolarization of the Membrane Potential of *Trypanosoma cruzi*. *Front. Pharmacol.* **2021**, *12*, 734127. [CrossRef] [PubMed]
22. da Silva, S.L.; Chaar, J.D.S.; Damico, D.C.S.; Figueiredo, P.D.M.S.; Yano, T. Antimicrobial Activity of Ethanol Extract from Leaves of *Casearia sylvestris*. *Pharm. Biol.* **2008**, *46*, 347–351. [CrossRef]
23. Urzúa, A.; Rezende, M.C.; Mascayano, C.; Vásquez, L. A Structure-Activity Study of Antibacterial Diterpenoids. *Molecules* **2008**, *13*, 882–891. [CrossRef] [PubMed]
24. Tavares, W.L.F.; Apolônio, A.C.M.; Gomes, R.T.; Teixeira, K.I.R.; Brandão, M.G.L.; Santos, V.R. Assessment of the antimicrobial activity of *Casearia sylvestris* extract against oral pathogenic microorganisms. *Rev. Ciênc. Farm. Básica Apl.* **2008**, *29*, 257–260.
25. Saputo, S.; Faustoferri, R.; Quivey, R.G. A Drug Repositioning Approach Reveals that *Streptococcus mutans* Is Susceptible to a Diverse Range of Established Antimicrobials and Nonantibiotics. *Antimicrob. Agents Chemother.* **2018**, *62*, e01674-17. [CrossRef]
26. Van Dijck, P.; Sjollema, J.; Cammue, B.P.A.; Lagrou, K.; Berman, J.; D'Enfert, C.; Andes, D.R.; Arendrup, M.C.; Brakhage, A.A.; Calderone, R.; et al. Methodologies for in vitro and in vivo evaluation of efficacy of antifungal and antibiofilm agents and surface coatings against fungal biofilms. *Microb. Cell* **2018**, *5*, 300–326. [CrossRef]
27. Marino, M.J. The use and misuse of statistical methodologies in pharmacology research. *Biochem. Pharmacol.* **2014**, *87*, 78–92. [CrossRef]
28. Bueno, P.C.P.; Passareli, F.; Anhesine, N.B.; Torres, R.B.; Cavalheiro, A.J. Flavonoids from *Casearia sylvestris* Swartz variety lingua (Salicaceae). *Biochem. Syst. Ecol.* **2016**, *68*, 23–26. [CrossRef]
29. Ribeiro, S.M.; Fratucelli, D.O.; Fernandes, J.M.; Bueno, P.C.P.; Cavalheiro, A.J.; Klein, M.I. Systematic Approach to Identify Novel Antimicrobial and Antibiofilm Molecules from Plants' Extracts and Fractions to Prevent Dental Caries. *J. Vis. Exp.* **2021**, *169*, e61773. [CrossRef]
30. Jeon, J.-G.; Rosalen, P.; Falsetta, M.; Koo, H. Natural Products in Caries Research: Current (Limited) Knowledge, Challenges and Future Perspective. *Caries Res.* **2011**, *45*, 243–263. [CrossRef]
31. Sims, K.R.; Maceren, J.P.; Liu, Y.; Rocha, G.R.; Koo, H.; Benoit, D.S. Dual antibacterial drug-loaded nanoparticles synergistically improve treatment of *Streptococcus mutans* biofilms. *Acta Biomater.* **2020**, *115*, 418–431. [CrossRef]
32. Rocha, G.R.; Sims, K.R., Jr.; Xiao, B.; Klein, M.I.; Benoit, D.S. Nanoparticle carrier co-delivery of complementary antibiofilm drugs abrogates dual species cariogenic biofilm formation in vitro. *J. Oral Microbiol.* **2021**, *14*, 1997230. [CrossRef]
33. Lobo, C.; Lopes, A.; Klein, M. Compounds with Distinct Targets Present Diverse Antimicrobial and Antibiofilm Efficacy against *Candida albicans* and *Streptococcus mutans*, and Combinations of Compounds Potentiate Their Effect. *J. Fungi* **2021**, *7*, 340. [CrossRef]
34. Cotter, P.D.; Hill, C. Surviving the Acid Test: Responses of Gram-Positive Bacteria to Low pH. *Microbiol. Mol. Biol. Rev.* **2003**, *67*, 429–453. [CrossRef]
35. Kuhnert, W.L.; Zheng, G.; Faustoferri, R.C.; Quivey, R.G. The F-ATPase Operon Promoter of *Streptococcus mutans* Is Transcriptionally Regulated in Response to External pH. *J. Bacteriol.* **2004**, *186*, 8524–8528. [CrossRef]
36. Vacca-Smith, A.; Venkitaraman, A.; Schilling, K.; Bowen, W. Characterization of Glucosyltransferase of Human Saliva Adsorbed onto Hydroxyapatite Surfaces. *Caries Res.* **1996**, *30*, 354–360. [CrossRef]

37. Koo, H.; Schobel, B.; Scott-Anne, K.; Watson, G.; Bowen, W.; Cury, J.; Rosalen, P.; Park, Y. Apigenin and tt-Farnesol with Fluoride Effects on *S. mutans* Biofilms and Dental Caries. *J. Dent. Res.* **2005**, *84*, 1016–1020. [CrossRef]

38. Falsetta, M.L.; Klein, M.I.; Lemos, J.A.; Silva, B.B.; Agidi, S.; Scott-Anne, K.K.; Koo, H. Novel Antibiofilm Chemotherapy Targets Exopolysaccharide Synthesis and Stress Tolerance in *Streptococcus mutans* to Modulate Virulence Expression In Vivo. *Antimicrob. Agents Chemother.* **2012**, *56*, 6201–6211. [CrossRef]

39. Zhang, Q.; Ma, Q.; Wang, Y.; Wu, H.; Zou, J. Molecular mechanisms of inhibiting glucosyltransferases for biofilm formation in *Streptococcus mutans*. *Int. J. Oral Sci.* **2021**, *13*, 30. [CrossRef]

40. Esker, M.H.V.D.; Kovács, T.; Kuipers, O.P. From Cell Death to Metabolism: Holin-Antiholin Homologues with New Functions. *mBio* **2017**, *8*, e01963-17. [CrossRef]

41. Ahn, S.-J.; Rice, K.C.; Oleas, J.; Bayles, K.W.; Burne, R.A. The *Streptococcus mutans* Cid and Lrg systems modulate virulence traits in response to multiple environmental signals. *Microbiology* **2010**, *156*, 3136–3147. [CrossRef] [PubMed]

42. Derr, A.M.; Faustoferri, R.C.; Betzenhauser, M.J.; Gonzalez, K.; Marquis, R.E.; Quivey, R.G., Jr. Mutation of the NADH Oxidase Gene (nox) Reveals an Overlap of the Oxygen- and Acid-Mediated Stress Responses in *Streptococcus mutans*. *Appl. Environ. Microbiol.* **2012**, *78*, 1215–1227. [CrossRef] [PubMed]

43. Hamilton, I. Biochemical Effects of Fluoride on Oral Bacteria. *J. Dent. Res.* **1990**, *69*, 660–667. [CrossRef] [PubMed]

44. Guha-Chowdhury, N.; Clark, A.G.; Sissons, C.H. Inhibition of purified enolases from oral bacteria by fluoride. *Oral Microbiol. Immunol.* **1997**, *12*, 91–97. [CrossRef] [PubMed]

45. Chen, H.; Tan, R.X.; Liu, Z.L.; Zhang, Y.; Yang, L. Antibacterial Neoclerodane Diterpenoids from *Ajuga lupulina*. *J. Nat. Prod.* **1996**, *59*, 668–670. [CrossRef]

46. Eloff, J. Quantification the bioactivity of plant extracts during screening and bioassay guided fractionation. *Phytomedicine* **2004**, *11*, 370–371. [CrossRef]

47. Pedraza, M.C.C.; Fratucelli, E.D.D.O.; Ribeiro, S.M.; Salamanca, E.J.F.; Colin, J.D.S.; Klein, M.I. Modulation of Lipoteichoic Acids and Exopolysaccharides Prevents *Streptococcus mutans* Biofilm Accumulation. *Molecules* **2020**, *25*, 2232. [CrossRef]

48. Lemos, J.A.; Abranches, J.; Koo, H.; Marquis, R.E.; Burne, R.A. Protocols to Study the Physiology of Oral Biofilms. In *Oral Biology*; Part of the Methods in Molecular Biology Book Series; Humana: Totowa, NJ, USA, 2010; Volume 6, pp. 87–102. [CrossRef]

49. DuBois, M.; Gilles, K.A.; Hamilton, J.K.; Rebers, P.A.; Smith, F. Colorimetric method for determination of sugars and related substances. *Anal. Chem.* **1956**, *28*, 350–356. [CrossRef]

50. Klein, M.I.; Duarte, S.; Xiao, J.; Mitra, S.; Foster, T.H.; Koo, H. Structural and Molecular Basis of the Role of Starch and Sucrose in *Streptococcus mutans* Biofilm Development. *Appl. Environ. Microbiol.* **2009**, *75*, 837–841. [CrossRef]

51. de Araujo Lopes, A.C.U.; Lobo, C.I.V.; Ribeiro, S.M.; Colin, J.D.S.; Constantino, V.C.N.; Canonici, M.M.; Barbugli, P.A.; Klein, M.I. Distinct Agents Induce *Streptococcus mutans* Cells with Altered Biofilm Formation Capacity. *Microbiol. Spectr.* **2022**, *10*, e00650-22. [CrossRef]

52. Zero, D.T. Dentifrices, mouthwashes, and remineralization/caries arrestment strategies. *BMC Oral Health* **2006**, *6*, S9. [CrossRef]

53. Florez Salamanca, E.J.; Klein, M.I. Extracellular matrix influence in *Streptococcus mutans* gene expression in a cariogenic biofilm. *Mol. Oral Microbiol.* **2018**, *33*, 181–193. [CrossRef]

54. Klein, M.I.; Scott-Anne, K.M.; Gregoire, S.; Rosalen, P.L.; Koo, H. Molecular approaches for viable bacterial population and transcriptional analyses in a rodent model of dental caries. *Mol. Oral Microbiol.* **2012**, *27*, 350–356. [CrossRef]

55. Stipp, R.N.; Gonçalves, R.B.; Höfling, J.F.; Smith, D.J.; Mattos-Graner, R.O. Transcriptional analysis of gtfB, gtfC, and gbpB and their putative response regulators in several isolates of *Streptococcus mutans*. *Oral Microbiol. Immunol.* **2008**, *23*, 466–473. [CrossRef]

56. Cury, J.A.; Seils, J.; Koo, H. Isolation and purification of total RNA from *Streptococcus mutans* in suspension cultures and bio-films. *Braz. Oral Res.* **2008**, *22*, 216–222. [CrossRef]

57. Yin, J.L.; Shackel, N.A.; Zekry, A.; McGuinness, P.H.; Richards, C.; Van Der Putten, K.; Mccaughan, G.; Eris, J.M.; Bishop, G.A. Real-time reverse transcriptase–polymerase chain reaction (RT–PCR) for measurement of cytokine and growth factor mRNA expression with fluorogenic probes or SYBR Green I. *Immunol. Cell Biol.* **2001**, *79*, 213–221. [CrossRef]

58. Bustin, S.A.; Benes, V.; Garson, J.A.; Hellemans, J.; Huggett, J.; Kubista, M.; Mueller, R.; Nolan, T.; Pfaffl, M.W.; Shipley, G.L.; et al. The MIQE guidelines: Minimum information for publication of quantitative real-time PCR experiments. *Clin. Chem.* **2009**, *55*, 611–622. [CrossRef]

Article

Valorization of Invasive Plant Extracts against the Bispecies Biofilm *Staphylococcus aureus–Candida albicans* by a Bioguided Molecular Networking Screening

Guillaume Hamion [1,*], Willy Aucher [1], Charles Tardif [2,3], Julie Miranda [2,3], Caroline Rouger [2,3,4], Christine Imbert [1,†] and Marion Girardot [1,†]

1 Laboratoire EBI, University of Poitiers, UMR CNRS 7267, F-86000 Poitiers, France
2 University of Bordeaux, UMR INRAE 1366, Bordeaux INP, OENO, ISVV, F-33140 Villenave d'Ornon, France
3 Bordeaux Sciences Agro, UMR INRAE 1366, Bordeaux INP, OENO, ISVV, F-33170 Gradignan, France
4 Bordeaux Metabolome, MetaboHUB, PHENOME-EMPHASIS, Centre INRAE de Nouvelle Aquitaine-Bordeaux, F-33140 Villenave d'Ornon, France
* Correspondence: guillaume.hamion@univ-poitiers.fr
† These authors contributed equally to this work.

Citation: Hamion, G.; Aucher, W.; Tardif, C.; Miranda, J.; Rouger, C.; Imbert, C.; Girardot, M. Valorization of Invasive Plant Extracts against the Bispecies Biofilm *Staphylococcus aureus–Candida albicans* by a Bioguided Molecular Networking Screening. *Antibiotics* **2022**, *11*, 1595. https://doi.org/10.3390/antibiotics11111595

Academic Editors: Ding-Qiang Chen, Yulong Tan, Ren-You Gan, Guanggang Qu, Zhenbo Xu and Junyan Liu

Received: 21 October 2022
Accepted: 9 November 2022
Published: 11 November 2022

Publisher's Note: MDPI stays neutral with regard to jurisdictional claims in published maps and institutional affiliations.

Abstract: Invasive plants efficiently colonize non-native territories, suggesting a great production of bioactive metabolites which could be effective antibiofilm weapons. Our study aimed to look for original molecules able to inhibit bispecies biofilm formed by *S. aureus* and *C. albicans*. Extracts from five invasive macrophytes (*Ludwigia peploides*, *Ludwigia grandiflora*, *Myriophyllum aquaticum*, *Lagarosiphon major* and *Egeria densa*) were prepared and tested in vitro against 24 h old bispecies biofilms using a crystal violet staining (CVS) assay. The activities of the extracts reducing the biofilm total biomass by 50% or more were comparatively analyzed against each microbial species forming the biofilm by flow cytometry (FCM) and scanning electron microscopy. Extracts active against both species were fractionated. Obtained fractions were analyzed by UHPLC-MS/MS and evaluated by the CVS assay. Chemical and biological data were combined into a bioactivity-based molecular networking (BBMN) to identify active compounds. The aerial stem extract of *L. grandiflora* showed the highest antibiofilm activity (>50% inhibition at 50 $\mu g \cdot mL^{-1}$). The biological, chemical and BBMN investigations of its fractions highlighted nine ions correlated with the antibiofilm activity. The most correlated compound, identified as betulinic acid (BA), inhibited bispecies biofilms regardless of the three tested couples of strains (ATCC strains: >40% inhibition, clinical isolates: ≈27% inhibition), confirming its antibiofilm interest.

Keywords: antibiofilm; invasive plants; natural products; molecular networking; *Staphylococcus aureus*; *Candida albicans*; betulinic acid

1. Introduction

Biofilms are complex structures in which microorganisms belonging to different species can grow, proliferate, interact, communicate and acquire original attributes allowing them to tolerate or resist numerous conventional antimicrobial agents [1].

In human health, biofilms have been the subject of numerous studies over the last few decades to understand the origin of therapeutic failures and relapses in case of infection related to biofilms. Indeed, many fungal and bacterial infections can be associated with a biofilm which develops on a medical device or a biotic surface [1]. The structure and architecture of biofilms are becoming increasingly known, particularly concerning those formed by *Staphylococcus aureus* bacteria and *Candida albicans* yeasts. These two important and ubiquitous species are among the most studied microorganisms because of their frequency of isolation in the case of infections. The retrospective study by He et al. indicated that *C. albicans* was the third most common organism isolated on central venous catheters, after *Acinetobacter* and *Staphylococcus epidermidis*, and was the third most common species causing central line-associated bloodstream infection, just after *Acinetobacter* and *S. aureus*. In this last case, the authors found close prevalence for *S. aureus* (13.1%) and *C. albicans* (12.1%) [2].

C. albicans is a commensal species of the human oral cavity, gastrointestinal and reproductive tract. It is also an opportunistic pathogen capable of causing superficial to systemic and hematogenously disseminated candidiasis, depending on the patient's immune status and other predisposing factors [3–5]. In addition, candidiasis is often associated with a biofilm [6]. The Gram-positive bacterium *S. aureus* is a commensal species, as well as an opportunistic pathogen, responsible for superficial to life-threatening diseases, often related to a biofilm [7,8]. *C. albicans* and *S. aureus* share numerous host niches contributing to their frequent coisolation [9].

Not surprisingly, numerous cases of mixed *C. albicans–S. aureus* infections have been reported, including bloodstream infections [10–12]. In this latter case, numerous risk factors have been recently identified, such as a prolonged stay in an intensive care unit, antimicrobial administration and the presence of two or more central venous catheters [11].

The propensity of *C. albicans* and *S. aureus* to grow together within interkingdom biofilms is well documented, as well as their respective ability to protect each other from antimicrobial treatments [13]. Thus, the resistance and tolerance inherent in the biofilm lifestyle are worsened by the protection induced by the polymicrobial character of the biofilm, in which communication and interaction processes further strengthen the microorganisms [14,15]. Unfortunately, no treatment is yet available to prevent the severe mortality and morbidity associated to interkingdom biofilms formed by these two infamous species. It is therefore necessary and urgent to search for new and original molecules capable of destroying or inhibiting mixed *C. albicans–S. aureus* biofilms, and thus to fight against associated infections [13–15].

Invasive alien plants (IAPs) are a source of compounds of interest and could therefore meet this expectation. Invasive alien species are defined as "alien species that reach the final stage of the invasion process and have the capacity to spread [...] with highly detrimental impact in the regions concerned, not only on local biodiversity and on the way ecosystems work, but also on socioeconomic parameters, including animal production and hence animal health, and lastly on public health" [16].

According to the International Union for Conservation of Nature (IUCN), invasive alien species are "one of the biggest causes of biodiversity loss and species extinctions" [17]. IAPs often have no natural predators in their new environments and display a high capacity of dispersion and of forming a dense monospecific population entering into competition with the native plants. It is known, for example, that invasive *Ludwigia* species, because of their capacity to rapidly cover the entire surface of a body of water, lead to a modification of the environment that is harmful to the local fauna and flora [18].

However, despite these features and the economic, ecological and health-related negative impacts of invasive plants, they also constitute a reservoir of molecules with great potential. Indeed, their ease of adaptation, control of the new habitat and resistance to predators involve their chemical machinery. Some studies mention their capacity to synthesize new or more concentrated allelopathic, defense or antibiotic biochemicals resulting in a different chemical composition than that of native plants [19,20].

Thanks to these compounds, some IAPs have previously demonstrated antioxidant, antimicrobial, antiviral, neuroprotective, antiproliferative and cytotoxic, anticholinesterase activities [20]. Among them, several aquatic species have been highlighted. For example, invasive *Ludwigia peploides* (*Onagraceae* family) previously demonstrated antimicrobial, antioxidant and antiproliferative activities [21], while invasive *Ludwigia grandiflora* also demonstrated antibacterial activities against Gram-positive and Gram-negative bacteria [22].

Invasive plant extracts are still poorly investigated for their curative activities against polymicrobial biofilms, despite their presumed interesting potentials. Thus, this work aims to demonstrate the interest of invasive plants, in particular aquatic IAPs, in the discovery of compounds active against polymicrobial *S. aureus–C. albicans* biofilms. This work focuses on five IAPs registered on the list of plants of concern in France: *Egeria densa*, *L. grandiflora*, *L. peploides*, *Myriophyllum aquaticum* and *Lagarosiphon major*. Thus, this study could provide a new treatment to reinforce the therapeutic arsenal against biofilm and biofilm-related infections.

2. Results and Discussion

2.1. Plant Extracts

The five plants (*E. densa*, *L. grandiflora*, *L. peploides*, *M. aquaticum* and *L. major*) were identified, sampled and dried. Then, they were successively extracted using four solvents of increasing polarity (MeTHF, EtOAc, EtOH and EtOH/W, respectively). Forty extracts were obtained, and their yields are reported in Table 1. For each plant, the best yield of extraction was obtained with the mixture EtOH/W (2.9–19.7%). The solvents MeTHF and EtOH led to intermediate yields: (1.1–11.2%) and (0.5–11%), respectively. The solvents MeTHF and EtOAc have a close polarity index (4 and 4.4, respectively), which largely explains the low yield associated to EtOAc. Indeed, MeTHF already drained the compounds in this polarity range. Thus, the EtOAc extracts would have a limited interest. The best total yields were obtained with *Ludwigia* species and *M. aquaticum*, especially their leaves. *E. densa* and *L. major* contained fewer compounds extractable by these solvents.

Table 1. Extraction yields of different invasive plant parts, obtained with four solvents.

Plants and Parts [2] Extracted		Extraction Solvents				Total per Plant	
		MeTHF [1]	EtOAc	EtOH	EtOH/W		
		Yield (%)				Yield (%)	Weight (mg)
E. densa	WP	1.09	0.58	1.39	4.59	7.67	2142
L. major	WP	1.19	0.06	0.56	2.89	4.69	940
M. aquaticum	S	4.85	0.95	8.52	19.30	33.64	6731
	L	6.59	1.05	11.09	19.75	38.49	7701
	AS	2.54	0.30	3.41	16.09	22.37	4476
L. peploides	SS	2.18	0.19	3.07	11.42	16.87	3376
	L	11.21	0.16	6.44	19.62	37.42	7496
	AS	2.73	0.12	2.84	13.20	18.90	3783
L. grandiflora	SS	2.35	0.13	1.48	8.63	12.60	2522
	L	5.33	0.56	6.92	16.50	29.32	5873

[1] The used solvents are listed in running order: 2-methyltetrahydrofuran (MeTHF), ethyl acetate (EtOAc), ethanol (EtOH) and ethanol/water 50:50 (EtOH/W). [2] Abbreviation of parts: WP: whole plant, S: stems, L: leaves, AS: aerial stems, SS: submerged stems with roots.

These results suggested that polar components were present in great quantity in the studied plants. Other studies also observed that, after extraction by solvents of increasing polarity, the highest yields were obtained with the most polar solvents [23]. This observation is not surprising given that numerous primary and secondary metabolites commonly present in plants are polar, including sugars, amino acids, organic acids or most components of the large category of phenolic compounds [24].

2.2. Antibiofilm Activities Screening

The forty extracts were first screened for their activity against bispecies biofilms of *C. albicans* and *S. aureus*, using three concentrations ranging between 50 and 200 μg·mL^{-1}. The ability of these extracts to reduce already-formed 24 h old biofilms was investigated. The activity varied according to the extracts (Figure 1a), but in general, the results suggested that the least polar solvents used (MeTHF and EtOAc) were the most active against biofilms, except for *L. grandiflora* leaves (Lg-L) and *L. peploides* leaves (Lp-L), with MeTHF ones especially being the most active of all. Several studies have already shown the weakly polar nature of many compounds active against biofilms. For example, essential oils have shown a great activity against biofilm adhesion [25]. Some lipids are also known for having antibiofilm activities, especially against mixed *C. albicans* and *S. aureus* biofilms [26]. Concerning *L. grandiflora*, more polar extracts were also highlighted (EtOH and EtOH/W) but they were not obtained from the same part, which suggested that all parts were of interest in this plant: polar compounds from leaves and less polar compounds from AS and SS parts.

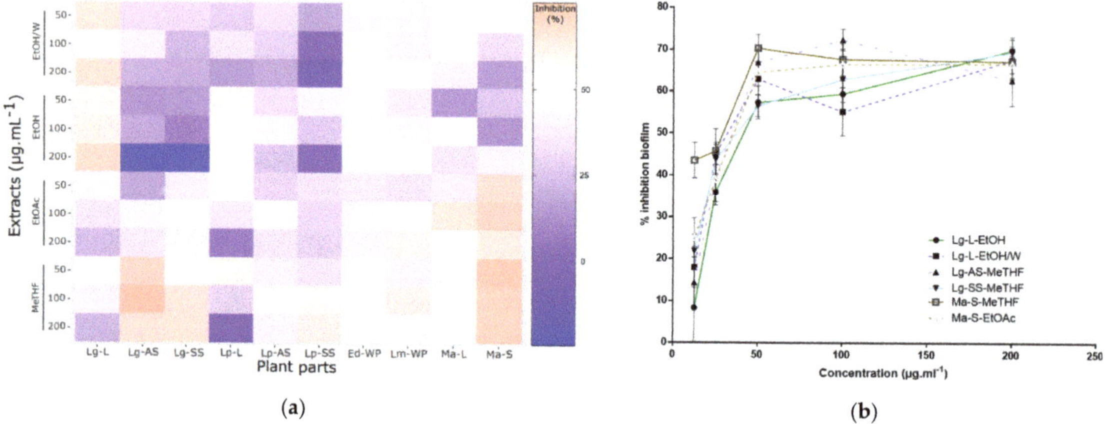

(a)   (b)

Figure 1. Plant extracts activity against *S. aureus–C. albicans* biofilm. Antibiofilm activities were assessed using crystal violet staining assay: (**a**) Heatmap displaying screening results of 40 plant extracts, with in ordinate the solvents of extraction and the final tested concentrations (50, 100, 200 µg·mL^{-1}), and in abscissa the extracted plant part; (**b**) Inhibition curves of the six most active extracts (6.25 to 200 µg·mL^{-1}), expressed in percentage of biofilm inhibition. Abbreviations: Lg (*Ludwigia grandiflora*), Lp (*Ludwigia peploides*), Ed (*Egeria densa*), Lm (*Lagarosiphon major*), Ma (*Myriophyllum aquaticum*), L (leaves), AS (aerial stems), SS (submerged stems with roots), S (stems), WP (whole plants).

Based on these results, completed by a Dunn statistical analysis, six extracts, from *L. grandiflora* leaves (Lg-L), aerial stems (Lg-AS), submerged stems with roots (Lg-SS) and *M. aquaticum* stems (Ma-S), demonstrated a significant antibiofilm activity compared to the nontreated control conditions and were therefore identified as promising: Lg-L-EtOH/W; Lg-L-EtOH; Lg-AS-MeTHF; Lg-SS-MeTHF; Ma-S-EtOAc; Ma-S-MeTHF. Their dose-dependency activity was then shown by testing concentrations above 50 µg·mL^{-1} (Figure 1b). Due to the closeness of the chromatographic profiles of the Ma-S-MeTHF and Ma-S-EtOAc extracts and the limiting amounts obtained for Ma-S-EtOAc extract, this last one was not further investigated. Thus, the antibiofilm activity of the remaining five selected extracts was then further detailed.

2.3. Characterization of Active Extracts

Through an FCM approach, we evaluated the activity of these five extracts specifically against the bacterial and fungal populations of bispecies biofilms. Indeed, the difference in cell size allowed us to distinguish these two populations, as we previously showed in another bispecies biofilm model [27]. The SYTO9 staining allowed counting the microorganisms obtained after biofilm scraping in order to compare the populations of the control biofilms (DMSO) to those treated by one of the five studied extracts. The number of bacterial cells counted was about 100 times as high as that of the fungal cells (Figure 2a,b). For the five extracts studied, only Lg-AS-MeTHF significantly reduced by a factor of three compared to the control for the bacterial ($p < 0.001$) and by a factor of two for the fungal ($p < 0.05$) population of the bispecies biofilms (Figure 2a,b). The Lg-SS-MeTHF and Lg-

L-EtOH extracts also reduced the *S. aureus* population ($p < 0.05$) but had no significant effect on the *C. albicans* one. Finally, the extracts Lg-L-EtOH/W and Ma-S-MeTHF were not active, regardless of the targeted population. Results obtained by CVS and FCM approaches may appear partially diverging. The large difference between the two populations can be explained by the shorter doubling time for bacteria than for yeast in the in vitro condition [28,29]. Moreover, differences between CVS and FCM approaches were not surprising, as these approaches targeted different constituents of the biofilm. The sonication performed before the FCM analyses eliminated the aggregates, resulting in single-cell suspensions. In addition, applied FCM settings allowed to provide quite strict microbial cell counts that excluded other constituents that may be present, such as matrix or cellular fragments and free components. In a different way, the CVS method required several successive washes that may detach the bacteria and yeasts less strongly attached to the biofilm. This method also tagged all constituents of the biofilm [30], thus giving a global view of the biofilm, wider than the microorganism's count. It is therefore possible that the extracts whose activity was not observed using FCM acted mainly on the matrix.

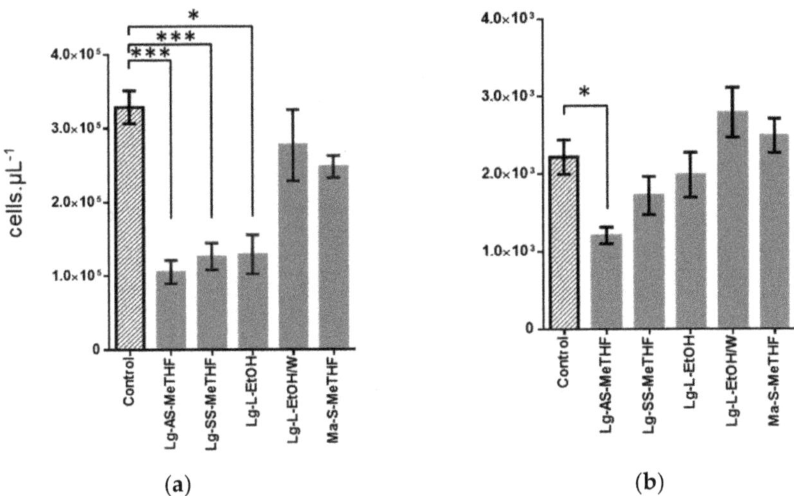

Figure 2. Inhibition activities of crude extracts at $100\ \mu g \cdot mL^{-1}$ on 48 h old bispecies biofilm of *C. albicans–S. aureus*: (**a,b**) enumeration of each population in bispecies biofilm (**a**) *S. aureus* and (**b**) *C. albicans* by FCM after 24 h of treatment with the five selected crude extracts, compared with control (DMSO 2% treatment) (error bars in SEM). *p*-value calculated by Dunn's test were given with *: $0.05 > p\text{-value} > 0.01$; ***: $0.001 > p\text{-value}$.

Addition of propidium iodide (PI) to the microbial suspensions analyzed by FCM allowed to evaluate the effect of the extracts on the cell membrane permeability. PI-labelled cells could be considered dead [27,31]. PI labelling did not reveal any difference between cells from treated biofilms and controls, regardless of the extract studied. This result suggested that the active extracts did not alter the membrane permeability, and thus that all microbial cells present in the cell suspensions analyzed by FCM were alive. This was consistent with the results of SEM observations of the bispecies biofilms treated or not with Lg-AS-MeTHF extract at $50\ \mu g \cdot mL^{-1}$ and $100\ \mu g \cdot mL^{-1}$. We did not observe any morphological modification of the cells after treatment, neither for bacteria nor for yeasts (Supplementary Materials Figure S1a,b). Unfortunately, this SEM approach did not allow any cell quantification to complete this result.

2.4. Bioactive Molecular Networking

In order to identify the largest possible panel of interesting compounds, the extracts both active with CVS and FCM methods most likely to have different compositions were selected for further investigation. As Lg-AS-MeTHF and Lg-SS-MeTHF presented similar HPLC profiles, suggesting a very close composition, the latter was not retained. The Lg-AS-MeTHF and Lg-L-EtOH extracts were finally considered for further investigation. Their fractionation resulted in seven and nine fractions, respectively (Table 2). The strongest antibiofilm activities (bispecies biofilm, CVS method) were associated with the fractions derived from the Lg-AS-MeTHF extract. Two of them exhibited comparable (F5) or even higher (F4) activities than the initial extract. Overall, the most active fractions of this extract were obtained with high percentages of acetonitrile (70–100%), suggesting the presence of moderately or highly apolar compounds. In the case of the Lg-L-EtOH extract, only two fractions (F3′ and F4′) exhibited activities comparable to that of the extract, while the others displayed lower activities. The Lg-AS-MeTHF extract was therefore considered the most promising. As fractions and extract were tested at the same concentration, it can be assumed that the fraction 4 was enriched in the compound(s) active against biofilm. A bioactive score was calculated for each fraction and extract. This score considered the variability of biological tests and was defined as the probability for a molecule of being bioactive [32]. These scores allowed composition–activity correlation studies that aimed to highlight compounds supposed to be responsible for, or involved in, the activity of the plant. For this purpose, the chemical profiles of the Lg-AS-MeTHF extract and all its fractions were analyzed, regardless of their bioactive score. Concerning the fractions from the Lg-L-EtOH extract, only the inactive ones (score ≤ 15) were considered so that the analysis was as discriminating as possible. This was made possible since the two selected extracts were prepared from the same plant.

Table 2. Yields, antibiofilm activity and bioactivity scores calculated for *L. grandiflora* Lg-AS-MeTHF and Lg-L-EtOH fractions and crude extracts.

		Yield	Activity (CVS, 50 µg·mL^{-1})		Bioactive Score
		(%)	Inhibition (%)	SD (%)	
Lg-AS-MeTHF (238 mg)	F1	7.11	0	20.03	0
	F2	1.50	0	27.50	0
	F3	2.59	30.82	9.89	37
	F4	1.10	63.96	6.89	100
	F5	1.77	50.83	8.16	74
	F6	1.87	39.19	13.63	44
	F7	14.54	25.61	12.55	23
	Extract		53.88	7.96	80
Lg-L-EtOH (600 mg)	F1′	15.71	0	21.20	0
	F2′	1.84	10.39	18.70	0
	F3′	9.13	33.85	13.96	34
	F4′	4.14	39.26	9.52	52
	F5′	1.50	27.78	8.94	33
	F6′	1.61	25.74	16.96	15
	F7′	4.05	0	24.28	0
	F8′	1.76	20.19	12.55	13
	F9′	7.13	15.56	20.20	0
	Extract		42.25	7.85	60

The samples (Lg-AS-MeTHF extract and its fractions 1–7, fractions 1′–2′ and 6′–9′ of Lg-L-EtOH) were analyzed by LC-HRMS^2 in positive and negative modes in order to cover as many compounds as possible. This approach also facilitated the identification of compounds detected in both active and negative modes. After processing the data on

MZmine2, a total of 993 positive precursor ions and 4735 negative ones were detected in the samples for MS[1]. The data obtained in positive and negative ionization were treated separately. Finally, an antibiofilm activity score was assigned to each sample and the area under the curve was calculated for each compound detected. As a result, nine compounds correlated with antibiofilm activity were found, seven in positive mode (LUg1 to LUg7) and two in negative mode (LUgA and B).

Molecular networks were constructed to partially or completely elucidate the structure of these nine ions, represented with their correlation scores and their identification from GNPS databases and/or from manual annotation using in silico fragmentation software MetFrag (Figure 3).

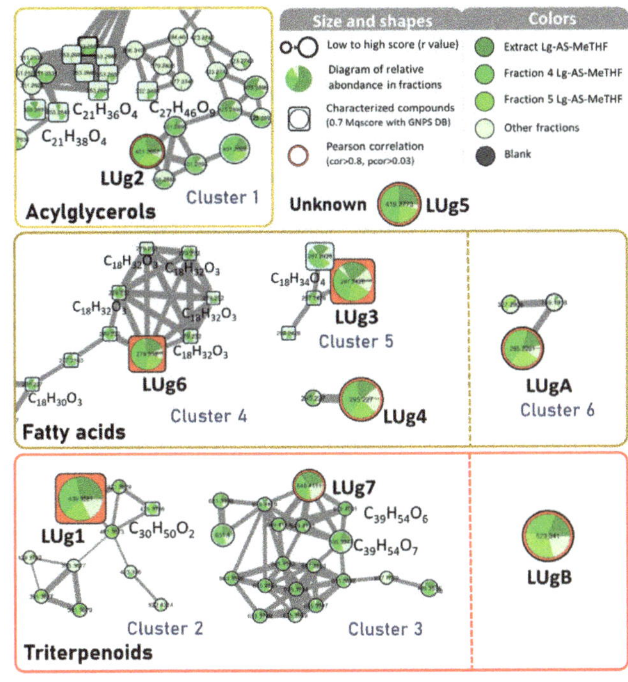

(a)

Detection	Name	r	*p*-value	*m/z*	RT (min)	Adduct	Predicated formula	Database identifications
Positive	LUg1	0.95	1.09×10^{-4}	439.36	18.62	$[M-H_2O+H]^+$	$C_{30}H_{48}O_3$	Betulinic acid
	LUg2	0.92	7.8×10^{-4}	401.27	16.95	$[M+Na]^+$	$C_{23}H_{38}O_4$	2-Arachidonoylglycerol
	LUg3	0.92	8.65×10^{-4}	297.24	16.89	$[M+H]^+$	$C_{18}H_{32}O_3$	12,13-Epoxy-9-octadecenoic acid
	LUg4	0.90	5.90×10^{-3}	295.23	15.77	$[M+H]^+$	$C_{18}H_{30}O_3$	C18:3 monohydroxylated fatty acid
	LUg5	0.89	6.47×10^{-3}	419.28	18.78	$[M+Na]^+$	$C_{23}H_{40}O_5$	Unknown
	LUg6	0.89	9.68×10^{-3}	279.23	15.72	$[M-H_2O+H]^+$	$C_{18}H_{32}O_3$	9,10-Epoxy-12-octadecenoic acid
	LUg7	0.87	2.32×10^{-2}	649.41	17.97	$[M+H]^+$	$C_{40}H_{56}O_7$	pentacyclic triterpenoid esterified with cinnamic acid derivated
Negative	LUgA	0.89	1.23×10^{-2}	295.23	15.44	$[M-H]^-$	$C_{18}H_{32}O_3$	9,10-Epoxy-12-octadecenoic acid
	LUgB	0.92	1.72×10^{-3}	523.34	18.62	$[M+HCOONa-H]^-$	$C_{30}H_{48}O_3$	Betulinic acid

(b)

Figure 3. Feature-based molecular networks corresponding to *L. grandiflora* fractions and extracts. (a) Molecular networks generated from positive and negative ionization mode, with legend of represented ions. (b) List of bioactive correlated ions including correlations scoring (r), *p*-value, *m/z* value, retention time (RT), adduct ions detected, predicted chemical formula and identification predicted by databases (GNPS, MetFrag).

Of the seven positive ions correlated with the bioactivity (Figure 3a), the most correlated one called LUg1 (correlation score r at 0.95, with *p*-value at 1.1×10^{-4}) was detected at *m/z* 439.3581 and a retention time (RT) of 18.62 min. GNPS databases identified it as an $[M-H_2O+H]^+$ ion with the calculated chemical formula $C_{30}H_{47}O_2$ corresponding to the pentacyclic triterpenoid betulinic acid (BA). In the same network as BA, another triterpenoid matched with the database search: the betulin ($C_{30}H_{50}O_2$), displaying a structure close to BA with a primary alcohol instead of carboxylic acid function on the 28th carbon (Figure S2). BA was available as commercial standard (Sigma-Aldrich, Saint Louis, MO, USA), and in order to confirm its identification, the HPLC-UV (210 nm) and targeted HPLC-MS/MS profiles of F4 and the Lg-AS-MeTHF extract were compared with the standard. Results showed a peak at the same retention time (Figure S3a) and a similar MS/MS fragmentation profile (*m/z* fragments and their relative intensity) (Figure S3b), which supported the BA identification hypothesis.

Two other putative bioactive compounds highly correlated, LUg3 (*m/z* 297.2426 at 16.89 min) and LUg6 (*m/z* 279.2319 at 15.72 min), were identified in the GNPS database as epoxidized fatty acids, respectively, as $[M+H]^+$ of 12,13-epoxy-9-octadecenoic acid and $[M-H_2O+H]^+$ 9,10-epoxy-12-octadecenoic acid. LUg3 was networked with three ions with the same *m/z* of 297.2426, including a hypothetical stereoisomer (RT at 17.1 min), $[M-H_2O+H]^+$ 9,10-dihydroxy-12-octadecenoic acid (RT at 16.95 min) and $[M+H]^+$ 9,10-epoxy-12-octadecenoic acid (RT at 13.63 min) (Table S1). Lug6 was also networked with identified compounds on GNPS: as probably a stereoisomer ($[M-H_2O+H]^+$ 9,10-epoxy-12-octadecenoic acid, RT at 15.44 min), five closed structures of linolenic acid ($[M+H]^+$ *m/z* 279.2320 at 17.97, 18.1, 18.35, 18.82 min), $[M-H_2O+H]^+$ 9-hydroxy-10,12,15-octadecatrienoic acid (*m/z* 277.2163 at 16.31 min) and two putative isomers of $[M+H]^+$ 9-oxo-10,12-octadecadienoic acid (*m/z* 295.2270 at 16.14 and 16.33 min). These networks of fatty acids and derivatives confirmed the nature of the LUg3 and LUg6 structures. However, the complexity of their determination and difficulties for synthetizing standards would require the isolation and investigation of their absolute structure.

The remaining bioactivity correlated ions were not identified within the GNPS database but through in silico fragmentation, either by direct comparison on MetFrag with their MS/MS spectra and/or by comparing the MS/MS spectra of their networked ions, thus helping to improve their identifications. For LUg2 (*m/z* 401.2667 at 16.95 min), with the chemical formula $C_{23}H_{38}O_4Na$ ($\Delta_{m/z}$ = 1.169 ppm), the postanalysis attributed 37 out of 59 similar fragments on its MS/MS spectrum with theoretical MS/MS spectrum of $[M+Na]^+$ 2-arachidonoylglycerol ($C_{23}H_{38}O_4$). On the LUg2 cluster, five putative stereoisomers of $[M+H]^+$ monolinolenin (*m/z* 353.2687 at 12.63, 13.07, 13.33, 14.53, 14.64 min), $[M+H]^+$ 1-linoleoylglycerol (*m/z* 355.2837 at 14.08 min) and $[M+NH_4]^+$ 9,12,15-octadecatrienoic acid, 3-(hexopyranosyloxy)-2-hydroxypropyl ester (*m/z* 532.3484 at 14.61 min) were identified with the GNPS database (Supplementary Materials Table S1). All these compounds were composed of a glycerol part esterified with a long chain unsaturated fatty acid. These assumptions are consistent with the identification of LUg2 as 2-arachidonoylglycerol (Figure S2).

LUg4 (*m/z* 295.2269 at 15.77 min) was found to be related to the in silico fragmentation of $[M+H]^+$ 17-hydroxyoctadeca-9,11,13-trienoic acid (47 out of 69 fragments in MS/MS spectrum). In parallel, a negative ion at the same retention time was detected (*m/z* 293.2124 at 15.76 min) (Figure 3b), and the MS/MS comparison gave $[M-H]^-$ 2-hydroxylinolenic acid (12/31 similar fragments), implementing the hypothesis that LUg4 was a C18:3 monohydroxylated fatty acid.

In the same way, LUg7 (*m/z* 649.4100 at 17.97 min) had a high number of similar peaks, with an in silico spectrum of eucalyptic acid (52 matches/79 peaks), 3-*O*-feruloyl-2-hydroxy-12-ursen-28-oic acid (52/79) and 11-hydroxy-10-{[3-(3-hydroxy-4-methoxyphenyl)prop-2-enoyl]oxy}-1,2,6a,6b,9,9,12a-heptamethyl-1,2,3,4,4a,5,6,6a,6b,7,8,8a,9,10,11,12,12a,12b,13,14b-icosahydropicene-4-carboxylic acid (58/79). These three molecules have $C_{40}H_{56}O_7$ as a molecular formula and consist of a pentacyclic triterpenic part esterified with a phenolic acid (ferulic or isoferulic acid). Two networked compounds with LUg7 were assimilated to similar structures: m/z 619.4001 at 17.75 min as 2-*O*-*p*-coumaroyl alphitolic acid (45/60) ($C_{39}H_{54}O_6$) and m/z 635.3947 at 16.98 min as 3-caffeoyloxy-2-hydroxyurs-12-en-28-oic acid (38/58) ($C_{39}H_{54}O_7$). Moreover, the negative ion of LUg7 at m/z 647.3964 showed spectral similarities with the 3-alpha-*O*-*trans*-feruloyl-2-alpha-hydroxy-12-ursen-28-oic acid and the *trans*-3-feruloylcorosolic acid (10/23 common fragments, molecular formula $C_{40}H_{56}O_7$) (Figure 3a). All these similarities converged towards the hypothesis of a pentacyclic triterpenic structure esterified with a cinnamic acid derivative for LUg7. Seven other networked compounds matched with the GNPS databases and were identified as triterpenoïds (Table S1).

LUg5 (m/z 419.2773 at 18.78 min) showed a low number of peaks in the MS/MS spectrum, but an intense ion at m/z 207.0995 (100% relative intensity) and another main one at m/z 335.2196 (7.5%). The calculated raw formula for the parent ion was $C_{23}H_{40}O_5Na$ ($\Delta_{m/z}$ = 0.965 ppm). Comparisons with databases gave no similarity with known natural compounds, suggesting the presence of a new compound.

Finally, the last two compounds correlated in the molecular network with the negative ionisation mode, LUgA (m/z 295.2281 at 15.44 min) and LUgB (m/z 523.3410 at 18.62 min), shown in Figure 3a, which exhibited similarities with the ions already described. LUgA was the $[M-H]^-$ adduct of LUg6. Its MS^1 calculated molecular formula ($C_{18}H_{31}O_3$) and the in silico comparison of its MS^2 spectrum resulted in its identification as $[M-H]^-$ 9,10-epoxy-12-octadecenoic acid (12/31 common fragments). LUgB showed the same retention time as LUg1 (BA), with a major MS^2 peak at m/z 455.3535, assuming that LUgB is the $[M+HCOONa-H]^-$ adduct of BA.

To summarize, the activity of Lg-AS-MeTHF is correlated with two families of molecules. The first one is lipids including acylglycerols and derivatives of fatty acids, especially hydroxylated and epoxidized derivatives. Epoxidized and hydroxylated fatty acids are often found in plants and are notably part of oxylipins biosynthesis (enzymatically oxygenated fatty acids), metabolites involved in intra/intercellular communication in plants [33]. The presence of free fatty acids as linoleic, linolenic and arachidonic acids was also confirmed in the close species *Ludwigia octovalvis* [34]. Several studies highlighted the antibiofilm activities of fatty acids such as linoleic acid and their derivatives on Gram-positive bacteria (including *S. aureus*) [35,36]. Moreover, fatty acids would have effects on intraspecies microbial communication as *C. albicans* quorum sensing. Indeed, they can, for example, mimic the effect of farnesol, an important signal molecule secreted by *Candida* yeasts, making the study of fatty acids interesting to target the yeast hyphal form [37,38]. The second chemical family of interest concerns pentacyclic triterpenes such as betulinic acid, a secondary metabolite of many plants involved in the biosynthesis of saponins, natural surfactant biomolecules [39]. BA and betulin were also identified in the *Ludwigia adscendens* extract [40]. Several pentacyclic triterpenoids, such as glycyrrhetinic acid, ursolic acid and BA, have been shown to display antibiofilm effects against Gram-negative bacteria *Acinetobacter baumannii* and *Pseudomonas aeruginosa* [41]; an activity of BA and several derivatives was also described on the *S. aureus* biofilm [42].

Among the nine correlated ions, BA appeared to be the best candidate molecule to be involved in the antibiofilm activity of the *L. grandiflora* extract. Its activity was investigated on the bispecies biofilm to confirm this hypothesis using commercial pure BA.

2.5. Antibiofilm Activity and Quantification of Betulinic Acid

BA effects on 24 h mature biofilms were assayed using the same protocol as for plant extracts, with a CVS measurement of the total biomass after treatment. Different concentrations, ranging from 6.25 to 50 $\mu g \cdot mL^{-1}$ (13.7 to 109.5 μM) (concentrations below BA's solubility limit), were tested on three couples of microorganisms: two were constituted of reference strains (Couple A: *C. albicans* ATCC 28367–*S. aureus* ATCC 29213; Couple B: *C. albicans* ATCC MYA2876–*S. aureus* ATCC 6538) and one of clinical isolates (Couple C: *C. albicans* Aca1–*S. aureus* SCO4). The results, shown in Figure 4, highlighted a significant inhibition of the biofilm of at least 12.5 µg·mL^{-1} for the reference strains (A and B). Biofilm inhibition reached up to 42% at 25 µg·mL^{-1} (55 µM) for pair A, with a loss of activity from 12.5 µg·mL^{-1} (27.4 µM). For couple B, BA showed activity at lower concentrations than for couple A, with still 34% inhibition at 6.25 µg·mL^{-1}, but showed similar activity at 25 µg·mL^{-1} (41–43% inhibition). For the clinical strains pair, we observed a lower inhibition than for the collection couples, with a maximum obtained inhibition of 27% at 25 µg·mL^{-1}.

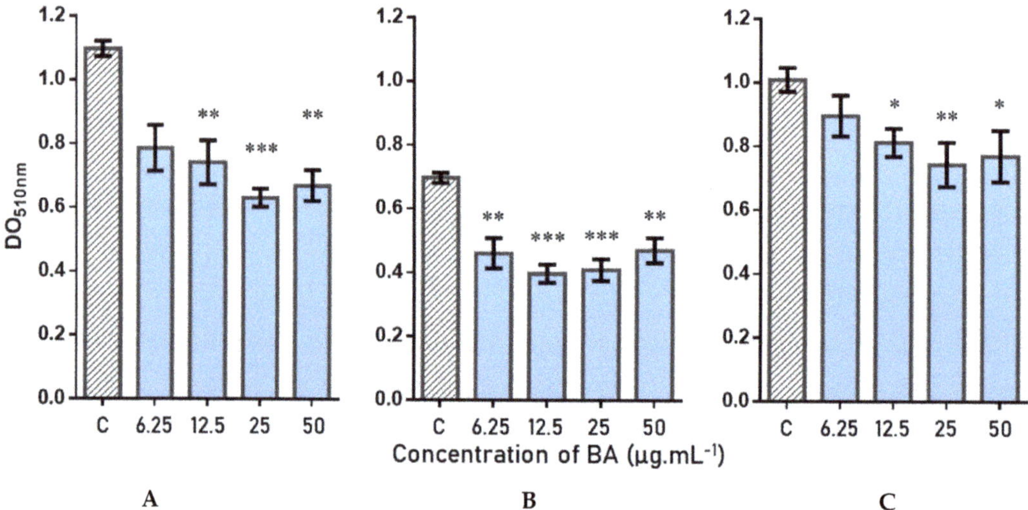

Figure 4. Betulinic acid (BA) activities on total biomass (CVS assay) of bispecies biofilm strains of *C. albicans–S. aureus* (error bars in SEM): (**A**) *C. albicans* ATCC 28367–*S. aureus* ATCC 29213; (**B**) *C. albicans* ATCC MYA-2876–*S. aureus* ATCC 6538; (**C**) *C. albicans* Aca1–*S. aureus* SCO4. *p*-values calculated by Dunn's test were given with *: 0.05 > *p*-value > 0.01; **: 0.01 > *p*-value > 0.001; ***: 0.001 > *p*-value.

Thus, these results showed an activity against the bispecies biofilm *C. albicans–S. aureus*, regardless of the tested couples (with a slightly lower activity for the clinical strains, generally more resistant to treatments), highlighting the nonstrain dependence and the interest of this compound.

BA was quantified by HPLC-UV within F4, F5 (most active fractions) and the Lg-AS-MeTHF extract in order to investigate the activity/BA quantity relationship (Figure S4). Analyses suggested that 1 mg of F4 and F5 contained 169.31 µg and 24.79 µg of BA, respectively, and that 1 mg of the extract contained 23.73 µg of BA. Thus, F4 has been strongly enriched in BA. The 64% and 50% of biofilm reduction induced by F4 and F5 at 50 µg·mL^{-1} corresponded to a concentration of 8.5 and 1.2 µg·mL^{-1} of BA in contact with the biofilm. BA alone did not demonstrate such high activity on the bispecies biofilm before, and efficient concentrations reached around 25 µg·mL^{-1} BA (43% biofilm inhibition), which suggested that BA was not the only compound responsible for the activity. Nevertheless, the highest concentration of BA in F4 compared to F5 seems to be correlated with a higher activity. A study of the other correlated compounds would allow to check the potential synergistic effect between the different active ingredients contained in the extracts of *L. grandiflora*.

BA is fairly well known, but its antibiofilm potential requires our full attention. A wide variety of actions has already been associated with BA (anticancer, anti-inflammatory, anti-HIV, antimalarial and anthelmintic activities), yet no cytotoxicity has been reported on healthy cells [43]. Some studies have described the mechanism of action of BA, particularly on mammalian myeloma cells. BA induced the apoptosis of cells by targeting mitochondria, generating an increase of reactive oxygen species in cytosol, and consequently an oxidative stress [44]. For all these applications, BA is still largely investigated in different studies, including its improvement of its bioavailability, due to its low solubility in aqueous medium that limiting its applications. One approach under consideration is its incorporation in nanoparticles, that permits to enhance its concentration and control its release during treatment [45]. The large applications of BA and actual studies of this molecule make it a promising treatment in human health.

3. Materials and Methods

3.1. Plant Material

Five aquatic IAPs: *E. densa*, *L. grandiflora*, *L. peploides*, *L. major* and *M. aquaticum* were collected in July 2020 in west region of France (Nouvelle-Aquitaine) (GPS localization: *E. densa* (46.569233, 0.640664); *L. grandiflora* (46.645098, 0.584291); *L. peploides* (46.910908, 0.247578); *L. major* (46.557362, 0.409080); *M. aquaticum* (45.645026, −0.053613)). Botanical identification was performed by the Conservatoire Botanique National Sud-Atlantique (CBNSA). Samples were washed in water baths and air-dried for one week. A voucher specimen of each plant was deposited at the Herbarium of the School of Pharmacy at the University of Poitiers (France).

3.2. Chemicals and Reagents

LC-MS grade acetonitrile (ACN) and methanol (Fisher Scientific, Waltham, MA, USA) were used for the UHPLC analysis. Extractions and fractionation were performed with 2-methyltetrahydrofuran (MeTHF), ethyl acetate (EtOAc), ethanol (EtOH), isopropanol (IPA) and cyclohexane (CHX) (E. Merck, Darmstadt, Germany). Deionized water (W) was purified by Milli-Q system (Millipore, Burlington, MA, USA). Betulinic acid analytical grade was purchased from Sigma–Aldrich Chemical Corporation (Saint Louis, MO, USA).

3.3. Extraction and Fractionation

Whole plants (WP) or parts of plants (stems (S) and leaves (L) for *M. aquaticum* leaves, aerial stems (AS) and submerged stems with roots (SS) for *L. grandiflora* and *L. peploides*) were reduced to a powder and 20 g were extracted by maceration assisted by sonication for 1 h at room temperature. Four solvents of increasing polarity were used successively on the same sample: MeTHF, EtOAc, EtOH, EtOH-W (1:1 *v/v*). After filtration by using a Büchner funnel, the extracts were evaporated under low pressure at 40 °C. A supplementary step of freeze-drying was added for MeTHF and EtOH-W extracts.

The most active extracts against bispecies biofilm (600 mg of EtOH leaves extract from *L. peploides* and 250 mg of MeTHF aerial stem extract from *L. grandiflora*) were fractionated by using flash chromatography (Puriflash® 4250 from Interchim (Montluçon, France) equipped with a diode array detector) on prepacked C18 columns (C18-HP 30 μm, 51 g for EtOH extract and 32 g for MeTHF extract, Interchim). Samples were solubilized in MeOH to perform a liquid loading, and compounds elution was monitored using UV detection at 220 and 265 nm. Compounds were eluted at 10 mL·min^{-1} with ACN/H$_2$O (5:95 to 30:70 in 30 min and then 30:70 to 100:0 in 5 min for EtOH extract; 5:95 to 60:40 in 15 min and then 60:40 to 100:0 in 5 min for MeTHF extract), and finally eluted for 30 min using, successively, 100% ACN, 100% IPA and 100% CHX to afford nine and seven fractions from EtOH and MeTHF extracts, respectively. The fractions were evaporated under low pressure at 40 °C and/or by lyophilization. Dried fractions and extracts were stored at −80 °C.

3.4. HPLC Analysis

All extracts and fractions were analyzed on DIONEX UltiMate 3000 UHPLC (Thermo Fisher scientific, Waltham, MA, USA) with a diode array detector (UHPLC-DAD) on a C18 analytical column (DIONEX, C18, 5 μm, 120 Å, 4.6 mm × 250 mm Acclaim®) protected by a Phenomenex® SecurityGuard (Torrance, CA, USA). The elution was performed with ACN/H$_2$O gradient complemented with 0.1% of trifluoroacetic acid (5:95 to 100:0 during

40 min and then 100:0 for 10 min, 0.8 mL·min^{-1}). The column oven temperature was set at 25 °C. UV detection was monitored at 210, 220 and 265 nm. Samples were injected from 1 to 10 mg·mL^{-1} in MeOH after centrifugation.

3.5. Organisms

Three reference strains of *C. albicans* and three reference strains of *S. aureus* were used: *C. albicans* ATCC® 28367™, *C. albicans* ATCC® MYA-2876™, *C. albicans* Aca1 (isolate recovered from venous catheter), *S. aureus* ATCC® 29213™, *S. aureus* ATCC® 6538 ™, *S. aureus* SC04 (isolate recovered from human lungs). *C. albicans* and *S. aureus* were grown for 48 h on Sabouraud glucose with chloramphenicol (0.05 g·L^{-1}) (SGC) (Sigma-Aldrich, Saint Louis, MO, USA) or brain heart infusion (BHI) (BD DifcoTM, Sparks, MD, USA) agar plates at 37 °C, respectively. Prior to biofilm assays, each strain was cultured in liquid BHI medium at 37 °C overnight, with agitation at 80 rpm only for *S. aureus*.

3.6. Biofilm Studies
3.6.1. Biofilm Growth

An amount of 25 mL of previously prepared BHI liquid cultures were centrifuged at 5000 g for 5 min. The pellet was washed with 10 mL of Phosphate Buffer Saline (PBS) (GIBCO, New York, NY, USA) and centrifuged again in the same condition. Cell concentration was determined by absorbance measurement at 600 nm for *S. aureus* and by using the previously determined equation:

$$1.4 \, DO_{600nm} = 23 \times 10^9 \, CFU/mL \tag{1}$$

A direct counting with a Fast-Read 102® counting chamber (Biosigma, Cantarana, Italy) was performed for cell concentration determination of *C. albicans*.

Single- and dual-species biofilms were cultured in 96-well polystyrene nontreated microtiter plates (Costar, Corning, NY, USA). For single-species biofilms, 200 μL of cultures at 10^6 CFU·mL^{-1} for *S. aureus* and 10^6 cell·mL^{-1} for *C. albicans* were inoculated in each well. For dual-species biofilms, 100 μL of each suspension at 10^6 CFU·mL^{-1}/10^6 cell·mL^{-1} were inoculated for a 1:1 ratio. After incubation at 37 °C for 2 h, culture medium was removed to eliminate nonadherent cells, and 200 μL of fresh BHI medium was added. After 24 h of total incubation at 37 °C, supernatants were discarded and biofilms were washed once with PBS. An amount of 196 μL of fresh medium and 4 μL of DMSO (control condition) or extract/fraction suspended at 10 mg·mL^{-1} in DMSO was added. Wells without treatment were preserved (negative control). Plates were incubated for 24 h at 37 °C.

3.6.2. Crystal Violet Staining (CVS) Assay

After 24 h of incubation, wells were washed with 200 μL of PBS. An amount of 200 μL of MeOH was then added in each well and left in contact during 10 min. After removing MeOH, 200 μL of crystal violet (0.3% in demineralized water) was added and the plates were incubated at room temperature for 5 min. Excesses of crystal violet were washed with demineralized water and 200 μL of acetic acid 33% was added and left in contact during 15 min under agitation (150 rpm). The absorbance was measured at 510 nm (Infinite M Plex absorbance reader, TECAN, Zürich, Switzerland).

The percentage of inhibition (*I*) of each sample was calculated with Formula (2) by comparing with intraplate controls (DMSO 2%).

$$I\,(\%) = \left(\frac{DO510_{control} - DO510_x}{DO510_{control}} \right) \times 100 \tag{2}$$

The inhibitory percentages and the concentration that inhibited 50% of the biofilm formation (IC$_{50}$) were determined for each tested sample by constructing a dose–response curve and selecting the closest tested concentration value above or equal to 50% inhibition.

3.6.3. Flow Cytometry (FCM) Assay

According to the adapted protocol of Kerstens et al. [46], after 24 h of incubation, each well was washed with 200 μL of PBS to remove residual nonadherent cells. An amount of 50 μL of PBS previously filtrated at 0.1 μm was added in each well and the biofilms were scratched vigorously with a sterile folded cone before pipetting. All the obtained suspensions were diluted one tenth in filtrated PBS, followed by 10 min of sonication and 30 s of vortex. Suspensions were double stained with 1 to 2 μL of 334 μM SYTO 9 (S9) and 1 μL of 2 mM propidium iodide (PI) (LIVE/DEAD™ Viability/Cytotoxicity Kit, Invitrogen, Carlsbad, CA, USA). Measurements were performed with a CytoFLEX flow cytometer (Beckman Coulter, Brea, CA, USA) equipped with a blue diode laser (excitation 488 nm) and a violet diode laser (excitation 405 nm) managed by CytExpert 2.0.0.153 software (Beckman Coulter) (SYTO 9 excitation filter, 525/40 nm; PI excitation filter, 610/20 nm). A compensation matrix was defined using unstained and single-stained heat-killed biofilm suspension (60 °C during 15 min) prior to sample measurements.

3.6.4. Cryo-Scanning Electron Microscopy

Bispecies biofilms were grown as described above with slight differences: 600 μL of each strain was inoculated at 1×10^6 cells·mL^{-1} on sterile polycarbonate coupons (diameter: 13 mm, thickness: BIOFOULING 3514 mm; BioSurface Technologies Corporation, Bozeman, MO, USA) deposed in 24-well microplates. After 2 h, the medium was substituted with 1.2 mL fresh BHI. After 24 h, extracts or 2% DMSO were added and the biofilm grew for an additional 24 h. Coupons were then recovered and dried few minutes before they were frozen with liquid nitrogen (Leica EM VCM Vacuum Cryo Manipulation system), sublimed and coated with platinum (Leica EM ACE600 High Vacuum Coater). Samples were then observed with a FEI Teneo Volume Scope (FEI Company, Hillsboro, OR, USA).

3.7. Molecular Networking

3.7.1. Data-Dependent LC-ESI-HRMS^2 Analyses

Selected extracts and fractions of *L. grandiflora* were suspended in MeOH at 1 mg·mL^{-1}. After vortex and sonication, samples were filtrated at 0.2 μm. The sequence was prepared by injecting the samples randomly, with a quality control sample every 10 samples analyzed, consisting in the mixture of an equal volume of all samples. A blank control with 100% MeOH was prepared and analyzed before and after the sample list. The UHPLC was performed on a Vanquish system (Thermo Fischer Scientific, Les Ulis, France) using a reverse-phase column (Zorbax RRHD SB-C18 1.8 μm, 2.1 mm × 100 mm, Agilent Technologies, Les Ulis, France) with a similar method to HPLC analysis, adapted for UHPLC (5:95 to 100:0 ratio of ACN/W gradient during 20 min and then 4 min at 100:0, 0.4 mL·min^{-1}). MilliQ W and ACN were acidified by 0.1% of formic acid. One μL per sample was injected. Column temperature was set at 30 °C.

Mass spectra were acquired in a Q-Exactive Plus™ mass spectrometer (Thermo Fischer Scientific) equipped with a heated electrospray ionization (HESI-II) probe. Acquisition was performed using both positive and negative ionization modes, setting spray voltage at 3.7 kV and 2.8 kV, respectively, capillary temperature at 310 °C and probe heater temperature at 280 °C. The MS^1 scan range was 150–1200 m/z with a resolution at 70,000 (full width at half maximum (FWHM) at m/z 200). Each full MS scan was followed by data-dependent acquisitions (DDA) selecting the 5 most intense ions and acquiring MS^2 between 50 and 1200 m/z, with a resolution of 17,500 and a normalized collision energy (NCE) of 20%, 35% and 50%. Data were acquired in centroid format.

3.7.2. MZmine 2 Data Preprocessing Parameters

The spectral features detection of MS/MS data was performed on MZmine 2.53 [47]. Each file of MS^2 analysis was imported in RAW format. Positive and negative ionization modes data were treated independently. Positive acquisitions were adjusted with asymmetric baseline corrector to 1E7. Peaks detection was set at 2E6 in MS^1 level and 0 in MS^2

level. Chromatograms were built for the detected ions with the "ADAP chromatogram builder" algorithm, with a minimum group size of 4 scans, a group intensity threshold of 2E6, a minimum highest intensity of 6E6 and an m/z tolerance of 0.0015 or 5 ppm. Peaks deconvolution was applied with the "local minimum search" algorithm with the following parameters: chromatographic threshold of 10%, search minimum in RT range of 0.5 min, minimum relative height of 10%, minimum absolute height of 6E6, min ratio of peak top/edge of 1 and peak duration range from 0 to 0.5 min. MS^2 scan pairing was set at the m/z range of 0.02 Da and the RT range of 0.1 min. Isotopic peaks grouper was applied with an m/z tolerance of 0.0015 or 5 ppm, with an RT tolerance of 0.2 min and a maximum charge of 2. Representative isotope was set on most intense. Features alignment step with Join aligner was performed with an m/z tolerance at 0.0015 or 5 ppm, with a weight for m/z of 75% and a weight for RT of 25%. The RT tolerance was set at 0.2 min. A final gap-filling step was performed with 10% intensity tolerance, 0.0015 m/z or 5 ppm tolerance, 0.2 min RT tolerance and with an RT correction. Exportation of data generated *.csv and *.mgf files with MS^2 data and MS^1 peaks area integration. The same parameters were applied to negative ionization data, with the exception of noise level, which was set to 5E5 for MS^1 peaks detection.

3.7.3. Molecular Networks Analysis

Molecular networks were created using Global Natural Products Social molecular networking (GNPS, http://gnps.ucsd.edu, accessed on 21 October 2022). The MGF and CSV files of processed data on MZmine 2 were uploaded on GNPS. Files were used to generate an MS/MS molecular network using the GNPS Feature-Based Molecular Networking workflow [48]. The precursor ion mass tolerance and the product ion mass tolerance were set to 0.02 Da. Networks were generated using 5 minimum matched peaks and a cosine score of 0.6. The library search options were set to minimum 5 matched peaks and a score threshold of 0.6 without search of analogs. In complement to GNPS databases, in silico fragmentation software MetFrag was used with several molecular databases (Pubchem, Human Metabolome Database (HMDB), Coconut).

Predictions of active compounds were realized with workflow from Nothias et al. [31] from an R-based Jupyter notebook available on GitHub, https://github.com/DorresteinLab oratory/Bioactive_Molecular_Networks (accessed on 21 October 2022). Briefly, bioactivity scores of antibiofilm inhibition at 50 μg·mL^{-1} from CVS tests were calculated to optimize disparity between samples with the Formula (3).

$$Bioactive\ score = \frac{(I - SD) \times 100}{I_{max} - SD_{Imax}} \tag{3}$$

where the standard deviation (SD) from each condition was subtracted to the percentage of inhibition (I). A bioactive score of 100 was attributed to the most active sample (I_{max}). For the other samples, scores were attributed proportionally relative to inhibition (%). A score of 0 was assigned to the samples with negative values.

Scores were added on the spectral features table obtained with MZmine 2 workflow. The Jupyter notebook applied 3 steps of analysis: (i) normalization of TIC intensity, (ii) calculation of Pearson correlation and its significance (p-value) between features and bioactivity scores and (iii) Bonferroni correction. Results were imported as a table in Cytoscape 3.8.2 [49] on the molecular network data from GNPS. Compounds which were clustered with correlated compounds, i.e., displaying MS/MS spectral similarities, were also studied to identify or confirm their molecular family. p-value and correlation value thresholds were respectively defined as ≤ 0.05 and ≥ 0.85.

3.8. Annotation and Quantification of Betulinic Acid

In order to confirm the identification and to quantify BA in the MeTHF aerial stems extract of *L. grandiflora* and its F4 and F5 fractions, commercial standard of BA was analyzed by HPLC-DAD and HPLC-MS/MS. HPLC-DAD analysis was performed with the same method as described before, adapted with a flow at 0.5 mL·min^{-1}. Ten μL of samples were injected, with BA at 0.5 mg·mL^{-1} in MeOH, and crude extract and fractions at 1 mg·mL^{-1} in MeOH. Each sample was injected separately and mixed in solution (extract or fractions mixed with BA). To quantify BA, 7 dilutions (0.5 to 0.01 mg·mL^{-1} in MeOH) of the standard were injected to realize a standard curve. UV detection was monitored at 210 nm, corresponding to the maximal absorbance of BA, as described in Zhao, Yan and Cao [50]. HPLC-MS/MS analysis was performed on a Waters system equipped with a time-of-flight XEVO™ G2 Q-TOF analyzer (Waters Corporation, Milford, MA, USA) with an ESI source in positive mode with the same chromatographic method as HPLC-DAD, and a 5 μL injection of the samples. MS^2 targeted acquisition was set at m/z 439.36 and 457.36, corresponding, respectively, to the [M+H−H$_2$O]$^+$ and [M+H]$^+$ adducts. Source was set at 120 °C at 3.7 eV, collision energy at 40 eV and acquisition range at m/z 50–1000 with centroid mode.

Data were analyzed using MassLynxTM software (V4.1, 2013) from Waters.

3.9. Statistical Analyses

All biological experiments were performed at least three times with triplicate for each condition.

The Kruskal–Wallis test with Dunn's multiple comparisons test was applied to determine the statistical significance between obtained measures with treated biofilms and controls, using GraphPad Prism® version 6.01 (GraphPad Software Inc, San Diego, CA, USA).

PCA and OPLS-DA tests on MZmine2 data including QCs were realized to check the method performance of LC-MS acquiring data, on SIMCA 14.1 software (Sartorius, Goettingen, Germany).

4. Conclusions

The aerial stem extract of *L. grandiflora* demonstrated the highest activity against the bispecies biofilm *C. albicans–S. aureus* among the 40 extracts prepared from five aquatic invasive plant species. Biochemometric studies have highlighted several families of compounds potentially responsible for this activity, including fatty acids and their hydroxylated and epoxidated derivatives, monoacylglycerols and pentacyclic triterpenoids. Antibiofilm tests confirmed the betulinic acid activity, a pentacyclic lupane-type triterpenoid, which opens interesting perspectives for the research of new treatment to fight multispecies biofilms. Its mechanism of action must now be further investigated and characterized, and further studies are needed to increase its solubility and bioavailability. Its original structure opens interesting perspectives for possible combinations with already available conventional antibiotic and/or antifungal agents that could be complementary by both destructuring the biofilm and killing microbial cells. Moreover, the isolation of the other compounds correlated to the antibiofilm activity in the pure state will be necessary to complete this study of the *L. grandiflora* extract antibiofilm potential. Finally, this work confirmed the interest of invasive aquatic plants in the discovery of compounds active against polymicrobial biofilms and encourages their further studies.

Supplementary Materials: The following supporting information can be downloaded at: https://www.mdpi.com/article/10.3390/antibiotics11111595/s1, Table S1: List of detected and analyzed LC-MS^2 data with bioactivity-based molecular networking for their identification and correlation with the antibiofilm activity; Figure S1: SEM approach of 24 h preformed polymicrobial biofilm C. albicans–S. aureus treated; Figure S2: Synthesis of structures putatively identified for detected ions; Figure S3: Identification of betulinic acid (BA) in Lg-AS-MeTHF extract and F4 fraction with comparison with standard: (a) HPLC-UV detection method at 210 nm; (b) MS^2 spectrum of MS targeted method (detection MS1 at m/z 439.36); Figure S4: HPLC-UV dosage of betulinic acid (BA) from extract Lg-AS-MeTHF and active fractions F4 and F5 of L. grandiflora compared to the relative bispecies biofilm inhibition: (a) quantification table of betulinic acid with quantity per mg, final concentration per wells and their respective biofilm inhibition activity; (b) standard range of betulinic acid curve area (absorbance at 210 nm) in function of concentration.

Author Contributions: Conceptualization, C.I. and M.G.; methodology, C.R., W.A. and G.H.; software, C.R., C.T. and G.H.; validation, G.H.; formal analysis, G.H.; investigation, G.H., C.R. and J.M.; resources, C.R., W.A., M.G. and C.I.; data curation, G.H.; writing—original draft preparation, G.H.; writing—review and editing, C.I., M.G., C.R., W.A. and G.H.; visualization, G.H.; supervision, C.I. and M.G.; project administration, C.I. and M.G.; funding acquisition, C.I. and M.G. All authors have read and agreed to the published version of the manuscript.

Funding: This research was funded by Region Nouvelle-Aquitaine (AAPR2020-2019-8408110—Research Grant), University of Poitiers and University of Limoges (AAP ARIC 2019 Research grant) (France). This study was also supported by the Bordeaux Metabolome Facility and the MetaboHUB project (ANR-11-INBS-0010), the 2015–2020 State-Region Planning Contracts (CPER), European Regional Development Fund (FEDER) and intramural funds from the Centre National de la Recherche Scientifique and the University of Poitiers (Poitiers, France).

Institutional Review Board Statement: Not applicable.

Informed Consent Statement: Not applicable.

Data Availability Statement: A voucher specimen of each plant was deposited at the Herbarium of the School of Pharmacy at the University of Poitiers (France).

Acknowledgments: The authors would like to thank the staff from CBNSA for the scientific discussions and identification of the plants. The authors also thank the staff of the Green space department of the university of Poitiers, Anaël Lachaise of the Joint Union of the Seugne Basin and the staff of the Technical Unit for Natural Environments of Grand Poitiers for providing the plants. The authors wish also thank the Direction of the green spaces of the city of Poitiers, the Regional Biodiversity Agency of New Aquitaine (ARBNA) for the scientific discussions. Authors also would like to thank the French society of medical mycology (SFMM) for the awarding of a prize for the part of Guillaume Hamion's Ph.D. work. The authors thank D. Debail and M. Maillet for revising the English text.

Conflicts of Interest: The authors declare no conflict of interest.

References

1. Donlan, R.M.; Costerton, J.W. Biofilms: Survival Mechanisms of Clinically Relevant Microorganisms. *Clin. Microbiol. Rev.* **2002**, *15*, 167–193. [CrossRef] [PubMed]
2. He, Y.; Zhao, H.; Wei, Y.; Gan, X.; Ling, Y.; Ying, Y. Retrospective Analysis of Microbial Colonization Patterns in Central Venous Catheters, 2013–2017. *J. Healthc. Eng.* **2019**, *2019*, 8632701. [CrossRef] [PubMed]
3. Mavor, A.L.; Thewes, S.; Hube, B. Systemic Fungal Infections Caused by *Candida* Species: Epidemiology, Infection Process and Virulence Attributes. *Curr. Drug Targets* **2005**, *6*, 863–874. [CrossRef] [PubMed]
4. Nobile, C.J.; Johnson, A.D. *Candida albicans* Biofilms and Human Disease. *Annu. Rev. Microbiol.* **2015**, *69*, 71–92. [CrossRef] [PubMed]
5. Bougnoux, M.-E.; Diogo, D.; François, N.; Sendid, B.; Veirmeire, S.; Colombel, J.F.; Bouchier, C.; Van Kruiningen, H.; D'Enfert, C.; Poulain, D. Multilocus Sequence Typing Reveals Intrafamilial Transmission and Microevolutions of *Candida albicans* Isolates from the Human Digestive Tract. *J. Clin. Microbiol.* **2006**, *44*, 1810–1820. [CrossRef]
6. Nett, J.; Andes, D. *Candida albicans* biofilm development, modeling a host–pathogen interaction. *Curr. Opin. Microbiol.* **2006**, *9*, 340–345. [CrossRef]
7. Van Dyck, K.; Pinto, R.M.; Pully, D.; Van Dijck, P. Microbial Interkingdom Biofilms and the Quest for Novel Therapeutic Strategies. *Microorganisms* **2021**, *9*, 412. [CrossRef]
8. Carolus, H.; Van Dyck, K.; Van Dijck, P. *Candida albicans* and *Staphylococcus* Species: A Threatening Two-some. *Front. Microbiol.* **2019**, *10*, 2162. [CrossRef]
9. Eichelberger, K.R.; Cassat, J.E. Metabolic Adaptations During *Staphylococcus aureus* and *Candida albicans* Co-Infection. *Front. Immunol.* **2021**, *12*, 797550. [CrossRef]
10. Bouza, E.; Burillo, A.; Muñoz, P.; Guinea, J.; Marín, M.; Rodríguez-Créixems, M. Mixed bloodstream infections involving bacteria and *Candida* spp. *J. Antimicrob. Chemother.* **2013**, *68*, 1881–1888. [CrossRef]
11. Zhong, L.; Zhang, S.; Tang, K.; Zhou, F.; Zheng, C.; Zhang, K.; Cai, J.; Zhou, H.; Wang, Y.; Tian, B.; et al. Clinical characteristics, risk factors and outcomes of mixed *Candida albicans*/bacterial bloodstream infections. *BMC Infect. Dis.* **2020**, *20*, 810. [CrossRef]
12. Klotz, S.A.; Chasin, B.S.; Powell, B.; Gaur, N.K.; Lipke, P.N. Polymicrobial bloodstream infections involving *Candida* species: Analysis of patients and review of the literature. *Diagn. Microbiol. Infect. Dis.* **2007**, *59*, 401–406. [CrossRef]

13. Harriott, M.M.; Noverr, M.C. *Candida albicans* and *Staphylococcus aureus* Form Polymicrobial Biofilms: Effects on Antimicrobial Resistance. *Antimicrob. Agents Chemother.* **2009**, *53*, 3914–3922. [CrossRef]

14. Schlecht, L.M.; Peters, B.M.; Krom, B.P.; Freiberg, J.A.; Hänsch, G.M.; Filler, S.G.; Jabra-Rizk, M.A.; Shirtliff, M.E. Systemic *Staphylococcus aureus* infection mediated by *Candida albicans* hyphal invasion of mucosal tissue. *Microbiology* **2015**, *161*, 168–181. [CrossRef]

15. Todd, O.A.; Fidel, P.L.; Harro, J.M.; Hilliard, J.J.; Tkaczyk, C.; Sellman, B.R.; Noverr, M.C.; Peters, B.M. *Candida albicans* Augments *Staphylococcus aureus* Virulence by Engaging the Staphylococcal *agr* Quorum Sensing System. *MBio* **2019**, *10*, e00910-19. [CrossRef]

16. Moutou, F.; Pastoret, P.P. Defining an invasive species. *Rev. Sci. Tech. l'OIE* **2010**, *29*, 37–45.

17. IUCN. Invasive Alien Species and Climate Change. 2021. Available online: https://iucn.org/resources/issues-brief/invasive-ali en-species-and-climate-change (accessed on 3 November 2022).

18. Reddy, A.M.; Pratt, P.D.; Grewell, B.J.; Harms, N.E.; Walsh, G.C.; Ndez, M.C.H.; Faltlhauser, A.; Ci-bils-Stewart, X. Biological Control of Invasive Water Primroses, *Ludwigia* spp., in the United States: A Feasi-bility Assessment. *J. Aquat. Plant Manag.* **2021**, *11*, 67–77.

19. Fan, P.; Marston, A. How Can Phytochemists Benefit from Invasive Plants? *Nat. Prod. Commun.* **2009**, *4*, 1407–1416. [CrossRef]

20. Máximo, P.; Ferreira, L.M.; Branco, P.S.; Lourenço, A. Invasive Plants: Turning Enemies into Value. *Molecules* **2020**, *25*, 3529. [CrossRef]

21. Smida, I.; Sweidan, A.; Souissi, Y.; Rouaud, I.; Sauvager, A.; Torre, F.; Calvert, V.; Le Petit, J.; Tomasi, S. Anti-Acne, Antioxidant and Cytotoxic Properties of Ludwigia Peploides Leaf Extract. *Int. J. Pharmacogn. Phytochem. Res.* **2018**, *10*, 271–278.

22. Smida, I.; Charpy-Roubaud, C.; Cherif, S.Y.; Torre, F.; Audran, G.; Smiti, S.; Le Petit, J. Antibacterial properties of extracts of *Ludwigia peploides* subsp. *montevidensis* and *Ludwigia grandiflora* subsp. *hexapetala* during their cycle of development. *Aquat. Bot.* **2015**, *121*, 39–45. [CrossRef]

23. Visweswari, G.; Christopher, R.; Rajendra, W.; Visweswari, D.G. Phytochemical Screening of Active Secondary Metabolites Present in *Withania somnifera* Root: Role in Traditional Medecine. *Int. J. Pharm. Sci. Res.* **2013**, *4*, 2770–2776.

24. Chinou, I. Primary and Secondary Metabolites and Their Biological Activity. In *Thin Layer Chromatography in Phytochemistry*; CRC Press: Boca Raton, FL, USA, 2008; pp. 60–74.

25. Sousa, S.N.B.; De Andrade, M.L.; Brito, R.D.V.D.D. Antibiofilm Activity of Natural Products: Promising Strategies for Combating Microbial Biofilms. *Ann. Public Health Rep.* **2020**, *4*, 92–99. [CrossRef]

26. Kim, Y.-G.; Lee, J.-H.; Park, J.G.; Lee, J. Inhibition of *Candida albicans* and *Staphylococcus aureus* biofilms by centipede oil and linoleic acid. *Biofouling* **2020**, *36*, 126–137. [CrossRef]

27. Bernard, C.; Renaudeau, N.; Mollichella, M.-L.; Quellard, N.; Girardot, M.; Imbert, C. Cutibacterium acnes protects *Candida albicans* from the effect of micafungin in biofilms. *Int. J. Antimicrob. Agents* **2018**, *52*, 942–946. [CrossRef]

28. Domingue, G.; Costerton, J.W.; Brown, M.R. Bacterial doubling time modulates the effects of opsonisation and available iron upon interactions between *Staphylococcus aureus* and human neutrophils. *FEMS Immunol. Med. Microbiol.* **1996**, *16*, 223–228. [CrossRef]

29. Mahto, K.K.; Singh, A.; Khandelwal, N.K.; Bhardwaj, N.; Jha, J.; Prasad, R. An Assessment of Growth Media Enrichment on Lipid Metabolome and the Concurrent Phenotypic Properties of *Candida albicans*. *PLoS ONE* **2014**, *9*, e113664. [CrossRef]

30. Christensen, G.D.; Simpson, W.A.; Younger, J.J.; Baddour, L.M.; Barrett, F.F.; Melton, D.M.; Beachey, E.H. Adherence of coagulase-negative staphylococci to plastic tissue culture plates: A quantitative model for the adherence of staphylococci to medical devices. *J. Clin. Microbiol.* **1985**, *22*, 996–1006. [CrossRef]

31. Alby, K.; Schaefer, D.; Sherwood, R.K.; Jones, S.; Bennett, R.J. Identification of a Cell Death Pathway in *Candida albicans* during the Response to Pheromone. *Eukaryot. Cell* **2010**, *9*, 1690–1701. [CrossRef]

32. Nothias, L.-F.; Nothias-Esposito, M.; da Silva, R.; Wang, M.; Protsyuk, I.; Zhang, Z.; Sarvepalli, A.; Leyssen, P.; Touboul, D.; Costa, J.; et al. Bioactivity-Based Molecular Networking for the Discovery of Drug Leads in Natural Product Bioassay-Guided Fractionation. *J. Nat. Prod.* **2018**, *81*, 758–767. [CrossRef]

33. Weber, H. Fatty acid-derived signals in plants. *Trends Plant Sci.* **2002**, *7*, 217–224. [CrossRef]

34. Mitra, S.; Sarkar, N.; Barik, A. Long-chain alkanes and fatty acids from *Ludwigia octovalvis* weed leaf surface waxes as short-range attractant and ovipositional stimulant to *Altica cyanea* (Weber) (Coleoptera: Chrysomelidae). *Bull. Èntomol. Res.* **2017**, *107*, 391–400. [CrossRef]

35. Lee, J.-H.; Kim, Y.-G.; Park, J.G.; Lee, J. Supercritical fluid extracts of *Moringa oleifera* and their unsaturated fatty acid components inhibit biofilm formation by *Staphylococcus aureus*. *Food Control* **2017**, *80*, 74–82. [CrossRef]

36. Yuyama, K.; Rohde, M.; Molinari, G.; Stadler, M.; Abraham, W.-R. Unsaturated Fatty Acids Control Biofilm Formation of *Staphylococcus aureus* and Other Gram-Positive Bacteria. *Antibiotics* **2020**, *9*, 788. [CrossRef]

37. Nigam, S.; Ciccoli, R.; Ivanov, I.; Sczepanski, M.; Deva, R. On Mechanism of Quorum Sensing in *Candida albicans* by 3(R)-Hydroxy-Tetradecaenoic Acid. *Curr. Microbiol.* **2010**, *62*, 55–63. [CrossRef]

38. Shareck, J.; Nantel, A.; Belhumeur, P. Conjugated Linoleic Acid Inhibits Hyphal Growth in *Candida albicans* by Modulating Ras1p Cellular Levels and Downregulating TEC1 Expression. *Eukaryot. Cell* **2011**, *10*, 565–577. [CrossRef]

39. Gauthier, C.; Legault, J.; Lavoie, S.; Rondeau-Gagné, S.; Tremblay, S.; Pichette, A. Synthesis of two natural betulinic acid saponins containing α-l-rhamnopyranosyl-(1→2)-α-l-arabinopyranose and their analogues. *Tetrahedron* **2008**, *64*, 7386–7399. [CrossRef]

40. Shilpi, J.A.; Gray, A.I.; Seidel, V. Chemical constituents from *Ludwigia adscendens*. *Biochem. Syst. Ecol.* **2010**, *38*, 106–109. [CrossRef]

41. Bhattacharya, S.P.; Mitra, A.; Bhattacharya, A.; Sen, A. Quorum quenching activity of pentacyclic triterpenoids leads to inhibition of biofilm formation by *Acinetobacter baumannii*. *Biofouling* **2020**, *36*, 922–937. [CrossRef]

42. da Silva, G.N.S.; Primon-Barros, M.; Macedo, A.J.; Gnoatto, S.C.B. Triterpene Derivatives as Relevant Scaffold for New Antibiofilm Drugs. *Biomolecules* **2019**, *9*, 58. [CrossRef]

43. Yogeeswari, P.; Sriram, D. Betulinic Acid and Its Derivatives: A Review on their Biological Properties. *Curr. Med. Chem.* **2005**, *12*, 657–666. [CrossRef] [PubMed]

44. Shen, M.; Hu, Y.; Yang, Y.; Wang, L.; Yang, X.; Wang, B.; Huang, M. Betulinic Acid Induces ROS-Dependent Apoptosis and S-Phase Arrest by Inhibiting the NF-κB Pathway in Human Multiple Myeloma. *Oxid. Med. Cell. Longev.* **2019**, *2019*, 5083158. [CrossRef] [PubMed]

45. Bocalon, L.G.; Tozatti, M.G.; Januário, A.H.; Pauletti, P.M.; Silva, M.L.A.; Rocha, L.A.; Molina, E.F.; Santos, M.F.C.; Cunha, W.R. Incorporation of Betulinic Acid into Silica-Based Nanoparticles for Controlled Phytochemical Release. *Anal. Lett.* **2022**, 1–15. [CrossRef]

46. Kerstens, M.; Boulet, G.; Pintelon, I.; Hellings, M.; Voeten, L.; Delputte, P.; Maes, L.; Cos, P. Quantification of *Candida albicans* by flow cytometry using TO-PRO®-3 iodide as a single-stain viability dye. *J. Microbiol. Methods* **2013**, *92*, 189–191. [CrossRef]

47. Pluskal, T.; Castillo, S.; Villar-Briones, A.; Orešič, M. MZmine 2: Modular framework for processing, visualizing, and analyzing mass spectrometry-based molecular profile data. *BMC Bioinform.* **2010**, *11*, 395. [CrossRef]

48. Wang, M.; Carver, J.J.; Phelan, V.V.; Sanchez, L.M.; Garg, N.; Peng, Y.; Nguyen, D.D.; Watrous, J.; Kapono, C.A.; Luzzatto-Knaan, T.; et al. Sharing and community curation of mass spectrometry data with Global Natural Products Social Molecular Networking. *Nat. Biotechnol.* **2016**, *34*, 828–837. [CrossRef]

49. Shannon, P.; Markiel, A.; Ozier, O.; Baliga, N.S.; Wang, J.T.; Ramage, D.; Amin, N.; Schwikowski, B.; Ideker, T. Cytoscape: A software environment for integrated models of Biomolecular Interaction Networks. *Genome Res.* **2003**, *13*, 2498–2504. [CrossRef]

50. Zhao, G.; Yan, W.; Cao, D. Simultaneous determination of betulin and betulinic acid in white birch bark using RP-HPLC. *J. Pharm. Biomed. Anal.* **2007**, *43*, 959–962. [CrossRef]

MDPI

St. Alban-Anlage 66

4052 Basel

Switzerland

Tel. +41 61 683 77 34

Fax +41 61 302 89 18

www.mdpi.com

Antibiotics Editorial Office

E-mail: antibiotics@mdpi.com

www.mdpi.com/journal/antibiotics